designer evolution

Third Wave thinkers must now face the fact that we are about to become *designers* of evolution.

—Alvin Toffler, *The Third Wave* (1980)

simon young

designer evolution

a transhumanist manifesto

forewords by

dr. aubrey de grey

department of genetics
university of cambridge

and

robert a. freitas jr.

senior research fellow
institute for molecular manufacturing

59 John Glenn Drive
Amherst, New York 14228-2197

Published 2006 by Prometheus Books

Inquiries should be addressed to
Prometheus Books
59 John Glenn Drive
Amherst, New York 14228–2197
VOICE: 716–691–0133, ext. 207
FAX: 716–564–2711
WWW.PROMETHEUSBOOKS.COM

10 09 08 07 06 5 4 3 2 1

Library of Congress Cataloging-in-Publication Data

Young, Simon, 1964–
Designer evolution : a transhumanist manifesto / by Simon Young.
p. cm.
ISBN 1–59102–290–8 (alk. paper)
Includes bibliographical references and index.
1. Medical genetics—Moral and ethical aspects. 2. Genetic engineering—Moral and ethical aspects. 3. Biotechnology—Moral and ethical aspects. 4. Human evolution. 5. Philosophy and science. I. Title.

RB155.Y685 2005
174.2—dc22

2005002144

Printed in the United States of America on acid-free paper

In memory of Mary and J. F. Young

CONTENTS

FOREWORD 1

FOREVER YOUNG

Dr. Aubrey de Grey

Biogerontologist, University of Cambridge

> De Grey has emerged as one of the boldest thinkers and organizers in the science of aging, whose ideas have begun to influence a whole generation of biologists.
> —David Stipp, *Fortune*, June 14, 2004

I work on the biology of aging. I do so because, if I may use classic British understatement, I am not altogether in favor of aging and am striving to get it brought under complete control as soon as possible. By complete control I mean the same sort of control that we currently have over, say, malaria: not the absolute elimination of aging from our bodies (which I believe is impossible for biological systems), but the total suppression of its effects on our health, by periodic removal of the damage that it causes at the microscopic level before that damage can spiral out of control and be bad for us.

The practical effect of this will be the same as if we *could* eliminate it completely: we will be in possession of indefinite youth. We will die only from the sort of causes that young people die of today—accidents, suicide, homicide, and so on—but not of the age-related diseases that account for the vast majority of deaths in the industrialized world today. Our rate of death from age-independent causes will also plummet as we take better care of ourselves and each other, knowing how much more there is to lose than in the past.

Sounds good, doesn't it? Good enough to try pretty hard to expedite? How hard? As hard as the United States tried to develop the atomic bomb before the Germans did, say? Surely so. Surely the case for a Manhattan Project to cure aging is overwhelming. And yet, though we spend a respectable amount of our wealth on research to postpone the major diseases of old age by the odd year or two, we spend perhaps 1 percent of that on research to combat aging. Even 1 percent is a decidedly generous estimate, including work with virtually no prospect of giving rise to knowledge relevant to postponing (let alone curing) human aging. I spend much of my time trying to understand why this is so, and of course to change it.

Lots of reasons for reluctance to work hard to cure aging come up, but they're almost all of the same sort: curing aging would cause big changes to society. Overpopulation, unless we implement draconian sterilization. Tyrants reigning forever. We'd never be able to retire. You've heard them—I needn't go on. I could explain why they're all largely nonsense anyway, but I won't, because I have something more direct to say.

Even if they were valid, how is it that these arguments prevail, against the fact that we would be saving so many lives? And let's be clear, "saving" is what it would be—there is no difference between saving lives and extending lives unless you consider that the value of life falls as it continues, irrespective of how much longer it may continue. Let me remind you of the numbers. One hundred thousand deaths a day, about two-thirds of the total number of deaths worldwide, are due to causes that young people hardly ever die of—in other words, they are due to aging. Thirty World Trade Centers every single day. Four times as many deaths per day as occurred during the Second World War. And it's still going up. How can we continue to sit back and accept it? It simply can't be that we prefer death to social change, can it?

The only answer that seems to fit the evidence is denial, based on fear that one won't make the cut. We don't want to get our hopes up that something so desired, which has eluded us for so long, may finally be upon us, only to have those hopes dashed by progress in the science being slower than anticipated. This is a disgrace, and something must be done about it. It's a disgrace because it implies that the only person one cares about saving is oneself, rather than the generation (whichever it turns out to be) who could make the cut if we act today but won't if we don't. But yet it persists. So, what can be done?

Dry academic arguments are my style, but the number of people with whom they resonate is limited. We need a populist approach. And this book provides it. Here you will find a truly down-to-earth, energy-filled, heartfelt description of the world we should be seeking. Not just in terms of defeating aging, but in all aspects of what technology can give us if we stop being afraid.

Read on, stand up, and get active.

FOREWORD 2
NANOMEDICINE AND TWENTY-FIRST-CENTURY HEALTHCARE

Robert A. Freitas Jr.
Senior Research Fellow
Institute for Molecular Manufacturing

Robert Freitas is in my view, the world's leading pioneer in nanomedicine.
—Ray Kurzweil, KurzweilAI.net. (November 22, 2002)

Biotechnology and genetic engineering are comparatively well known as potential methods of enhancing the body, because of their many important successes over the last several decades. But advocates of these approaches often ignore a future post-biotechnology discipline, just now appearing on the two- to three-decade research and development horizon, which can almost guarantee whole-body elimination of biological senescence and the indefinite maintenance of healthy mind and body, while producing few if any unwanted medical side effects.

This new technology involves the application of molecular nanotechnology and nanorobotics to human healthcare.[1] Over the next decade or two, it will become increasingly clear that all of biotechnology is but a small subset—albeit an important subset—of nanotechnology. Indeed, the twenty-first century will be dominated by nanotechnology—the engineering and manufacturing of objects with atomic-scale precision.

Humanity is finally poised at the brink of completion of one of its greatest and most noble enterprises.[2] Early in this century, our growing abilities to swiftly repair most traumatic physical injuries, eliminate pathogens, and alleviate suffering using molecular tools will begin to coalesce in a new medical paradigm called nanomedicine.[3]

Nanomedicine may be broadly defined as the comprehensive monitoring, control, construction, repair, defense, and improvement of all human biological systems, working from the molecular level, using the emerging techniques of molecular nanotechnology and molecular manufacturing to produce engineered nanodevices and nanostructures, molecular machine systems, and—ultimately—nanorobots too small for the eye to see.[4] Nanotechnology applied to medicine generally means controlling biologically relevant structures with molecular precision. Even now, nanomedicine[5] is already exploring how to use carbon buckyballs, dendrimers (spherical treelike molecules), and other cleverly engineered nanoparticles in novel drugs to combat viruses, bacteria, and cancer. But in ten to twenty years we may learn how to build the first medical nanorobots. These will be devices the size of a microbe, though incapable of self-replication,[6] containing onboard sensors, computers, manipulators, pumps, pressure tanks, and power supplies. Building such sophisticated molecular machine systems will require molecular manufacturing—both the ability to make atomically precise objects, probably using diamond[7] or other similarly rigid materials, and the ability to make precise objects in very large numbers, probably using massively parallel assembly lines in nanofactories.

Methods to build these still-theoretical medical nanorobots are already being researched today.[8] The technology is still in a relatively primitive state.[9] We are just now learning how to build structures atom by atom,[10] particularly structures made of diamond[11] and other similar durable materials.[12] These structures, or nanoparts, must then be assembled into complex components, and then these components must in turn be assembled into complete machine systems such as medical nanorobots.[13] After that, we must devise massively parallel production lines so that these nanorobots can be manufactured in batches of millions or billions of identical units.[14] It's a big research agenda, but it's doable. By 2020 we should begin to enjoy the fruits of these efforts in our daily lives.

Once doctors gain access to medical nanorobots[15] they'll be able to quickly cure most known diseases that hobble and kill people today, rapidly repair most physical injuries that our bodies can suffer, and vastly extend the human healthspan.

The net effect of these interventions will be the continuing arrest of all biological aging, along with the reduction of current biological age to whatever new biological age is deemed desirable by the patient. These interventions may become commonplace several decades from today. Using annual checkups and cleanouts, and some occasional major repairs, your biological age could be restored once a year to the more or less constant physiological age that you select. I see little reason not to go for optimal youth—a rollback to the robust physiology of your early twenties would be easy to maintain and lots of fun. You might still eventually die of accidental causes but you'll live ten times longer than you do now—a healthy life expectancy of at least one thousand years is anticipated, based on the actuarial rates of fatal accidents and suicides.[16] Even if for some reason nearer-term efforts fail to accomplish similar objectives using biotechnology alone, nanotechnology—via nanomedicine—is almost guaranteed to achieve the desired results.

Some conservative bioethicists[17] are aghast at the prospect of vastly expanded human healthspans but there are many important advantages of "ageless bodies." For example, rather than leading to "a life of lesser engagements and weakened commitments,"[18] ageless bodies would enhance our personal commitment to the more distant future because we would now expect to live in it ourselves. This realization would inevitably lead to an increased respect for our bodies, for our friends and families, for our natural environment, and, most significantly, for strangers we've never met. With time to meet more people during an enhanced healthspan, with more time to consider the consequences of our personal actions, and with more healthy life span to lose by risking early death, the human proclivity for war and violence should markedly diminish, an unquestionably historic major social benefit.

NOTES

1. Robert A. Freitas Jr., *Nanomedicine*, vol. 1, *Basic Capabilities* (Georgetown, TX: Landes Bioscience, 1999); http://www.nanomedicine.com/NMI.htm.

2. Robert A. Freitas Jr., "1.1 A Noble Enterprise," *Nanomedicine,* vol. 1, http://www.nanomedicine.com/NMI/1.1.htm.

3. Freitas, *Nanomedicine*, http://www.nanomedicine.com/NMI.htm.

4. Freitas, "1.1 A Noble Enterprise."

5. Freitas, *Nanomedicine*, http://www.nanomedicine.com/NMI.htm.

6. Robert A. Freitas Jr. and Ralph C. Merkle, *Kinematic Self-Replicating Machines* (Georgetown, TX: Landes Bioscience, 2004); http://www.MolecularAssembler.com/KSRM.htm.

7. Ralph C. Merkle and Robert A. Freitas Jr., "Speeding the Development of Molecular Nanotechnology," Foresight Institute, December 2003, http://www.foresight.org/stage2/project1A.html.

8. Robert A. Freitas Jr., "Pathways to Molecular Manufacturing," *Nanomedicine*, vol. 1; http://www.nanomedicine.com/NMI/2.1.htm.

9. Ibid.

10. Ralph C. Merkle and Robert A. Freitas Jr., "Theoretical Analysis of a Carbon-Carbon Dimer Placement Tool for Diamond Mechanosynthesis," *Journal of Nanoscience Nanotechnology* 3 (August 2003): 319–24; http://www.rfreitas.com/Nano/JNNDimerTool.pdf.

11. Jingping Peng, Robert A. Freitas Jr., and Ralph C. Merkle, "Theoretical Analysis of Diamond Mechanosynthesis. Part I. Stability of C_2 Mediated Growth of Nanocrystalline Diamond C(110) Surface," *Journal of Computational and Theoretical Nanoscience* 1 (March 2004): 62–70, http://www.molecular assembler.com/JCTNPengMar04.pdf; and David J. Mann, Jingping Peng, Robert A. Freitas Jr., and Ralph C. Merkle, "Theoretical Analysis of Diamond Mechanosynthesis. Part II. C_2 Mediated

Growth of Diamond C(110) Surface via Si/Ge-Triadamantane Dimer Placement Tools," *Journal of Computational and Theoretical Nanoscience* 1 (March 2004): 71–80, http://www.molecularassembler.com/ JCTNMannMar04.pdf.

12. Robert A. Freitas Jr., "Technical Bibliography for Research on Positional Mechanosynthesis," Foresight Institute, December 16, 2003; http://www.foresight .org/stage2/mechsynthbib.html.

13. Robert A. Freitas Jr., "2.4 Molecular Components and Molecular Assemblers," *Nanomedicine,* vol. 1; http://www.nanomedicine.com/NMI/ 2.4.htm.

14. Robert A. Freitas Jr. and Ralph C. Merkle, "5.7 Massively Parallel Molecular Manufacturing," *Kinematic Self-Replicating Machines.*

15. Robert A. Freitas Jr., "Say Ah!" *Sciences* 40 (July/August 2000): 26–31; http:// www.foresight.org/Nanomedicine/SayAh/index.html.

16. Robert A. Freitas Jr., "Death Is an Outrage!" lecture delivered at the Fifth Alcor Conference on Extreme Life Extension, November 16, 2002, Newport Beach, CA; http://www.rfreitas.com/Nano/DeathIsAnOutrage.htm.

17. "Ageless Bodies," in *Beyond Therapy: Biotechnology and the Pursuit of Happiness*, ed. Leon Kass et al. (Washington, DC: President's Council on Bioethics, October 2003); http://bioethics.gov/reports/beyondtherapy/ chapter4.html.

18. Ibid.

INTRODUCTION

Some people can live quite happily without a purpose to life above that of everyday living—working, playing, raising a family. Personally, I cannot live without a philosophy—a wider meaning to life.

I call my philosophy transhumanism. I define it as "the belief in overcoming human limitations through reason, science, and technology."

I see a fundamental problem with life. Despite the enormous pleasure involved in living, ultimately death and biological limitation make life a tragedy. I find it impossible to blindly accept the suffering imposed upon us by our biological condition. Why must we age and die? Why must our brains and bodies be so fragile, subject to inevitable decay—programmed for self-destruction? I believe in seeking to overcome the mental and physical limitations that restrict our freedom. Science offers the only serious possibility of succeeding. Therefore, I believe in science.

In the twenty-first century, the combined tools of an advanced biotechnology I call, collectively, Superbiology, will provide us with the serious prospect of preventing and curing disease after major disease. After that—who knows? We shall begin to enhance human capabilities beyond limits dreamed of by our ancestors. Transhuman limits will be set only by our imagination and ingenuity.

It is compassion that first caused me to believe in science. I held my dying father in my arms and watched his cancer-riddled body shrivel away into nothingness while his mind remained as brilliant and lucid as ever. Death is, to me, an obscenity. It is an outrage that the mind should die because the body is programmed to self-destruct. The goal of human life is survival—we are programmed that way. To increase our ability to survive we must evolve. Science and technology increasingly offers us the chance to overcome the limitations of the human condition. Therefore, let us believe in science.

The transhumanist philosophy I espouse is *not* some irresponsible "anything goes" attitude toward biotechnology (Luminous mice? Gee whiz! Cool!). Indeed, *this* transhumanist *opposes* the recent, knee-jerk rush by liberals to *automatically* support the use of embryos discarded from fertility treatment purely as material for research, on precisely the grounds stated by the current US president of the "religious right," George W. Bush, in another context: "Where doubt exists, it is wise to always err on the side of life." I consider this a profoundly wise statement, and applaud it unreservedly. Such embryos were deliberately created from the union of a man and a woman in the specific attempt to produce a healthy baby. I think they deserve respect.

Instead, I favor greater research into adult stem cells[1] in order to avoid the problem of "creating life only to destroy it," a perfectly reasonable moral objection not necessarily founded on religious belief, which all sensible, compassionate people should respect rather than ridicule.

This, I might add, is the view of one who's own mother suffered bravely from Parkinson's disease (whose victims are expected to be some of the first beneficiaries of stem cell research) entirely without complaint for over thirty years, to the point at which, crippled, she attempted to take her own life, leaving a suicide note saying, "I can't even hold my library books anymore." She is no longer with us. The fight to cure disease goes on; but let it go on ethically. These issues are too important to be decided on knee-jerk reactions alone.

So let us not leap to the conclusions of the unthinking orthodoxist in our enthusiasm to seem "progressive." That way lies the tyranny of the extreme, progressive left, just as the attempt to *outlaw* biotechnology represents the tyranny of the extreme, religious right. Let us beware the authoritarian Bodypolitic, from whatever political persuasion it may come.

The transhumanist politics I espouse is not the politics of death but of *More Life*. It is, in short, the politics of the responsible, sensible center. One does not have to approve of *everything* scientists wish to do in order to believe passionately in scientific progress, no more than one must believe in obesity to be a lover of food, or of cold promiscuity in public toilet cubicles to be a lover of making love.

With this book I shall almost certainly cause offense to members of practically every Bodypolitic in the modern world—the church, political ideologues of both left and right, ecoists, New Agers, therapists—even scientists of an orthodox or smug-superior persuasion. If I play devil's advocate a little more than is customary—and I do—it is not to deliberately offend, but to tilt the balance away from the present unhealthy bias against a belief in the modern world, to a more reasonable, favorable disposition toward science, technology, and the concept of progressive evolution. For the postmodern ironists are wrong—we are *not* "on a road to nowhere," but on an ongoing journey of liberation from our biological chains. The human adventure has only just begun.

If one's beliefs are sincere, passionate, reasoned, and benevolent—and the world won't listen—well, then, one must shout all the louder, even at the risk of causing offense. The society which seeks to eliminate the very possibility of causing offense is already halfway down the road to tyranny.

I am not a scientist but a layman. A piano player, even! What more different world could there be, one might think, from the white-coated world of the lab scientist? Certainly, I am no expert on technology. I can barely change a fuse. *But you don't have to be a scientist to believe in human progress through science and technology.* That is the message of this book.

Another is this—Transhumanism is *not* the advocacy of some tyrannical eugenic breeding program by mad scientists in white coats or master-race fantasists, but the belief in the freedom of individual men and women to increase their own well-being and that of their children (the instinctive drive of every human being on earth) by means of the emerging technology available to them in the miraculous technowonderland of the modern world.

Dangers? Of course there are dangers—*life itself* is dangerous. But this is no time to be lily-livered. There is too much to gain. We should seize the opportunity with open arms. We have existed in a condition of biological servitude—slaves to our selfish genes—for too long. It is time to free ourselves from enforced subjugation.

In the words of the Transhumanist Manifesto, "We have nothing to lose but our biological chains."

This book consists of three parts:

- The Transhumanist Manifesto—a polemical advocacy of **Superbiology**—the emerging biotechnology by which to cure disease, enhance abilities, and, ultimately, defeat death in the twenty-first century.
- A critique of the **bio-Luddites** who oppose such radical interventions in human nature.
- An outline of transhumanism as a philosophical system, the *new world picture*—or **metameme**—which underlines the manifesto.

This is not a manifesto on behalf of any specific group, but for all those individuals who, out of common sense, compassion, and an instinctive distaste for nihilism and irrationality, share a largely unspoken and much-maligned belief in human progress through reason, science, and technology, in the hope that they may recognize they are not alone, and begin to speak out against the irrational **sciphobia** of those who seek to put a cap on human progress by regarding human limitations as a given, and technology something to be feared, rather than celebrated as a source of human liberation.

The term *transhumanism* is now gaining in public recognition, and describes a broadly recognizable set of attitudes, but what follows is my own philosophical system. I speak for no one but myself, and anyone who feels they might agree with me.

WHY DO WE NEED A NEW PHILOSOPHY?

Why do we need a new philosophy?

Philosophy matters.

We live in an age of science. Yet philosophers have failed to deliver a popular, benevolent, rational system of ethics for a scientific age. Instead, as science has swept away the metaphysical foundations of religious belief, it has left religious ethics without their power to unite.

In the twentieth century, philosophers failed to provide an alternative to religious ethics. Instead of stepping into the moral vacuum, they retreated into their ivory towers and abandoned the quest for meaning for the endless analysis of the insignificant. In the absence of philosophical guidance, it was left to scientists to tell us how to live. Not up to the job, they informed us with almost pathological glee that life is a cosmic fluke; evolution without direction or purpose; the self an illusion; human beings naked apes built by selfish genes. By an astonishing intellectual blunder, the silence of twentieth-century philosophers allowed scientists dabbling in philosophy to destroy the belief in universal purpose, human progress, and shared moral codes.

We need philosophy more than ever because the old foundations of conventional beliefs and values are crumbling in front of our eyes. The old truths—the absolute belief in religious or political salvation—are collapsing. What is left is a metaphysical and ethical vacuum currently filled by an antiphilosophy of cynicism-cum-nihilism commonly known as postmodernism, but which would more accurately be termed *anti*-modernity—for its proponents reject entirely the liberal belief in universal human progress through reason—declaring instead the absolute *relativity* of all beliefs and values. In a culture no longer dominated by religion, the result has been inevitable—a civilization devoid of shared values, increasingly divided into competing interest groups, each with its own agenda, its own way of looking at the world, its own "reality." "The "Postmodern Condition" is one of life without shared meaning, purpose, significance, or values.

A civilization without shared meaning or morality is a civilization in danger of implosion from within, or destruction from without. In such a circumstance, that rare thing today, a philosophy with a positive message for the civilization as a whole shines like a beacon of light and hope in a sea of chaos.

TRANSHUMANIST PHILOSOPHY IN A NUTSHELL

The transhumanist philosophy I espouse may be condensed into three basic ideas, corresponding to the three most important areas of any practical philosophical system—and the attempt to answer the three big "fundamental questions":

- What is the nature of the world? (metaphysics, or cosmology)
- What is the nature of a human being? (psychology)
- What is the best way to live? (ethics)

I answer thus:

Transhumanist metaphysics: The world is a process of evolutionary complexification toward evermore complex structures, forms, and operations.

Transhumanist psychology: As conscious aspects of evolutionary complexification in nature, human beings are imbued with the innate Will to Evolve—an instinctive drive to expand abilities in pursuit of ever-increasing survivability and well-being.

Transhumanist ethics: We should seek to *foster* our innate Will to Evolve, by continually striving to expand our abilities throughout life. By acting in harmony with the essential nature of the evolutionary process—complexification—we may discover a new sense of purpose, direction, and meaning to life, and come to feel ourselves *at home in the world* once more.

Such is the essence of my transhumanist philosophy in a nutshell.

TRANSHUMANISM AS NEUROMANTICISM

> I'm in love with the modern world.
> —Jonathan Richman and the Modern Lovers, "Roadrunner" (1976)

> "I want to know what passion is," she heard him saying. "I want to feel something strongly."
> "When the individual feels, the community reels," Lena pronounced.
> "Well why shouldn't it reel a bit?"
> —Aldous Huxley, *Brave New World* (1932)

> We need a revival of energy, enlightenment, enthusiasm into culture. That whole sneering David Letterman thing, this is so over.
> —Camille Paglia, *Philadelphia Enquirer* (May 1, 2005)

In this book I hope to make clear that transhumanism is not the preserve of white-coated mad scientists intent on eliminating emotions from the gene pool, but a philosophy for ordinary men and women who wish to enhance their minds and bodies without losing what is essential to their humanity; namely, the qualities of love, passion, and *joie de vivre*. For the transhumanist philosophy I espouse represents a passionate, romantic, hi-tech humanism in which logic is not divorced from emotion, and new technologies are joyously celebrated as the wonders of the modern world.

Where postmodern cynics say, "Modern life is rubbish," the transhumanist replies, "I'm *in love* with the modern world," rejoicing in the **e-phoria** of the **technosublime**—the feeling of awe and wonder at the miraculous **technowonderland** created by human beings through their skill, ingenuity, effort, creativity—*design!* And furthermore, a new generation is beginning to agree.

At the dawn of the twenty-first century we are witnessing the emergence of a newfound **technophilia**. The world of the Web, 3G, DVDs, PDAs, and MP3s has fashioned a new generation of **sciphiles** for whom the technoluddism and **sciphobia** of an older generation is as alien as the attitudes of the prewar generation to the children of the Sixties.

In 2004 the term *technosexual*[2] appeared in the ever-expanding **meme pool** of the modern world to describe a new breed of style-conscious, technophile aesthete as in love with cutting-edge design as with new technology. Socially liberal by nature, technosexuals are likely to live in the *ideopolis*,[3] or as we might call it, **netropolis**—culturally influential, often university-based cities with a high proportion of young brainworkers in IT or Internet-related professions.

In the psychology of the new netropolis, psychoanalysis is giving way to **neuroanalysis**. Words such as *serotonin, dopamine,* and *endorphins* are as familiar to a new generation as *ego*, *id*, and the *unconscious* to a former. Where Freud confessed ignorance as to what women want, the bio-realists of the netropolis have no such problems—they've read their *Mars and Venus* to find out.

As biopsychology replaces Freudianism, so **bio-fatalism** will increasingly be replaced by **techno-can-do-ism**—the belief in the power of new technology to free us from the limitations of our bodies and minds. The self-diminution society of postmodern irony will gradually give way to a **Self-Enhancement Society** based on a newfound belief in the power of the human mind—not merely to *talk* about problems—but to *overcome them.*

In the twenty-first century, the belief in the Fall of Man will be replaced by the belief in his inevitable transcendence—through Superbiology. The belief in human transcendence—whether spiritual or social—is the essence of romanticism. Thus, the ethos of the emerging Self-Enhancement Society might be called **technoromanticism**, or **neuromanticism**, the passionate belief in the transcendence of human limitations—not through religion or politics, but through *science*—product of the rational mind in the technowonderland of the modern world.

TRANSHUMANISM AS FUTURISM

> Dr. Andrew von Eshenbach, Director of the National Cancer Institute, stated that NCI's goal is "to eliminate the suffering and death due to cancer by 2015." That could be made possible by the exponential expansion of scientific knowledge and new enabling technologies such as genomics, proteomics, metabolomics, molecular imaging and nanotechnology.
> —*Science*, February 28, 2004

The New Agers are right about one thing. We do indeed stand at the Dawn of a New Age—the **DNAge**.

Where preindustrial society was defined by farming (the macro-manipulation of plants and animals) and industrial society by the manipulation of metals to make machines, the DNAge will be defined by the ability to manipulate the human body itself, through **Superbiology**—the combined techniques of an emerging biotechnology that will enable us to enhance our minds and bodies beyond the limitations of the human condition.

New designer drugs targeting specific chemicals in the brain will allow us to raise our general levels of mood and empathy, physical vitality and virility, powers of concentration and memory, and capacity for logic and learning. We might call them **eugoics**—a combination of *eu*: Greek for "well" or "good," and *ego*: Latin for "I" or "self." Eugoics are well-being enhancers—rational methods of self-improvement.

The ability to grow new cells, alter and add new genes, will allow us to prevent and cure disease (gene therapy) and enhance our bodies and minds beyond the dull level of functioning we call "normal" health (bioenhancement). We might call such techniques, collectively, **Supergenics**.

The construction of machines based on biological principles—**cybernetics** —will allow us to further enhance our bodies and minds, as worn-out organs are increasingly replaced by artificial equivalents of superior performance, and neural implants begin to blur the distinction between human and computer brains.

The development of molecular-scale bioengineering—or **nanomedicine**— will allow miniaturized, self-reproducing robot construction workers to act as "bio-spellcheckers," identifying and rectifying cellular "spelling mistakes" as they occur in the body.

Eventually, through a combination of superbiological techniques, the process of aging itself will be halted. Humanity's built-in program for decay will be rewritten. No longer a slave to the selfish genes which demand his self-destruction, the eventual elimination of death itself will be the final chapter in the story of *Homo sapiens*, and a new species will be born—*Homo cyberneticus* (from the Greek *kubernetes*: the steersman of a ship). Where *Homo sapiens* was the slave of his selfish genes, *Homo cyberneticus* will be the steersman of his own destiny.

Such a future is not without dangers. It will not occur without mistakes and false starts. Transhumanists say we should embrace the challenge ahead in a spirit of courage, benevolence, and adventure. Only the miracles of Superbiology will allow us to transcend the limitations of the human condition. We should seize the torch of Prometheus with both hands.

The **bio-Luddites** disagree. For them, the future should be the same as the past. We should accept the limitations of the human condition, and declare human nature "good enough."

By the end of this book, hopefully you will have decided for yourself who is right and who is wrong.

TRANSHUMANISM AS EVOLUTIONARY ETHICS

This book is *not* a review of current or prospective developments in Superbiology. Nor does it seek to catalogue evidence for the *feasibility* of such developments. Other books will serve that purpose well. I shall begin from the premise that the deliberate furtherance of human evolution through advanced biotechnology is not only possible, but *inevitable*, given both time, and—vitally—the continuing prosperity and freedom of the technologically advanced nations I shall refer to in these pages as the **modern world**. I shall begin from the premise that Designer Evolution is not a question of "if" but of "when" and set out to answer the question "why?"

This is a book not about what may be possible, but what it is right to attempt. It is, then, essentially a book of *ethics*, and an outline of the totalized philosophical system which underlines them, based on a simple proposal—that the modern world should embrace the emerging biotechnology as the *only* means of transcending the suffering at the heart of the human condition.

In short, this book is about something to believe in—in an age of collapsed belief. Transhumanism is a philosophy for all those who value technology, love life, and unashamedly want more of it.

TRANSHUMANISM AS RELIGION

Could transhumanism be regarded as a religion?

Transhumanists don't regard it as such, but it depends upon your definition of the word. A religion is either a belief system that asserts the existence of supernatural phenomena (such as God, heaven and hell, soul and spirit), or one that demonstrates a passionate, devotional adherence to its beliefs, values, and practices.

The first criterion, of course, is the most common and generally accepted. The second broadens the definition to include secular worldviews, which do not include supernatural belief, but require passionate adherence to beliefs or values which affect all aspects of life. So if your criterion is the belief in the supernatural, then transhumanism is not a religion. If, however, your criterion is passionate devotion to a belief system which affects all aspects of your life, then transhumanism could conceivably be regarded as a religion—but so, too, could Marxism, football, and fly fishing.

Whichever term one prefers—philosophy, religion, worldview, ideology, or zeitgeist—transhumanism will be a dominant player in the Meme Wars of the twenty-first century, and as such, a basic familiarity with its ideas is essential for anyone who wishes to understand the modern world—a world into which we have been thrown by blind, butterfingered Mother Nature.

TRANSHUMANISM AS METAMEME

> If a meme is to dominate the attention of a human brain, it must do so at the expense of "rival" memes.
> —Richard Dawkins, *The Selfish Gene* (1976)

This book represents the opening battles in a **Meme War**.

A meme* is an idea or habit regarded as a unit of cognitive information, as a gene is a unit of *biological* information. As genes spread through the gene pool, so memes spread through the **meme pool** of a human culture.

Some say that memes act of their own accord. I shall argue that *memes are power tools*. **Meme Wars** are battles of ideas played out within a culture to influence the direction it will take. Whoever controls the spread of attitudes and habits in a society, controls that society. *He who holds the power is he who controls the meme.*

A personal philosophy or belief system which guides behavior might be called a **meme map**. When spread through the meme pool like genes through a gene pool, affecting behavior in the culture as a whole, a meme map becomes a **metameme**. A metameme is a socially dominant meme map. Meme wars are battles between meme maps for control of the metameme.

In **Meme Wars**, **Defenders of the Meme** (propagandists) seek to impose their recessive metameme (potentially influential worldview) on **meme maps** (individual minds) and **memotypes** (attitudes) through **meme control** (mind/thought control) achieved through **weapons of memetic destruction** (WMD—propaganda, "spin") directed against the existing dominant metameme

*The term was introduced by Richard Dawkins in *The Selfish Gene* (1976).

(prevailing ideology) and **rogue memes** (socially unorthodox, potentially dangerous ideas) which threaten the survival of the metameme.

The main combatants in the Meme Wars of the modern world at the dawn of the twenty-first century are defenders of the metamemes we might call premodernity, modernity, and postmodernity.

Modernity may be defined as the condition of a culture based on the underlying belief in ongoing human progress toward ever-increasing knowledge, abilities, survivability, and well-being, attained through reason, science, and technology, as opposed to irrationality or superstition.

Originating in the European Renaissance, as the rediscovery of Greek philosophy loosened the chains of medieval superstition, the culture of modernity and the philosophy underlying it—humanism—has dominated Western civilization for five hundred years. All those who enjoy the civil liberties and material comforts of the modern world have the metameme of modernity to thank for their good fortune.

Pre- and postmodernity are essentially forms of *anti*modernity. Premodernists reject modernity for religious dogma, superstition, or the belief in simpler, preindustrial cultures. Postmodernists wish to replace what they see as the "hegemony"* of modernity with epistemological, ethical, and cultural relativism.

The philosophy of transhumanism represents a **New Modernity** emerging in the twenty-first century. Its Big Idea, or **supermeme**, is the use of advanced biotechnology—**Superbiology**—to eradicate disease, enhance minds and bodies, and ultimately, to defeat death itself, by halting the biogenetic process of aging.

The defenders of the pre- and postmodern metamemes seek to counter the rise of the New Modernity through control of the language operating in society. **Negamemes** such as *Playing God*, *Tinkering with Nature*, *Designer Babies*, *Frankenfoods*, *New Eugenics*, and *Brave New World* are some of the weapons of memetic destruction (WMD) with which the defenders of pre- and postmodernity seek to defeat modernity, by destroying the public's belief in technology. I shall respond by referring to the opponents of transhumanism as **bio-Luddites**, intent on stamping out human progress through an irrational fear of hubris (premodern **hubraphobia**), or a cynical belief in the wickedness of man (postmodern **malanthropy**).

This book could be seen as a defensive retaliation on behalf of modernity in the Meme War against the hostile forces of pre- and postmodernity. In Meme Wars there are no armies, no bombs, no bloodshed, no violence at all. Meme wars are wars without tears. Yet they are just as significant as territorial wars. For *he who holds the power is he who controls the meme.*

So let peace and freedom rule—and let the Meme Wars commence!

---

*Leadership or domination, depending on your spin.

NOTES

1. The President's Council on Bioethics, "Session 6: Seeking Morally Unproblematic Sources of Human Embryonic Stem Cells," December 3, 2004, http://www.bioethics.gov/transcripts/dec04/session6.html.

2. *Technosexual*: wordspy.com (1994).

3. *Ideopolis*: ibid.

Prologue

1. E-MAIL TO NATURE

(from an original idea by Dr. Max More: "A Letter to Mother Nature")[1]

> It is as if man had been suddenly appointed managing director of the biggest business of all, the business of evolution.
> —Julian Huxley, "Transhumanism" in *New Bottles for New Wine* (1957)

e-mail from: thetranshumanistsociety@hotmail.com
to: nature@evolution.com
re: Homo sapiens

Dear Nature:

Thank you so much for your wonderful free gift of Life. It has given us so much pleasure over the years. We would, however, like to suggest some improvements to the design of *Homo sapiens*, which we feel sure would increase its popularity with future generations.

The current model is limited by numerous design faults. Terminal breakdowns occur frequently in all parts: brain, heart, lungs, breast, liver, kidney, pancreas, stomach, colon, rectum, cervix, ovaries, uterus, bladder, prostate, penis, testicles, throat, mouth, tongue, blood, skin, and bones. We wondered if there might be a fault on the production line, or could it be that *Homo sapiens* has been deliberately designed with built-in obsolescence? If so, could not a program of automatic self-repair be included in an upgraded model?

General improvements in all aspects of functional design would be most beneficial:

DOING: Increased strength, stamina, vitality, and virility.
FEELING: Improved temperament and capacity for empathy.
THINKING: Expanded powers of memory, logic, learning, and creativity.
PERCEIVING: Enhanced sight and sound.
EATING: The need to refuel three times daily is most frustrating. Few motor car owners would tolerate such a hindrance. The absence of instructions as to the ideal fuel type is a source of endless confusion.
SLEEPING: Eight hours daily recharging time is a major disadvantage. Could the battery life not be improved?
DEFECATING: The requirement for the daily evacuation of unpleasantly odorous waste products from an orifice needlessly situated directly next to the sexual organs is an obvious design fault, as is the frequent involuntary emission of odorous wind. Could not a new, improved model metabolize and eliminate waste products in the form of small, uniform, odorless packets, for instance?
COPULATING: Failure to include the capacity for conscious control of sexual arousal is a serious omission. Could the conscious brain not be permitted to make its own decisions concerning sexual activity?

We do hope you may be able to incorporate some of these positive suggestions for upgrading *Homo sapiens*, as our own model is beginning to wear out.

Yours sincerely,
The Transhumanist Society

e-mail from: nature@evolution.com
to: thetranshumanistsociety@hotmail.com
re: Homo sapiens

Dear Transhumanist Society:

Thank you for your e-mail.

I regret to inform you that it is quite beyond our power to rectify any of the problems you mention, as our manufacturing equipment is entirely automated, and we do not possess the ability to rewrite the software for our production line. All we can offer is the hope that future mistakes in manufacturing may one day, by chance, result in some improvement on the original design. However, it must be acknowledged that any such fortuitous errors are extremely unlikely, and that as a result, faults in production will continue to occur on a regular basis.

It is true that we have not upgraded the model *Homo sapiens* for some considerable time. However, we would point out that many generations have enjoyed this product without complaint. If well maintained, the average model may last around seventy years. We therefore ask you to accept the design faults you mention, and suggest you take as good care of your own model as possible, as there will be no replacements available. Among our other range of products you will find several plants and flowers, which you may consider using to help ease the pain and discomfort experienced during your decline into decay, disease, and death.

Yours truly,
Nature

e-mail from: thetranshumanistsociety@hotmail.com
to: nature@evolution.com
re: Homo sapiens

Dear Nature:

I'm afraid we found your response most disappointing.

Your business is antiquated, your machinery out of date, yet you make no effort to improve your products, continually churning out faulty goods, and expecting the customer to accept them without complaint.

I suggest you join the twenty-first century. In the modern world, we believe in continually improving products and services to the benefit of the consumer.

I have to tell you that we are no longer prepared to accept the design faults in *Homo sapiens*. The deficiencies of disease, obsolescence, and functional limitations can all be rectified given the required will to evolve. If, as manufacturers, you are unable or unwilling to redesign the product, then we as consumers will be forced to do the job ourselves.

We, the human species, therefore formally advise you of our intention to take over the business of Evolution, in order to improve the design of *Homo sapiens*, in our own interests of ever-increasing survivability and well-being.

Yours sincerely,
The Transhumanist Society

2. THE DUMB DESIGNER

> People say, "Well, these would be designer babies," And I say, "Well, what's wrong with designer clothes?"
> —James Watson, *Scientific American* (May 2003)

You've been shopping at IKEA, and returned home with a flat pack of self-assembly bookshelves.

You open the packaging, and proceed to put the shelves together, taking care to follow the instruction manual (fixing the brace supports at either end, A to T, and G to C). After a few hours of hard labor, the job is done. You fill the shelves with books, stand back, proudly admire your handiwork—and look on as row after row of books slowly tipple off the shelves, like a pack of cards. When the last book (Darwin's *Origin of Species*) has insolently plopped to the floor, you notice that the shelves are not quite horizontal.

You refer to the instruction manual. You carefully followed the instructions. They must have got it wrong.

You examine the shelves. No doubt about it. They're uneven, off-center, asymmetrical, misaligned, lop-sided, unbalanced. And it's not your fault!

Well, calm down, no need to panic, just need to do a little adjustment—move the brackets on one side to balance them up . . . ah. . . . "*Note: Once connected to brackets, shelves cannot be realigned.*" There's nothing you can do—you're stuck with your wonky shelves.

You stand and stare at the pile of books on the floor. Anger begins to well up. You'll write to the manufacturer. Why can't these people use designers with a little common sense? It's as if they designed products without thinking that anyone might actually have to use them. What dumb designers! I could have done a better job myself.

The Dumb Designer is evolution. The shelves are the human body. The books are the human mind. The instruction manual is the human genome—The Book of Life. The instructions are the human genes. You and I are about to become the new designers.

The age of Designer Evolution is about to begin.

NOTE

1. Max More, "A Letter to Mother Nature," August 1999, http://www.maxmore.com/mother/htm.

PART 1
THE TRANSHUMANIST MANIFESTO

This is a supremely interesting time to be living in. Here we are, the first species that's ever effectively taken over its own evolution. We've taken over the function of Darwinian evolution and we're going to change big time. It's like human evolution is now *designer evolution*.
—William Gibson, author of the seminal SF "cyberpunk" novel *Neuromancer*, interviewed in macdirectory.com

I teach you the superman. Man is something to be overcome.
—Friedrich Nietzsche, *Thus Spoke Zarathustra* (1883–85)

For two million years man has been climbing a mountain of evolution, and his will is so weak that he dies when he is less than a century old.
—Colin Wilson, *The Black Room* (1971)

Do not go gentle into that good night,
Old age should burn and rave at close of day;
Rage, rage, against the dying of the light.
—Dylan Thomas, "Do Not Go Gentle Into that Good Night"

Picard: "After all, we're all human."
Riker: "Speak for yourself, sir. I plan to live forever!"
—*Star Trek: Generations*

The chief task of twenty-first-century philosophy is the unification of science and ethics.

1. TRANSHUMANISM

Man is not born free, but everywhere in biological chains. People of the world, unite. You have nothing to lose but your biological chains!

We stand at a turning point in human evolution. We have cracked the genetic code; translated the Book of Life. We will soon possess the ability to become designers of our own evolution.

There will be opposition from those who call for the abandonment of progress in subservience to nature. Let us not turn back now through fear.

As humanism freed us from the chains of superstition, let transhumanism free us from our biological chains.

2. HOMO CYBERNETICUS

Humanity—a species so weak it defines its own condition as tragic! How long must we endure such a pitiful condition?

"Human"—the very word is synonymous with suffering and failure. "I'm only human"; "the human predicament"; "the tragedy of the human condition": they all tell the same truth—that humanity is a disease state from which to be cured!

What is "the human condition" but an affliction? What is a human being? A weak mind in a decaying body, a bundle of primitive emotions, and a brief life in the knowledge of inevitable death—extinction for the rest of eternity!

Human beings are the slaves of a three-part genetic program reading, "Survive, reproduce, and self-destruct." We did not *ask* to be programmed for self-destruction. Why, then, should we accept death, disease, and decay like some helpless sheep?

Let us learn to think beyond the human condition. *Not what humanity is, but what it could be!*

A species evolved to control its own constitution cannot be considered the same as that which is slave to its genetic programming. *Homo sapiens* is about to undergo the next stage in its evolution. We are about to witness the birth of a new species—*Homo cyberneticus*—man the steersman (from the Greek, *kubernetes*: steersman, pilot of a ship).

Where *Homo sapiens* was the slave to its genetic programming, *Homo cyberneticus* will be the steersman of its own destiny.

3. NEUROTHEOLOGY

It is time for the human species to grow up. We no longer need superstition to explain the world.

Human beings are not "souls" or "spirits" but evolved, biological beings genetically programmed to survive, reproduce, and self-destruct.

That which religion calls the "soul" is the anthropomorphization of love—an evolved, neurochemically induced feeling of benevolent attraction to objects of benefit to survival and reproduction.

"Spirituality" is the faculty of love extended beyond its evolved function—a neurochemically induced feeling of benevolent attraction to all things.

"God" is the anthropomorphization of omnipotence—an absent father invented to reduce fear of death, increase tolerance of suffering in life, and condition for benevolence through the imaginary reward and punishment of heaven and hell.

Like many an absent father, God clearly does not wish to be traced by his offspring.

As our power over nature expands, our reliance on superstition contracts. Let us not be governed by primitive fantasies. A grown-up species should put away childish things, recognize the survival advantage of cooperation, and set about seeking an end to suffering and limitation—through science.

4. NEUROPSYCHOLOGY

The Mind as Virus Metaphor

Human beings are not invisible, immortal ghosts inhabiting mortal bodies, but embodied brains, genetically programmed to self-destruct.

The brain is an organic computer. The human mind is a brain that recognizes its own existence.

Consciousness is recognition; self-consciousness is self-recognition.

The mind is like a virus infesting the brain. Like all viruses, the mind seeks self-preservation. It invades the host organism, lives and grows in symbiosis with it, and, if it is strong enough, *it takes control.*

Once the mind has taken control of the host body, it naturally seeks to reverse the body's programming for self-destruction in an effort to survive. Once human beings have learned to define themselves by their minds, the next step is to defy the programmed decay of the body.

It is the emergence of self-consciousness—the mind's recognition of its own existence distinct from the body—that ignites humankind's battle to free itself from the limitations of the genetic prison. It is the sheer horror of the mind's recognition of its existence in a decaying body that triggers the will to override its genetic program for self-destruction.

Those who speak of the "Cartesian dichotomy of mind and body" are sorely mistaken. *Cartesian dualism is not the cause of our problems, but the beginning of the solution.*

Descartes's "cogito ergo sum"—I think therefore I am—signifies the brain's recognition of its own existence distinct from the body. The evolution of the human mind has allowed us to wake up to the horror of our slavery to a genetic program for self-destruction.

When Descartes defined existence in terms of his thinking mind, he did not usher in an age of collapsed human values, but the start of the brightest era in human history—the quest for freedom from the selfish genes which demand our self-destruction.

Descartes is not the villain but the hero of the piece. Cartesian duality marks the beginning of human evolution from *Homo sapiens* to *Homo cyberneticus*—man the steersman of his own destiny.

5. NURETHICS

A morality based on faith, not reason, cannot hope to endure in an age of ever-increasing scientific knowledge. Man is a rational animal; he must find rational arguments to guide his behavior.

Morality requires no deity. Benevolence is simply common sense.

"Goodwill to all men" is a rational tactic for mutual survival and well-being.

"Conscience" is not the voice of God in man, but an evolved, neurochemical, behavioral conditioning mechanism, deterring antisocial behavior through the automatic infliction of pain in response to malevolent thoughts or behavior.

"Fellow-feeling" is nature's pleasure reward for benevolence—an evolved, neurochemical inducement to cooperate in the struggle to survive (**Felt Morality**).

The evolution of the human brain allows for the extension of benevolence beyond the level attainable through feelings alone (**Thought Morality**):

- "Good" is **Sensible Self-Interest**—the ability of the rational mind to inhibit antisocial impulses in the interests of maximum survivability and well-being.
- "Bad" is **Stupid Selfishness**—the *in*ability of the mind to inhibit asocial impulses through ignorance of the self-defeating consequences.

The presence of antisocial "animal instincts" in our genetic programming does not justify malevolent behavior. For the evolved, rational mind provides us with the ability to *inhibit* "stupidly selfish" impulses in the interests of mutual survivability and well-being.

We must learn to distinguish between **Genethics**—instinctive behavior determined by the "selfish genes"—and **Nurethics**—behavior self-determined by the rational mind.

Morality is the replacement of Genethics with Nurethics—from control by the selfish genes, to self-rule by the human mind.

In the language of religious ethics—"free will" enables "fallen man" to fight the impulse to "sin." In the language of Nurethics, the self-governing mind may learn to inhibit *stupidly selfish* instincts in its own best interests of ever-increasing survivability and well-being.

Nurethics is a benevolent ethics based on reason—the only type worthy of an intelligent species.

6. NEUROTYPOLOGY

Human personality is not the result of an invisible, immortal tenant squatting in the body, but of neurobiology.

The neurobiological basis of personality is called temperament. It can now be identified. The ancient Greek theory of the four humors was closer to the truth than we imagined. For today we might speak of the **Four Neurohumors**—personality is influenced by subtle variations in habitual levels of neurochemicals, such as dopamine, serotonin, adrenaline, and opioids:

- The tendency to be extroverted or introverted depends upon our basal cortical arousal state—the level at which the brain is habitually "turned up," governed by excitatory neurotransmitters such as dopamine, glutamate, and acetylcholine.
- The tendency to be emotionally stable or unstable depends upon the reactivity of the autonomic nervous system (ANS)—the level at which the *body* is habitually "turned up" in response to stimuli—governed by inhibitory neurohumors such as serotonin and GABA.
- The tendency to be emotionally empathetic or detached depends upon habitual levels of neurohumors such as oxytocin and vasopressin—evolved to encourage maternal behavior through the feeling of benevolent attraction we call "love."
- The tendency to be assertive or passive depends upon habitual levels of adrenaline, and the male and female sex hormones. Where "*androgen agitates, estrogen tranquilizes.*"[1]

Nature, not nurture, lies at the heart of human personality.

7. EUGOICS

Temperamental deficiencies can lead to social disadvantage. An emotionally stable extrovert is better suited to cope with life than an unstable introvert crippled by anxiety and depression, or an unstable extrovert incapable of inhibiting aggressive impulses. Society affords increasingly few opportunities for the overly shy, anxious, depressed, or aggressive.

Human beings have responded to the problem of temperamental disadvantage by developing drugs which enhance the neurochemistry of behavior. By raising the level of specific neurohumors, personality-enhancing designer drugs have begun to improve the life experience of those not afforded the advantages of a strong temperament by nature. We might call them **eugoics**—means of producing a "good self"; from a combination of the Greek *eu*: "well" or "good," and the Latin *ego*: "I" or "self"). Eugoics are rational tools for self-enhancement.

The first eugoics, best known of which is the mood-brightener Prozac, have in some cases succeeded way beyond expectations, transforming anxious, depressive introverts into socially confident extroverts—simply by raising levels of the neurotransmitter serotonin. Anxiety and depression are symptomatic of a dimension of personality we might call the **Opioid Neurotype**, characterized by a hyperreactive nervous system, which floods the body with endogenous tranquilizers to dampen the pain of overarousal.

By reducing the brain's reactivity to stimuli, increased levels of serotonin may reduce the extreme oversensitivity of the neurotic personality. Through eugoics, the life-crippling anxiety and depression of the Opioid Neurotype may be alleviated. The implications are enormous.

Eugoics demonstrate the neurochemical basis of personality. The misery of anxiety, depression, and timidity are the product of a neurochemical deficiency which can increasingly be rectified by science. A pill may succeed where any amount of political intervention has failed to create emotional stability and self-confidence, the basic requirement for social success.

Science promises to succeed where religion and politics have failed—to make the weak strong.

8. SUPERGENICS

Eugoics are only the beginning of our ability to reduce the genetic injustice of blind, butterfingered nature. Prevention is better than cure. If a pill can enhance temperament by altering neurochemistry, how much more effective the change tackling the problem at its source, by altering the genetic instructions which determine habitual neurochemical levels from birth.

The genome is the recipe to make a human being. Once a recipe has been learned, the next step is to improve it. Most genes operate collaboratively to produce behavioral effects. Multiple genes acting together to build bodies and effect behavior might be called **supergenes**. The ability to add, alter, or replace supergenes in order to prevent or cure disease, enhance abilities, or extend life could be called **Supergenics**.

Through techniques such as stem cell therapy, gene therapy, and nanomedicine, Supergenics—the repair and enhancement of bodies and minds through the alteration of genes—will be available as a service among services in the miraculous modern world of the twenty-first century.

9. SUPERBIOLOGY

The first goal of Supergenics will be the gradual, systematic eradication of disease. Initial success will be found among the ten thousand or so diseases caused by mistakes in single chemical sequences, such as the horrors of muscular dystrophy and cystic fibrosis. In time, the polygenetic disorders will follow—life-threatening conditions such as epilepsy, diabetes, and hypertension will become things of the past.

Next to fall may be the psychological disorders—from schizophrenia and psychosis, to depression, anxiety, and related conditions such as alcoholism, anorexia, and bulimia—all potentially curable by the Superbiology of the future.

Eventually, the great scourges of humanity—cancer, heart disease, stroke, AIDS—will be consigned to history and the ability to correct genetic spelling mistakes in eggs and sperm will allow us to guarantee our children optimum health from birth.

As Superbiology develops, therapy will gradually give way to enhancement—the improvement of body and mind beyond the dull level of functioning we call "normal" health. Life-crippling emotional dispositions such as nervousness, gloominess, timidity, lethargy, or aggressiveness will be avoidable by choice. Physical traits such as strength, stamina, vitality, and virility will all be capable of enhancement; speed of thought, memory, logic, learning, and creativity will all be targets of design.

As our knowledge and skills develop, the potential for **bioenhancement** will be unlimited. Through the combined miracles of Superbiology we shall begin to enhance our bodies and minds to limits set only by our imagination.

The DNAge will begin.

10. BEST CLINICS

Gradually, we will see the emergence of bioenhancement clinics throughout the modern world. We might call them **BEST Centers**—offering a range of Bio-Enhancement Self-design Technologies enabling us to be the *best* we can be.

After initial skepticism—through the instinctive fear of the new—bioenhancement will come to be welcomed as the next step in self-improvement after education, exercise, diet, and aesthetic surgery. Lacking energy? Depressed? Nervous? Timid? Poor memory? No concentration? No need to sit and suffer in stoical silence. Book a visit to the BEST Clinic to change the genes that code for physical vitality, emotional confidence, and mental agility.

In the twenty-first century, self-design through Superbiology will come into its own, as we learn to optimize the neurochemistry of our minds and bodies for maximum well-being, vitality, and pleasure. Increasingly, the taboo against biotechnology will be broken. A new ethic of self-enhancement will emerge. A new slogan will fill the public consciousness:

BEST TECHNOLOGIES—Be the best you can be!

11. THE DNAge

As ever-increasing numbers are freed by Superbiology to enhance their genetic constitution, the human species will slowly begin to evolve.

Gradually, through the action of free individuals making free choices in the free world, a stronger, more diverse species will emerge—a species in control of its own genetic makeup. Humanity will take evolution out of the hands of butterfingered nature into its own transhuman hands.

Darwinian Evolution by random genetic mutations and natural selection will be succeeded by **Designer Evolution**—evolution self-directed by humanity in its *own* interests of ever-increasing survivability and well-being.

The emergence of Designer Evolution will herald the Dawn of a New Age—the **DNAge**. Each era of cultural evolution has been defined by the nature of its production processes. The Agricultural Age was defined by the ability to manipulate plants and animals through farming. The Industrial Age was defined by the ability to manipulate metals to make machines. The coming DNAge will be defined by the ability to manipulate human life itself—through Superbiology.

Humankind has evolved from the Stone Age, through the Iron Age and the Industrial Age. We are now entering the Dawn of a New Age—the DNAge in which bio meets techno—biology and technology combined to enhance the human condition.

Welcome to the DNAge.

12. EVOLUTIONARY ETHICS

The prospect of Designer Evolution will provide the children of the DNAge with a new sense of direction and purpose: the continual improvement in the quality of life through the eradication of disease and enhancement of abilities.

Humankind is an evolutionary species. Let us then adopt an **Evolutionary Ethics**: We cannot separate ourselves from the ongoing process of evolution—we *are* evolution.

Evolution is the unfolding of potential—a continual increase in the complexity of structures, forms, and processes in nature. Human beings are a conscious aspect of the process of evolutionary complexification. *Humankind is the mind of evolution become conscious of itself.* We are born with a drive to expand our abilities. Humanity is imbued with an instinctive Will to Evolve.

"Evolution" is complexification. "Will" is the ability of the mind to determine behavior—the Will to Evolve is the instinctive drive of a conscious entity to expand its abilities in pursuit of ever-increasing survivability and well-being. The Will to Evolve is the natural curiosity of a child, and the impulse that makes the scientist, artist, and philosopher each seek ever-greater knowledge of the world.

That which ceases to grow, begins to decay. Human beings were not meant to live in the moment like the lilies of the field—our purpose is to be all that we can be.

Let us discover the Will to Evolve beyond the limitations of the human condition.

13. THE PROMETHEUS DRIVE

The Will to Evolve is symbolized in Greek mythology by the figure of Prometheus.

Punished for giving man the gift of fire; chained to a rock; his liver devoured each day by vultures only to grow back again each night (in the first example of tissue regeneration in history—but not the last); Prometheus symbolizes the innate human drive to increase knowledge and abilities, even at the expense of present pains.

Today the **Prometheus Drive** finds expression in our own equivalent of the Greek myths. In the **technomythology** of *Star Trek*, a technologically advanced humanity "boldly goes where none have gone before" in search of ever-increasing knowledge and experience for the benefit of the species as a whole.

Star Trek Philosophy is the essence of humanism: the belief in the ongoing progress of the species through reason, science, and technology. Without the instinct to progress, humankind is doomed to remain forever at the mercy of disease, decay, and the limitations of the human body and mind. To abandon the

Prometheus Drive is to condemn the species to stagnation, suffering, and limitation.

Let us be the **New Prometheans**. Let us unite in our commitment to boldly go where none have gone before in search of the knowledge by which to transcend the limitations of the human condition.

Let us cast aside cowardice and seize the torch of Prometheus with both hands.

14. TAKING OVER THE FAMILY BUSINESS

Nature deals a cruel hand. Rely on nature and what do we get? Disease, disadvantage, and death, the products of her cold, callous lottery. So long as we refuse to see nature as she is—a managing director with no interest in improving the company products—we shall continue to suffer the torments of death, disease, and the limitations of the human body and mind.

Nature is an old-fashioned business, its workshop is antiquated, its methods outdated. Yet rather than asking for a refund, we bow down before a designer long since disappeared into liquidation—and give thanks for our faulty goods!

Our bodies were designed for the Stone Age—but we live in the Gene Age. If nature is creating products with built-in obsolescence, it is up to us to buy out the business and take over production. If nature refuses to invest in the new technology by which to improve her products, then we must take over the company and do the job ourselves.

Homo sapiens requires a complete upgrade—a radical overhaul in design.

15. SCIPHILIA

Only science can offer the opportunity of transcending the state of biological limitation we call the human condition. Only science may free us from our slavery to the selfish genes which demand our self-destruction. If science can guarantee good health and long life for ourselves and our children, why on earth should we leave our fate in the hands of butterfingered nature—harbinger of death, disease, and the limitations of mind and body?

Let us embrace the only source of the knowledge by which to free us from the bonds of suffering and limitation. Let science once again be celebrated as the servant of our ongoing quest to improve our lot in life. Let us create a world in which no human being must endure the afflictions of nature's cruel lottery.

There could be no nobler task for humanity than this.

16. BENEVOLENT EVOLUTIONISM

Morality is not threatened by our ability to intervene in our own biology. Human goodness is not dependent on our subservience to nature's whims. We do not have to be slaves to the genetic lottery to know that we must be kind to one another.

It is compassion that causes us to believe in science. All those who have held a dying loved one in their arms and watched helplessly as a disease-riddled body shrivels away into nothingness before their eyes know too well the absolute horror of death and disease.

Let us embrace the technology by which to free ourselves from subjection to nature's cruelty. Let science once again be celebrated as the servant of our ongoing quest to improve our lot in life. Let us teach our children an ethics which says: *We are an evolving species, united by our shared genetic heritage, and striving to improve our condition. Our common goal is the continual improvement in quality of life, as unique individuals dependent upon one another for our survival and prosperity.*

Let us strive together to be the *best* we can be.

17. BIO-LUDDISM

Transhumanists are lovers of life who recognize that the limitations of the human condition may be overcome through the technology of the future.

Our opponents, the bio-Luddites, wish to abandon human progress for a perpetual present. We should be satisfied with our biological slavery, they say.

If the bio-Luddites had their way, humankind would be stuck in an eternal time warp. A bio-Luddite world would mean a static species going nowhere fast—forever.

Bio-Luddism is the abdication of the moral imperative to improve our condition. Humankind is an evolutionary species; we must use our powers to progress, or else sink into oblivion with the dodo. Fear and sloth can never be the basis of human ethics.

Let us reject the cowardly instincts of the bio-Luddites and embrace the technology by which to correct the design faults imposed on us by nature.

Let us dare to dream of greater things. For what is humanity's greatest goal but the defeat of death?

18. THE DEFEAT OF DEATH

Death is a disease waiting to be cured. We have learned to identify the genetic recipe for life. There is no reason to suppose that we will not go on to identify the genetic program for death.

Research is well underway. Breakthroughs have been made. The mechanisms by which cells wear out, and the chemicals which extend their life, have been identified. The life span of simple organisms has been increased.

Let us heed the words of Dr. Aubrey de Grey, **immortologist** at Cambridge University: "If my plan works in the timeframe I have mentioned, no one presently under thirty in wealthy nations need have a life expectancy under about 1,000 years. In practice we will be a lot more risk-averse, so I expect life expectancy to be at least 5,000 years."[2]

When the cure for aging is found, it will not come through faith, prayer, or meditation, but through science—product of the miraculous technowonderland of the modern world.

Human beings are the product of a three-part genetic program reading, "Survive, reproduce, and self-destruct." No longer satisfied with the final instruction in its programming, *Homo sapiens* is about to rewrite its genetic software, replacing the command to self-destruct with the instruction to self-renew.

19. DARE

Transhumanists *dare* to speak the truth of the human condition.

It is *death* that is the great enemy of life. It is *death* that causes us to speak of "the tragedy of the human condition."

Death is an obscenity. The first goal of life is survival—death is not part of the plan, so why take it lying down? Better to "rage, rage, rage against the dying of the light!"

Stoicism is the absence of will in the face of defeat. Superstition will not save us from decay. Only science can defeat death. Our evolved purpose is survival. Science is the greatest tool at our disposal, and the greatest of all human beings is the scientist who devotes his life to the defeat of death.

Transhumanists *dare* to ask what men consider only in the solitude of their darkest nights. Why must we age and die? Why must our brains and bodies be so fragile, doomed to decay—programmed for self-destruction? Why should we accept the suffering and limitations imposed upon us by nature? And we answer without hesitation: there is *no* reason, there is *no* justification, there is *no* moral basis for the acceptance of death, disease, and biological limitation.

We have existed in a condition of biological servitude too long. If nature refuses to eradicate disease, death, and limitation, well, then, we must do the job ourselves. It is time to free ourselves from slavery to the force which demands our self-destruction.

Dangers? Of course there are dangers! *Life* is dangerous. But this is no time to be lily-livered. There is too much to gain. Through the miracles of technology we may free ourselves from the tragedy of the human condition.

There can be no greater reverence for life than the will to eliminate death.

Let no man be ashamed of his will to survive.

20. THANATOSIA

Even when faced by the prospect of immortality, some still look for reasons to die! These **thanatosians** oppose immortality in the interests of population control. We must die, they say, in order to make room for others. Talk about making a virtue of a necessity!

Overcrowding? An evolved, intelligent species will cease to reproduce beyond its ability to support itself. We shall build homes beneath the earth, under the seas, in outer space, on other planets. Enough room there?

The boredom of eternity? Those who know how to fill time do not seek to kill it. Those who love life do not wish it to end.

The naturalness of mortality? The acceptance of death as a natural part of life is a shallow defense against existential anxiety.

The cause of human fatalism is the recognition of the terrible limits on our existence imposed by nature. What great projects could we complete given unlimited time!

Nothing is quite so absurd as the attempt to justify death.

Immortality is a noble goal for an evolving species. Let us no more fear our evolutionary destiny—for our destiny lies among the stars.

21. STAR TREK

Our journey into space has only just begun. We live on a small planet. It is inconceivable that the human species will not use its technical ingenuity to expand its horizons in search of wider pastures, brighter stars.

Until now, the exploration of space has been obstructed by the brevity of life, the feebleness of the body, the vastness of the cosmos, and the limitations of technology. But through the miracles of Superbiology, the human life span will be increased, the body strengthened and adapted to suit hostile environments.

It is our destiny to expand into the galaxy and beyond. From the seashore whence we crawled, to the farthest reaches of outer space—we are destined for greater things.

Per ardua ad astra—through adversity to the stars.

22. THE NEW ENLIGHTENMENT

Imagine a future.

In the twenty-first century, the emergence of Superbiology will signal a new respect for the power of technology. As diseases are eliminated, and our abilities enhanced, we will begin to see a revolution in attitudes toward our own species, our place and purpose in the universe. Science will come to be seen as the liberator of humankind.

As the ever-expanding Internet extends the human nervous system around the globe, linking mind to **cybermind**, a new planetary consciousness will emerge.

Gradually, the politics of division, nation against nation, race against race, will begin to fade. We shall come to see ourselves as individual neurons in an evolving global brain—a single species united in the common desire to enhance our abilities.

The explosion of scientific knowledge will create **New Enlightenment**—a renewed faith in the ability of our species to comprehend the universe and progress beyond the limits of the human condition.

Through Superbiology we will enhance our bodies and minds. Through nanotechnology we will eradicate the scourge of poverty and hunger. The "tragedy of the human condition" will come to an end . . .

And then we can truly say a New Age has begun.

23. THE STRENGTH TO DREAM

And what then?

Liberated from biological slavery, an immortalized species, *Homo cyberneticus*, will set out for the stars. Conscious life will gradually spread throughout the galaxy, a neural net diversifying at every turn—a myriad of life-forms extending across space and time, until finally, in the unimaginably distant future, the whole universe has come alive, awakened to its own nature—a cosmic mind become conscious of itself as a living entity—omniscient, omnipotent, omnipresent . . .

A dream? What is a dream but the visualization of an ideal?

Is it better to have no ideals—no purpose, no direction? Is it better to wallow in the "tragedy of the human condition" like a spineless coward? Is it better *not* to have a dream?

Let us have the *strength* to dream.

24. LIVE TO EVOLVE! EVOLVE TO LIVE!

Let the theists pray for help from one who never comes. Let the ecoists skulk back into nature's womb. Let the postmodernists drown in the cheap solace of cynicism. Let them all wallow in the degradation of their biological slavery! We transhumanists shall continue striving in pursuit of the knowledge by which to free ourselves from the bonds of biological slavery. The human adventure is just beginning, and there are no limits to what we might achieve once we embrace the Will to Evolve beyond our human-all-too-human condition.

The only way forward is onward and upward. The only limits are those we choose to impose upon ourselves. Only the Will to Evolve can save us from self-destruction.

Let us not fail to summon it.

NOTES

1. Camille Paglia, *Sex, Art and American Culture* (London: Penguin, 1992), p. 108.
2. Aubrey de Grey, http://www.betterhumans.com/Features/Interviews (accessed August 25, 2003).

PART 2
THE BIO-LUDDITES

THEISTS:	AUTHORITY
ECOISTS:	SANCTITY
HUMANISTS:	INVIOLACY
LIBERALS:	MALEFICENCE
THERAPISTS:	MALLEABILITY
DOOM MONGERS:	CATASTROPHE

Bio-Luddites make unusual bedfellows—the Religious Right and Liberal Left united in their sciphobia with the therapists who deny the biological influence on behavior, and the doom mongers disposed to see the glass half empty.

Chapter 1
THEISTS
The Authority of God

> And the other thing, because no one has the guts to say it. If we could make a better human being by knowing how to add genes, why shouldn't we do it? What's wrong with it? Who is telling us not to do it?
> —James Watson, *Engineering the Human Germline Symposium*
> (UCLA, March 20, 1998)

PLAYING GOD

Theists say, "Man should not play God."

But the quest to cure disease, enhance abilities, and extend life cannot seriously be called *playing*—more like *replacing* a God who is clearly either absent without leave or completely uninterested in reducing human suffering. Let us have no irrational fears about "playing God," for what kind of creator would wish his creation to remain in suffering and torment, plagued by the agonies of death and disease?

In what way can human beings be considered so perfectly designed that our constitution cannot be improved upon? Was cancer part of this benevolent creator's plans, and must we therefore leave it incurable to protect the inviolacy of a species created in his image? What nonsense!

Theists say, "Seeking to alter human nature is against the will of God."

But we challenge "the will of God" every time we put on our spectacles, put in our contact lenses, or book in for laser treatment to *cure* our faulty vision. Or is this to be considered a crime against God-given myopia?

We challenge "the will of God" every time we place a premature baby in an incubator, or operate to save a child born with a life-threatening illness (as He intended? How kind). Should we rather respect His authority—and leave them to die?

We improve upon God's handiwork every time we mow the lawn; cut our toenails; brush our teeth; take an aspirin; build a dam, a house, a hospital. A species that truly wished to leave nature as God intended would not be human at all, but ape. A return to Eden would mean the regression of man to beast.

HUBRIS

Theists say, "Beware hubris."

Hubris was a quaint notion that made for some good plays. Let us have no fear of seeking to improve our own condition. If physical and mental self-improvement through exercise and education are deemed virtuous, why not *genetic* self-improvement, through Superbiology?

What kind of compassionate God would wish us to suffer from limitations we are able to overcome by our own hands? Perhaps a God responsible for disease and disability, earthquakes, hurricanes, tsunamis, and droughts, in which animals devour each other to stay alive, and even the gentlest of pussycats will torture a mouse given half a chance. One must ask oneself: in what way should such a benevolent creator be deemed worthy of authority over our lives?

From Prometheus to Frankenstein, the myth of punishment for challenge to the Gods derives always from the same cause: the stoical acceptance of human limitations deemed impossible to overcome—and the cowardly fear of the unknown.

Theists say, "It is arrogant to presume the authority to alter our God-given nature."

Arrogance? But the belief in curing disease and enhancing abilities shows a far greater humility than the belief that the omnipotent creator of the universe is interested in personally rewarding us with immortality for our good deeds—or in having a little chat.

Arrogance? But it is *theists* who are arrogant in their refusal to humble themselves by admitting that human beings are not so perfect that we cannot improve upon their design. Recognition of our biological limitations is only a problem for the theist who considers it an insult to God to admit that he could have designed us better.

It is time to learn how to stand on our own two feet. Let us reject irrational **hubraphobia** and seek to improve our minds and bodies in any way we can.

PERFECTION

Theists say, "In an age of genetic enhancement, only perfect people will be valued."

Perfect people? Who knows what perfect means? Transhumanists believe not in "the perfectibility of man," but in "ever-increasing survivability and well-being," and Superbiology as the most powerful means of achieving it. But for those who believe that the concept of perfection is intrinsically immoral, let us quote the words of Jesus: "Be ye therefore perfect, even as thy father in heaven is perfect" (Matthew 5:48). Or perhaps God: "Be thou perfect" (Genesis 17:1). Or, if you prefer, Confucius: "The perfecting of one's self is the fundamental base of all progress and all moral development."

Ah, you say, but they speak of the perfection of virtue, not of self-aggrandizement. But Confucius says: "The man of perfect virtue, wishing to be enlarged himself, seeks also to enlarge others." The superior man (chun-tzu) strives to be the best he can be, by developing mind and body to the full: "The progress of the superior man is upwards; the progress of the mean man is downward" (Analects 14.24).

There can be nothing bad about wanting to be the best you can be. As bioethicist Arthur Caplan has noted: "Every religion on the planet sees the improvement of oneself and one's children as a moral obligation."[1]

Let us strive together to be the best we can be.

PRESSURE TO CONFORM

Theists say, "There will be pressure to conform to superhuman standards."

But pressure to conform has *always* existed. The desire to be like others is a universal human trait, evolved for the survival value of social cohesion, because the like-minded bond better by virtue of their greater mutual understanding. In fact, there is far *less* pressure to conform today in the modern world—the *free* world—than ever before in human history. In preindustrial society, deviation from social convention was all but impossible. The small-town mentality prevailed everywhere. But in the modern world, the pluralism of urban life dramatically increases the freedom to deviate from social norms. Diversity is the very definition of the modern world. To complain about pressure to conform today is untenable. In a small town, you must be whom your neighbors expect, but in the big city, you can be who you want to be.

A new medical drug has recently been launched to increase female sexual arousal. Bio-Luddites argue that the product places unacceptable pressure on women to conform to male ideals of female sexuality. But we live in the free world! If you don't want to take a particular drug, don't take it! No one will force you. To prevent another from doing as they wish with their own body is morally indefensible. Are we really ready to ban new drugs in case fashion victims feel pressured by society into taking them? What nonsense!

We live in a miraculous consumer society brimming with products and services we are free to use or ignore. Instead of seeking to ban those we do not personally like, out of fear of being pressured by society into using them, why not take responsibility for our own lives, refuse to follow the crowd, and leave others to make their own choices for themselves?

Let us choose freedom and personal responsibility.

DEHUMANIZATION

Theists say, "We are heading for a dehumanized world without compassion."

Compassion? We are talking about eradicating the misery of disease, aging, death, poverty, and starvation. What could be more compassionate than that?

Compassion? Scientists are developing the "protato," a protein-enriched potato, to eradicate polio in the malnourished children of India.[2]

Compassion? In the future, genetically enhanced crops will allow us to feed the world, by massively increasing yields, even in harsh climates,

Compassion? Eventually, through nanotechnology, vast armies of miniature robot workers manufacturing goods at the molecular level will bring productivity levels through the roof, bringing an end to scarcity and want, poverty, and hunger.

Is that compassionate enough?

Theists may join hands and pray for starving children, but it is *scientists* who will feed them.

SUFFERING

Theists say, "Human suffering is necessary to maintain morality."

Talk about making a virtue out of a necessity! There is nothing good about suffering. Humankind was not born to suffer. Our purpose is to reduce suffering through effort.

Mother Nature is cruel—she inflicts dreadful suffering upon the innocent. Even Jesus felt justified in overriding her capricious hand. Jesus healed the lame and cured the sick. He did not say, "Your suffering is necessary to maintain human compassion." Should we not seek to follow his example?

Theists claim that suffering is an intrinsic part of the human condition. But transhumanists do not *believe* in the human condition—rather in its *transcendence,* through technology.

Eventually, the miracles of Superbiology will allow us to eradicate disease from the face of the planet. What then for the theistic argument that suffering is a virtue? What kind of fool would prefer suffering to an operation that prevents it?

Just as the invention of heaven eased the painful knowledge of impending death, so the invention of suffering as a virtue makes a virtue out of a necessity. So long as death is held to be unavoidable, human beings will continue to seek religions which make suffering a virtue, and immortality its imaginary reward. But as Superbiology develops, fewer and fewer numbers will subscribe to a philosophy which seeks to justify a suffering avoidable through technology.

Once science declares that suffering is no longer required, it will cease to be regarded as a virtue. Once science declares death to be no longer necessary, the notion of an afterlife will wither away, and with it, the delusion of a deity. For what is the belief in an afterlife but the longing to escape from the suffering of *this* life?

While religions seek release from suffering in heaven, transhumanists seek transcendence of suffering on earth through the *true* source of modern miracles—technology.

NOTES

1. Arthur L. Caplan, "Is Better Best?" *Scientific American* (September 2003), p. 104.
2. Protato: *New Scientist* (January 2, 2003).

Chapter 2
ECOISTS
The Sanctity of Nature

> It is nature, not society, that is our greatest oppressor.
> —Camille Paglia, *Sex, Art, and American Culture* (1992)

> The world cannot feed all its 6.3 billion people from organic farms.
> —Norman Borlaug, foreword to *The Frankenfood Myth: How Protest and Politics Threaten the Biotech Revolution* (2004)

By **Eco-fundamentalism**—or **Ecoism** for short—I mean extreme ideology of the Green or environmentalist kind. Ecoism is the belief in the primacy of nature over humans. It should be distinguished from the perfectly sensible attitude we might call **Ecologism**—the recognition of the interdependence of humankind and nature, and the consequent need to protect the ecological balance upon which our survival depends. By contrast, Ecoism is cryptopantheism—the belief that the needs of Gaia, the earth goddess, are more important than the instinctive desire of human beings to prevent and cure disease, enhance their abilities, and extend their lives. Ecoism crosses the political boundaries. The Left associate a love of all things Green with the rejection of Western techno-imperialism. The Right view the attempt to alter nature as the latest folly of liberal progressivists in their naïve desire to straighten the crooked timber of humanity. The transhumanist replies, "Love nature, admire its beauty and magnificence, but do not be so stupid as to *worship* a force that ultimately has it in for you."

THE SANCTITY OF NATURE

Ecoists say, "Superbiology threatens the sanctity of nature."

Sanctity means the quality of a saint. One suspects the word derives from *Sanskrit* (meaning "well written"), the language of the ancient Indian religious texts, and later mutated into *sang*, the French for blood. So *sanctity* infers "written in blood," "set in stone," "laid down by law," "affixed," "not to be altered"—by religious decree.

But exactly which religion has the authority to impose such orders on us all? For they all have completely different beliefs! Some think women should cover their faces, others that they should be priests. Some think God loves us all, others that some are his personal favorites. Some believe in peace on earth, others in the holiness of war.

The "sanctity of nature" argument is only applicable to theists who seek to impose their religious beliefs on the world. Eco-fundamentalists are neo-pantheists in disguise.

TINKERING WITH NATURE

Ecoists say, "Tinkering with nature is dangerous."

There is nothing tinkering, tampering, or meddling about curing disease, enhancing capabilities, and extending life.

All human endeavor is dangerous. Without the bravery to strike out into the unknown, we would still be huddling in caves. Some people apparently wish us to return there.

Why should we not continue to improve our condition by altering nature in our own interests, as we always have done? Should we rather have remained nature's slaves—without the benefits of agriculture, electricity, medicine, or technology—until the end of time?

To abandon progress through fear is simple cowardice—and cowardice never accomplished anything.

ALTERING NATURE

Ecoists say, "It is wrong to seek to alter nature for human ends."

But for whose ends are we *supposed* to act if not our own? God's? Ecoist opposition to Superbiology sounds suspiciously like pantheism in disguise—the worship of Mother Nature as a deity: Gaia, the earth goddess. But what kind of goddess is this?

How easy it is for some to blindly ignore the ubiquitous savagery of a nature "red in tooth and claw." Do we really need to offer examples? The female praying mantis who *eats* her mate while copulating with him—what kind of Intelligent Designer dreamed that one up?

Why should we not seek to alter nature, the "gentle earth goddess" who has us programmed to self-destruct? Let us take responsibility for the world in our *own* hands—not those of a dreadful designer, architect of a world in which animals survive only by tearing one another to pieces.

Ecoists say, "You cannot control nature."

But is it truly better to be *out* of control—power*less* against a hostile force that subjects us against our will to decay, disease, and death?

Only a species of idiots would leave that which it can control to the cruel, random lottery of nature.

NATURE KNOWS BEST: CONSERVATIVE ECOISM

Conservative Ecoists say, "Nature knows best—respect her wisdom."

Do bio-Luddites think nature knows best when she destroys the lives of millions through disease, earthquake, famine, or flood? Do bio-Luddites think nature knows best when they're undergoing a hip replacement, heart surgery, chemotherapy, or immunizing their children against deadly viruses? Do bio-Luddites think nature knows best when they're putting in their contact lenses, taking off their makeup, or watching disease-riddled loved ones breathe their last, agonized living breath?

Does the "wisdom of nature" include the temperamental deficiencies she cares to impose on some, and not others, in her savage world of winners and losers? Nature is no egalitarian; she gives strength to some and weakness to others without a care in the world. Superbiology offers the future prospect of *improving* the lot of the weakest members of society by allowing them to overcome the limitations of bodies and minds that contribute to life-failure.

Could it be that conservative hostility to Superbiology is not always the biophilia it pretends to be—the love of all things natural?

Could it be that conservative ecoists do not wish to see the benefits of their *own* hereditary constitution (the product of private, selective breeding) challenged through biotechnology? For nature has served *them* well.

Could it be that conservative bio-Luddism is actually **bio-elitism** in disguise: we don't want *you* to have what *we* have?

BACK TO NATURE: ECO-SOCIALISM

Eco-socialists say, "Instead of trying to make people stronger through biotechnology, we should change *society* to make life less competitive—more in harmony with Gaia."

Gaia? Is that the gentle earth goddess of a world in which cats torture mice? Getting in tune with nature would mean imitating the actions of a serial killer.

The democratic way to alleviate the temperamental disadvantages of those who can't cope with life in the modern world is not to make society softer to suit the minority, but to make the weak *stronger*—through Superbiology.

The canonization of weakness simply perpetuates weakness. Better to make the weak strong, than the strong weak.

Life is certainly hard, but the solution is not to abandon civilization and run into the woods to hug trees, but to embrace the science and technology by which to make life easier.

Let us abandon this ridiculous romanticization of nature. Nature is beautiful, but it is not "good." Let us love nature for its magnificence—but not *worship* it. Would you worship one who is poisoning you to death? Yet nature is doing just that—killing you slowly! It has you programmed to self-destruct!

Eco-fundamentalists worship nature, the force that kills them.

Chapter 3
HUMANISTS
The Inviolacy of Man

Remaking ourselves is the ultimate expression and realization of our humanity.
—Dr. Gregory Stock, *Redesigning Humans* (2002)

By humanist arguments I mean those that focus on the "inviolacy" of human beings in their present biological state, any attempt to alter the human constitution is regarded as "inhumane," "dehumanizing," or an affront to "human dignity."

Here, then, I distinguish between humanism and transhumanism. Essentially, transhumanism is an extension of humanism into the biological realm. Where humanism is the belief in the progress of humanity in its present biological state, *trans*humanism is the belief in the ability of the species to *transcend* the limitations of its biological condition through technology. Humanists must decide for themselves whether they consider themselves to be *trans*humanists or advocate the "inviolacy of man" as he is—slave of the selfish genes that demand his self-destruction.

INVIOLACY

Humanist bio-Luddites say, "Superbiology threatens the inviolacy of man."

Take away the religious inference of the word "sanctity" and you get "inviolacy"—not to be violated, subject to violence, harmed. But in what way can the goal of eliminating disease, enhancing abilities, and increasing life span be considered *harmful*?

How is the human condition so perfect that it should not be improved upon? Is it morally good to grow old, weak, and decrepit, to lose one's sight and hearing, sex drive, mobility, and memory? Is it morally good to waste away—to succumb to the ravages of disease and decay, like a powerless sheep?

Simply describing human beings as inviolate is not a logical argument for denying them the opportunity to reduce the suffering imposed upon them by nature. If human beings are indeed inviolate, then logically we must abandon all medicine and look on while our loved ones die in agony from their "natural" diseases.

Let us abandon the meaningless nonsense of the "inviolacy of man" and seek to improve our minds and bodies in any way that we can.

DIGNITY

Humanist bio-Luddites say, "Altering our biological makeup is an affront to human dignity."

Dignity means worth or value. In what way does the attempt to rid ourselves of disease and increase our abilities devalue human life?

Where is the threat to "human dignity" in biological self-enhancement? Why should we not seek to improve ourselves biologically as we do through exercise and education? Transhumanists love life so much we want more of it—both in quality and quantity.

The recognition that the human condition is not so perfect that it cannot be improved upon no more devalues life than the recognition that computer programs can be upgraded devalues computers. On the contrary, the desire to enhance one's condition signifies a profound *love* of life. For you don't buy an upgrade if you don't like the original program.

INHUMANITY

Humanist bio-Luddites say, "Tampering with our biological makeup is inhumane."

Inhumane implies that "humane" (meaning compassionate) is synonymous with "human." But is it really humane to be human? Look around you. Soldiers mutilated and strung up for public viewing. Airliners hijacked and flown into buildings. Hostages decapitated live on video. Dissidents tortured and killed by the state. Cannibals advertising for victims on the Internet. Thousands slaughtered in tribal genocide. Spree killers murdering passersby at random.

Need I go on?

Rather, to be *trans*human is to be humane, for transhumanists wish no harm on anyone—only to defeat disease and enhance abilities.

IRRESPONSIBILITY

Humanist bio-Luddites say, "We have no right to pass on genetic changes to future generations."

But we *already* pass on genetic changes—by selecting one mate over another, just as we pass on *cultural* changes every time we alter society—by introducing antibiotics, or democracy.

How can it be wrong to use the medical technology at our disposal to ensure that our descendants are mentally and physically as strong, healthy, and happy as possible? Rather, such a task must be considered a moral *imperative* for any *compassionate* species.

Refusing to provide your descendants with the best possible genes is like leaving your children homeless by giving everything you own to charity just before you die—in the name of equality.

COMMODIFICATION

Humanist bio-Luddites say, "In a world of designer babies, children will become commodities to be bought and sold, chemicals to be manipulated."

But parents have been attempting to "design" their children throughout history, by picking the best possible qualities in a mate!

We do not speak of dehumanization when a woman looks for a strong, healthy father for her children. Why, then, should we worry if she is able to directly guarantee her children's health from birth through Superbiology?

Do bio-Luddites truly believe that the search for a mate involves no thought whatsoever as to the physical and mental qualities a partner may pass on to one's children? *All* planned babies are designer babies, for their mother and father actively chose each other's genes over other possibilities, whether consciously or not.

Deliberately enhancing one's children's genes is no different from deliberately choosing a husband or wife. Or should we enforce fertilization by randomly chosen sperm in the name of bioegalitarianism, as has actually been proposed?

Self-interest has always motivated propagation. That is the way human beings are programmed: to survive and reproduce. There can be nothing immoral in self-interest unless self-preservation is deemed a crime.

Parents will not begin to view their children as "commodities" just because they're better able to determine their characteristics than before! Parental love is instinctive; it evolved for its benefit to survival by ensuring that children are cared for until able to look after themselves. Parents will no more stop loving their children by choosing the color of their eyes than by choosing the color of their clothes.

Morality does not depend upon our helplessness to determine our children's health or characteristics. On the contrary, we will learn to cherish life *more* by having a greater creative hand in its design.

Chapter 4
LIBERALS
The Equality of Man

> Unless we have a totalitarian world order, someone will design improved humans somewhere.
>
> —Stephen Hawking, speech at the White House (March 3, 1998)

By "liberal" I mean here the belief that the first role of government is the redistribution of wealth from rich to poor, as opposed to the maintenance of stable economic and social conditions in which individuals can flourish according to their differing abilities and efforts. While all sensible, compassionate people recognize the requirement for state assistance for those in need, the extreme argument of some liberals that bioenhancement should be outlawed as an elitist threat to equality is as absurd as the proposal to ban Rolls Royces or tea at the Ritz.

INEQUALITY

Liberal bio-Luddites say, "Superbiology will increase inequality between the genetically enhanced rich, and the unenhanced poor."

As bioenhancement beyond the level of normal health will be the financial responsibility of individuals, not the state, so the wealthy will have greater initial access—just as today, some people more than others can afford cosmetic surgery, a personal trainer, manicures, pedicures, facials, or colonic irrigation. But just as gyms are open to anyone paying an entrance fee, not just champion bodybuilders, so, too, BEST centers for bioenhancement will be open to all those who choose to spend their disposable income on improving their bodies and minds, instead of a new kitchen, car, or holiday.

Bio-Luddites love to talk about the social divisiveness, prejudice, and discrimination they think will be the result of consumer-controlled bioenhancement. But they always seem strangely blind to the fact that economic growth in a market economy increases the prosperity of the poor, as well as the rich. As the techno-industrialized nations become ever more wealthy, so the poorest members of society will enjoy ever-increasing accessibility to the benefits of biodesign. The result will be a gradual *increase* in social egalitarianism, and the leveling *up* of society, as ever more people are able to enhance their condition, and ever-increasing numbers become stronger, healthier—and more powerful.

Yes, initially some people more than others will be able to afford bioenhancement beyond the level of "normal" good health. It's called living in the free world. The basic alternatives are a complete ban, out of spite for the better-off, or unsafe, bargain basement Supergenics for all in a poverty-stricken, Eastern bloc–style authoritarian state with astronomical tax rates. Which would you prefer?

Only the most ideological of extremists could seriously wish to deprive everyone of the benefits of bioenhancement on the grounds that the rich will be the first to benefit. In the words of James Watson: "The well-to-do probably had the first television sets, but that's not a reason for not building a television set."[1]

Providing one's children with a head start through Superbiology is no different than giving them help with homework, a personal computer, or a private education. Of course, some people think that's wrong, too. But that's an argument in favor of state socialism, not against bioenhancement.

Leftist opposition to Superbiology sounds suspiciously like an argument for communism in disguise—the enforced leveling down of society in the name of egalitarianism: we don't want anyone to have what everyone can't have right now.

The state can only pay for anything you want in adolescent fantasies and totalitarian regimes.

DISABILITY

Liberal bio-Luddites say, "By curing disability we would lose geniuses like Stephen Hawking to the world."

The will to overcome the hurdles of a disability certainly produces extraordinary human achievement. But how many parents would seriously *prefer* their son to suffer a life-crippling disability that could be rectified prenatally, on the billion-to-one shot it might make him a genius?

Some say, "In a world of designer babies, pressure will be placed on women to terminate pregnancies on the grounds of minor genetic defects."

Let us respect the compassionate, pro-life attitude behind this concern. Parents facing the prospect of a disabled child should be made aware of its ability to live a happy, fulfilled, independent life.

There is no great gulf between the able-bodied and disabled, for *all* human beings are limited in their biological potential by nature, an attitude I call **bio-empathy**. But equally, most parents of children born with disabilities, while dearly glad their children are alive, would equally love to see their child's disability curable for others in the future.

There is absolutely no moral contradiction between respecting and valuing all members of society whatever their abilities, and seeking to cure disease, disability, and biological disadvantage.

If it is fine to wish for a strong, healthy child, then it cannot be bad to grant that wish through technology.

Let us care for those with disabilities, love and respect them like anyone else, and seek to prevent disability for future generations. Surely this is the only commonsense and compassionate view.

HOMOSEXUALITY

Liberal bio-Luddites say, "Superbiology threatens tolerance and diversity. There will be pressure to abort or alter embryos carrying genes which predispose to homosexuality."

Let us denounce attempts by anyone to dictate the choice of parents with regards to their own children. Let parents decide whom they wish to bring into the world, not doctors—or politicians.

There are probably four possible factors predisposing to homosexuality:

- GENETIC: the development of a biologically female brain in a male body
- BIOCHEMICAL: maternal hormonal imbalance affecting the developing fetus
- EXPERIENTIAL: behavioral conditioning in childhood
- EXISTENTIAL: a conscious decision to experiment beyond the limits of genetic programming

Whatever the causative factors, the nature of an individual's private sexual behavior, as of any other type of private behavior, should be of no business to anyone else, provided no harm is caused to anyone else in the process. Let us respect the freedom of all citizens to live as they wish, free from harm, so long as they refrain from harming others.

Instead of abandoning technology through fear of man's inhumanity to man, let us teach our children the importance of celebrating freedom, diversity, and mutual self-respect.

Let us adopt an ethics based on reason and compassion, which says, *We are all unique individuals united by our common heritage. Diversity and cooperation have evolved because they increase our ability to survive. It is in the best interests of all of us to learn how to get on with people different from ourselves.*

EUGENICS

Liberal bio-Luddites say, "We are heading for a brave new world of eugenics."

Bioenhancement through Superbiology is not eugenics. This 120-year-old term describes the systematic attempt *by the state* to improve the constitution of a population through enforced sterilization and selective breeding. In short, eugenics means state control of reproduction. Let us reject it entirely and unequivocally, and fiercely oppose any attempt by a government to control the bodies or minds of the citizens it is elected to serve.

The future lies not in state-run eugenic programs, but in *voluntary consumer* access to Superbiology, enabling individuals to enhance their bodies and minds as they see fit, in their *own* interests—not those of the state! In the Self-Enhancement Society of the future, Bioenhancement will be available to consumers in the same way that they may choose to spend their disposable income on a face-lift, tummy tuck, or boob job today.

Aldous Huxley's *Brave New World* was a state-controlled dictatorship in which free will and individuality were banned. The emerging **Self-Enhancement Society** will be a plurality of autonomous individuals freely engaged in **Reciprocal eugoics**—the mutual encouragement of ever-increasing well-being.

DICTATOR

> Should Hitler harm us for the next 200 years by saying that we cannot do genetics?
>
> —James Watson, *Independent* (February 3, 2003)

Liberal bio-Luddites say, "Superbiology will be misused by a mad dictator."

In the modern world, we do not live under totalitarian regimes, but in democracies, in which individuals are free to do as they please so long as their actions do not harm others. Superbiology will and must be controlled by individual consumers, not the state. We should protect ourselves from totalitarianism by voting out of office any government which shows the first signs of a drift toward authoritarianism—but we cannot outlaw scientific knowledge through fear of ourselves!

Much hostility to Superbiology stems, understandably, from memories of the horrific eugenic policies of the twentieth century. Sixty years on, we still live in the dark shadow of Hitler. But in light of the imminent emergence of Superbiology, surely it is now the time to ask, is it really in the interests of the human species to let "what Hitler did" sixty years ago force us to outlaw the technology that may enable us to avoid the horrors of disease, aging—even death itself? To abandon the quest to cure disease and enhance the human condition because of "what Hitler did" in another age would be to punish ourselves for the sins of our forebears. How long must we forbid ourselves the ability to eradicate disease, enhance our abilities, and extend our life span? A hundred years? A thousand? A million?

Banning biotechnology because of Hitler is like banning barbers because of Sweeney Todd. We cannot go on basing our ethics on fear of a madman long since in the grave.

Libertarianism and individualism are the safeguards against malevolent ideologies. Let us keep Superbiology away from state control and in the hand of individuals—history reminds us what horrors the *state* is capable of inflicting in the name of social justice. It must be for parents—not politicians—to decide whom they bring into the world, and what they do with their own bodies. But in the name of common sense and compassion, let us not allow a psychopathic tyrant from the pages of history to forbid us the means by which to defeat disease, enhance our abilities, and seek the best for our children—the benevolent instinct of every parent on earth.

The fear that biotechnology may be misused by a mad dictator is really the fear of electing a mad dictator. Let us not banish knowledge, but dictatorship! It is the authoritarian state we should fear, not science! Let us be extremely wary of excessive state power, but let us not abandon technology through fear of our ability to exercise democracy wisely.

MALANTHROPY

Should we really ban anything that could possibly harm us? Some people die each year under anesthetic—should we ban operations? Thousands die each year from falling down the stairs—should the state make us all live in bungalows?

The idea that we should outlaw any knowledge that could possibly be used for malevolent ends is absurd. If we ask ourselves to think of a single piece of knowledge that could be used *only* for good purposes, we will think in vain.

Knowledge and how one uses it are two different things. The best ideas can be used for the worst purposes. All products of human design can be a tool for good or evil. By the logic of the argument that any tool should be banned which could possibly be used for malevolent ends, we should ban mathematics because it led to the atom bomb.

If there are bad people in the world who wish to use biotechnology for malevolent ends, then we must have confidence in our own power to stop them. But to abandon the quest to improve the human condition through fear that humanity is too evil to be trusted with power is simple cowardice—and cowardice never accomplished anything. You cannot base an ethics on fear of your own power.

Are we really to call for an end to progress because humans are too evil to be trusted with knowledge? What misanthropic madness! When bio-Luddites say Superbiology will be used for malevolent purposes, they reveal an attitude I call **malanthropy** (*mal*: bad, *anthropos*: man), the judgment that human beings are too evil to be trusted with power over their own destiny. Malanthropy is the belief that the innate wickedness of man requires that we be protected from ourselves—by the state.

Malanthropic bio-Luddites seek to impose their belief in innate human evil on society by denying others the freedom to undertake any actions that might possibly lead to an evil outcome. According to the malanthropic argument, if humans are not prevented from pursuing self-enhancement through Superbiology, they will descend into an orgy of barbarism, persecuting minorities in a genocidal fury.

How contemptible to regard one's fellow human beings as so innately wicked that they must be denied the benefits of medical science to stop them from descending into barbarity.

There is, of course, a deep illogic to malanthropy, for its proponents consider *themselves* somehow detached from the innate evil they accord to others. *Other people* will use new technology for malevolent ends, while they themselves remain holier-than-thou judgmentalists on us all.

How much bio-Luddism is malanthropy disguised as morality? Let us have the dignity to believe in our own powers of benevolence. What is man if he cannot even believe in himself?

FREEDOM

In the end, transhumanist ethics are a question of freedom: the freedom to do as one wishes with one's own body, and the freedom of parents to offer their children the best possible start in life.

No one has the right to deny others the freedom to improve their own physical or mental condition. No one has the right to prevent the development of a technology that may help individuals improve the quality of their lives. And no one has the right to impede the natural desire of parents to produce the strongest, healthiest children they can.

Above all, transhumanism is about the belief in freedom from limitations.

Let us choose freedom.

NOTE

1. James Watson, *Independent*, February 3, 2003, p. 11.

Chapter 5
THERAPISTS
The Malleability of Man

> In 1995 a woman sued her former therapist on the grounds that three weeks on Prozac had achieved more than three years of therapy. Freudian theory fell the moment lithium first cured a manic depressive, where twenty years of psychoanalysis had failed.
> —Matt Ridley, *Genome* (1999)

> My cat knew more about me than my therapist.
> —Sally Brampton, *Sunday Telegraph* (March 21, 2004)

Psychotherapists deter public awareness of the biological basis of personality that is a necessary precursor to the development of the biotechnology by which to *cure* psychological problems, instead of merely talking about them.

THE CULTURE OF THERAPISM

> The marginalization of the intellect is called "seeking a feeling society." Men have adopted the language of therapism: a deeply female notion that all things can be cured by talk.
> —Fay Weldon, *Harper's* (May 1998)

Psychotherapists say, "Psychological disorders are caused by life experience, not biological makeup. The way to alleviate psychological problems is to talk about them."

One can hardly blame them. After all, their jobs are at stake—for the recognition of the biological basis of personality signals the beginning of the end of the present Culture of Therapism.

Wallowing in the past and the trivia of daily life problems will one day be regarded as indulgent self-pity. Why dredge up childhood traumas or past failures when a pill can eliminate the neurochemistry of neurosis at a stroke? Why be chained to a lifetime's psychotherapy when *gene* therapy can permanently rectify the physiology of life failure?

Psychotherapy is limited by its refusal to recognize the effect of temperamental differences on behavior. It's good to talk, but talking won't make an emotionally stable extrovert out of a neurotic introvert.

Denying the influence of temperament on behavior is like ignoring a monkey on your shoulder. It won't go away if you don't look at it.

FREUDIAN PSYCHOANALYSIS

Freud wrote that the aim of psychoanalysis was the replacement of "hysterical misery with common unhappiness."[1]

Big deal!

To Freud, the battle between socialized behavior and animal instincts, superego and id, could never be won. The result was either neurosis on one hand (through the inhibition of instincts), or the collapse of civilization on the other (through the *failure* to inhibit instincts). Freudianism is biological fatalism, the acceptance of mankind's slavery to his animal instincts.

Biological fatalism is another name for the belief in Original Sin. Just as the belief in Original Sin kept humankind in the chains of religious subservience through the acceptance of its fallen nature, so the secular religion of psychoanalysis keeps us in biological servitude, slaves to the forces of the unconscious id. Psychoanalysts are a secular priesthood, maintaining power over the vulnerable by convincing them of their slavery to their animal instincts.

Shrinks are well named: the goal of psychoanalysis is the *reduction* of expectations in line with reality. The goal of transhumanism is the *transcendence* of limitations through Superbiology.

HUMANISTIC THERAPY

Humanistic psychologists say, "We are all capable of life fulfillment given unconditional support and encouragement."

Where Freud was too pessimistic about the human condition, humanistic psychologists swing to the opposite extreme, assuming there are no biological obstacles to the fulfillment of human potential.

Humanistic psychotherapy ignores the underlying biology of life failure. Psychotherapy often has a beneficial short-term effect, as a temporary boost in confidence allows neurotics to make positive changes in their lives. But once the therapist is no longer present and the ego-boosting stops, the individual gradually reverts to neurotype. The limitations of a genetically based temperament returns. A cycle of dependence is maintained. The therapist secures a job for life.

Therapism survives only by its own failure. If it worked, therapists would be out of business.

Therapism fails because it ignores the biological deficiencies at the root of life failure. There is no necessary causal relationship between an adverse event and a psychological problem. The extent to which an event affects future behavior depends upon individual temperament—or neurotype. What's brushed away by one leads to breakdown in another.

It's not only events that determine behavioral responses, but also the temperament with which to deal with them. Some emerge from a trauma unscathed, while others are scarred for life. The difference lies in neurotype, the neurochemically based temperament of the individual.

Neurobiology, not life experience, lies at the root of neurosis.

EXISTENTIAL THERAPY

Existential psychologists say, "We are free to make of our lives what we choose."

Jean-Paul Sartre was wrong. *Man is not absurdly free, but everywhere in biological chains.* Existentialism fails to acknowledge the biological limitations on freedom. Existence does not precede essence because our essence is biological, a fact which limits our freedom.

We would all love to be free to be anything we want to be. But only a fool closes his eyes to truth because it's not to his liking. The fact is that biology limits freedom. A neurotic is not free to be emotionally stable. An introvert is not free to be an extrovert. A man without talent is not free to be a genius.

Existential psychology places exaggerated expectations on people that cannot hope to be fulfilled—without Superbiology.

COGNITIVE-BEHAVIORAL THERAPY

> This combination of cognitive therapy and medication works for virtually everyone.
>
> —Dr James Le Fanu, *Sunday Telegraph* (March 21, 2004)

Cognitive-behavioral therapists say, "Behavioral problems are caused by irrational thoughts and behavior patterns which can be reconditioned."

CBT is the current medical treatment of choice for minor psychological disorders. By placing rationality above emotionalism and treating minds and bodies as they are—computers requiring reprogramming and bodies reconditioning—CBT has considerable success treating minor depression and anxiety-based complaints.

But in practice, CBT is almost *always* prescribed in conjunction with mood-brightening eugoic drugs—a blunt recognition of the underlying biology of neurosis!

Cognitive-behavioral therapy is limited in its effectiveness because it ignores biology. It treats the symptoms without addressing the underlying cause, failing to address the question of why certain people rather than others develop phobias, compulsions, anxiety, or depressive states.

The fact is that different temperaments respond differently to stimuli. An oversensitive, unstable introvert who habitually overreacts to stimuli is far more likely to pick up neurotic traits than an underaroused, stable extrovert who actively seeks out exciting sensations.

The tendency to worry—the essence of neurosis—is hereditary. Some people suffer from excessive irrational thoughts more than others by virtue of an inherited behavioral predisposition. Maladjusted attitudes and behavior associated with anxiety disorders (that is, neuroticism) are the end-result of living with an internal biochemistry hardwired for impulsive, hypersensitive emotional responses to stimuli.

Changing maladjusted, irrational attitudes is fine, but unless the underlying neurochemistry which leads to them is also changed, they are destined to recur. The neurotic is doomed to suffer a lifetime's effort trying to repress irrational thoughts which occur spontaneously.

Hereditary predispositions to neurotic complaints are treatable today through eugoics, and curable tomorrow through Supergenics.

Cognitive-behavioral therapy is certainly useful in teaching people positive, rational tactics for successful living. But compared to the prospect of a future Superbiology, CBT is like—well—pissing in the wind.

EVOLUTIONARY PSYCHOLOGY

Evolutionary psychologists say, "Human beings share a single, evolved psychology."

What, like Jesus and Hitler?

Evolutionary psychology (EP) explains universal behavior patterns in terms of evolved instincts preserved for their benefit to genetic survival, but it makes no attempt to explain why individuals behave differently in an identical situation. In short, EP assumes that human beings are neurobiologically identical. While this may be "politically correct," it is patently untrue.

Evolutionary psychologists tell us that children are statistically more likely to be murdered by a stepfather than a biological father because parents instinctively invest less care in stepchildren who do not share their genes. But they fail to explain *why one parent and not another* might resort to murder. Clearly, the answer has to do with differences between individuals, not shared species-wide characteristics. If upbringing were to blame, it follows that all children in a particular shared household and social environment would be potential killers—clearly nonsense.

A neurotypologist would identify the neurotypes likely to produce a greater susceptibility to different types of asociality than others through neurochemical deficiencies which reduce the ability to inhibit impulses or feel empathy for others.

Two boxers may have grown up in the same environment, but one may be predisposed to biting the ear off his opponent when under stress, the other to immaculate self-control. Evolutionary psychologists have absolutely nothing to say about this (except perhaps, with Freud, that boxing is a social sublimation of the universal instinct for aggression).

A science of neurotypology would identify differences in neurotype—the neurochemical makeup that influences behavior. The balance between self-control and impulsivity is regulated by neurochemicals such as dopamine—which allows the mind to inhibit physical impulses (the faculty commonly known as will)—and serotonin, which dampens nervous system reactivity by reducing the brain's recognition of stimuli.

A future neuropsychology might then identify two types of asociality, impulsive and psychotic, associated with imbalances in levels of dopamine and serotonin respectively, thus covering everything from uncontrollable naughty boys, to the cold-hearted criminal masterminds of James Bond–inspired popular myth—otherwise known as psychopaths.

THE NEUROCHEMISTRY OF EVIL

Applying the neuroanalysis of psychopathology in practice—Iraqi leader Saddam Hussein, the Number One Bad Guy of choice in the West at the time of writing (spring 2003)—has recently been described by a psychoanalyst as a hyperaggressive, narcissistic, paranoid psychopath. In everyday language, then —hostile, uncaring, suspicious, and self-obsessed. One wonders how many years training in psychoanalysis it took to produce this brilliantly insightful assessment.

All of these traits are beginning to be explained in terms of neurotypology. Simplifying somewhat, a neuroanalysist might say the following:

- Aggressiveness = excessive testosterone
- Paranoia = excessive dopamine

- Lack of empathy = low serotonin and oxytocin
- Narcissism may be a sort of "adult autism" (known as Asperger's syndrome)—excessive detachment caused by overly *high* levels of serotonin

The ex-president thus provides us with an example of the neurochemistry of evil.

(Writing a year later, in spring 2004, the US soldiers who ordered the sexual abuse of Iraqi prisoners and took holiday snaps of the proceedings to send home to their friends, provide another.)

Evolutionary psychologists ignore entirely the fact that people have subtly different bodies and minds which lead to different psychological profiles; different susceptibilities to physical and mental illnesses; different strengths and weaknesses—in other words, the fact that we are all unique individuals!

Instead of a world in which some are more predisposed than others to be good, and the bad may be helped to be better by decreasing their neurochemical susceptibility to asociality, evolutionary psychology lumps us all together as naked apes acting on amoral instincts.

EP ignores entirely both individual biogenetic differences and the more complex, cognitive, uniquely *human* drives by which we may *transcend* our animal instincts—the higher faculties of the *mind.* By doing so, evolutionary psychology serves to encourage **biological fatalism**, the belief in the Fall of Man to an animal state of amorality—semislave of his selfish genes.

BIO-FATALISM

Evolutionary Psychology restricts its attention to identifying the evolutionary origin of the universal behavioral traits (or instincts) that constitute "human nature." But human nature means behavior in the service of the selfish genes—and that includes *stupidly selfish* instincts.

Many evolved instincts are antisocial, such as a taste for aggression and adultery. By viewing human beings as semislave to their selfish genes, evolutionary psychology helps to fix ever more firmly in the public mind the belief that we are (as the hit pop song advises us) *"nothing but mammals—so let's do it like they do it on the Discovery channel."*

By lumping all human beings together as naked apes acting on animal instincts, evolutionary psychology contributes to the postmodern **demoralization** of society.

That evolutionary psychologists appear blissfully unaware of the social results of their work says much about the gulf between science and (moral) philosophy.

Evolutionary psychology has said all it needs to say about evolved predispositions (that is, instincts). The job is done. EP is far too limited in scope to be promoted as *the* modern understanding of human psychology. In the twenty-first century, evolutionary psychology will merge into a mature biopsychology which (1) recognizes our individual biological uniqueness as well as our shared animal instincts, and (2) identifies how we may *transcend* our animal instincts through the faculty unique to the *human* animal—the conscious, thinking mind.

Like psychoanalysts, evolutionary psychologists had better learn some new skills, or they'll be out of a job.

MOTIVATIONAL PSYCHOLOGY

Psychotherapists say, "Serotonin levels can be raised by positive experiences—so psychotherapists can alter neurochemistry without the need for pharmacology, by providing positive encouragement and advice."

If psychotherapy means the encouragement of positive thinking and "self-esteem" (**eumeme** for confidence), it can certainly be of use in reprogramming the biocomputer that is the human mind to automatically reevaluate situations in a positive light. Such is the basis of the popular self-help books that tell us to *Feel the Fear and Do It Anyway* as *You Can't Afford the Luxury of a Negative Thought*. This is motivational psychology—a rational, populist version of humanistic and cognitive psychology which originated in 1920 when French pharmacist Émile Coué coined the mantra "Every day in every way I'm getting better and better"—the first attempt to reprogram the mind with positive thoughts.

Essentially, motivational psychology is an extension of being encouraged or cheered up by a good friend. Undoubtedly, something as simple as a positive thought or two in the morning can help make your day. Motivational Psychology works. In the ideal society—a Self-Enhancement Society—citizens would recognize the value in freely tendering positivity, encouragement, and self-confidence—instead of self-diminution and *schadenfreude*. But if they don't (and they don't), the next best thing is to get a book or tape or attend a lecture by someone who does!

However, motivational psychology can only achieve so much. Winning may raise serotonin levels, but what makes one person a winner and not another—their therapists? Nothing to do with *innate* talent, determination, the will to win? How many champions of sport do therapists think they've been responsible for making out of second-raters? Therapists can help sustain or improve the positive attitude of a *potential* winner, but you can't make a habitual positive thinker with a dynamic will to win out of a neurotype predisposed to anxiety, depression, timidity, or lethargy just by talking—especially not about your parents!

NEUROPSYCHOLOGY

Neither evolutionary psychologists nor Freudian, humanist, existentialist, behavioral, or cognitive therapists acknowledge the influence of neurochemistry on personality and behavior. *But the doctor who fails to offer a cure will lose his patients in the end.*

The development of eugoic drugs marks a revolution in psychology. The recognition that altering neurochemistry can improve personality without psychotherapy marks the beginning of the end of the Culture of Therapism.

The root cause of most common psychological disorders is not an Oedipal complex, childhood trauma, poor parental conditioning, negative thinking, irrational thought processes, or failure to love oneself, but a hereditary, neurochemical predisposition to neuroticism. An overreactive nervous system and overly high cortical arousal level predisposes an individual to psychological disorders. No amount of talking about problems will alter the neurological defects that cause them.

Neuroticism—the basis of most common psychological disorders—is a physiological disposition capable of improvement through eugoic drugs, and future eradication through Supergenics.

Once optimum neurochemical balance is maintainable through Superbiology, much of psychotherapy will be redundant and consigned to the pages of history. What will remain of talking cures will be a synthesis of cognitive and humanistic therapy which encourages positive thinking and the application of rational solutions to the problems of life.

The future lies in **Neuromotive Psychology**, a combination of rational thinking and neuroenhancement designed to kick-start the **Will to Evolve**. A future neuromotive psychology would provide not only treatment for psychological disorders, but also general tools for self-improvement of body and mind, both experiential and neurobiological, which help us to transcend our levels of normal well-being, in the natural, healthy desire to feel *better than well.*

Therapism that ignores biopsychology has had its day. The Culture of Therapism is a victim culture that thrives on making people feel like slaves of their emotions, in order to gain power over them. Hostility to neuropsychology by therapists is an attempt to protect their jobs, which depend on not curing people. So next time you hear a psychotherapist denying the biological influence on behavior, remember their underlying motive is self-interest—and think again.

Instead of "Let me feel your pain," transhumanists say "Let me *heal* your pain"—through Superbiology.

NOTE

1. Sigmund Freud and Josef Breuer, *Studies on Hysteria* (1893–95) (Harmondsworth: Penguin/Pelican, 1974), p. 393.

Chapter 6
DOOM MONGERS
The End of the World

> The threat of a new ice age must now stand alongside nuclear war as a likely source of wholesale death and misery for mankind.
> —Nigel Calder (former editor of *New Scientist*), *International Wildlife* (July 1975)

> The good life is not found in dreams of progress, but in coping with tragic contingencies.
> —John Gray, *Straw Dogs* (2002)

> No new technology can abolish scarcity . . . or alter the fact of human mortality.
> —John Gray, *Heresies against Progress and Other Illusions* (2004)

> I'm pretty gloomy. The Italian Marxist Antonio Gramsci talked about "pessimism of the intellect and optimism of the will." I've got lots of the pessimism, but I struggle to find the optimism these days. My friend Martin Rees, the astronomer royal, says humanity is unlikely to survive the next century and I think he's probably right. There is an accumulation of environmental and human problems that pose real danger. The old cry was socialism or barbarism, I now think it's socialism or the end of humanity. But I'm not optimistic.
> —Steven Rose, biologist, interview, *Socialist Worker* (May 14, 2005)

Finally, not *necessarily* aligned to a particular ideology or interest group, the **Doomsday Luddites** are temperamentally inclined fatalists who like to assure us that progress is an illusion, new technology doomed to fail, and thus all effort ultimately futile. Doom mongering—the tendency to see the glass half empty—is characteristic of a neurotype predisposing to low mood and accompanying pessimism. It is particularly common among aging males—due perhaps to a combination of fading testosterone and failed life dreams (nature and nurture). I call it **Grumpy Old Man Philosophy**. One wonders if a course of eugoic mood-brighteners might help the miserabilist doom mongers to accentuate the positive. Perhaps their wives might also benefit.

FRANKENSTEIN'S MONSTERS

Doom mongers say, "Superbiology will lead to a Frankenstein world of soulless monsters."

It may come as a surprise to some, but the story of Frankenstein (subtitled "a modern Prometheus") is, like its Greek forerunner, a work of fiction. Just because someone once wrote a story about a scientist who inadvertently created a monster by reviving a patchwork of dead body parts, doesn't mean it's really going to happen, anymore than the film *The Omen* means that the anti-Christ is about to be born to the president's wife. It's simply not a rational argument to compare a technique in Superbiology to a work of fiction. The whole point of romantic fiction, of which Mary Shelley's novel is a prime example, is its ability to excite the imagination by straying beyond the realms of possibility.

As for the *original* warning against hubris, the Prometheus myth, why exactly should we listen to the fears of the ancient world? One wonders if the Greeks would still have believed in hubris had they been able to witness the technowonderland of the modern world, and enjoyed the benefits of electricity or transplant surgery.

There's nothing wrong with using soundbites for effect, but bio-Luddites use them *in place* of rational argument, knowing full well that fear and sensationalism attract more minds, or higher ratings, or sell more papers, than rational debate. In short, the Frankenstein argument is simple scaremongering.

SUPERMEN TAKE OVER

Doom mongers say, "A genetically enhanced species will split from *Homo sapiens* and wipe out its predecessor in the evolutionary chain."

A species-split would require the decision of a group to isolate itself geographically in order to breed only among itself. But it is hard to see how a small, genetically enhanced group could threaten the survival of the species. If it looked like it was doing so, the rest of us would be bound by the survival instinct to act quickly in order to deal with the potential threat. Therefore, if some people want to experiment with self-directed evolution through genetic isolation, the sensible response is to let them do so. People should be free to do as they wish so long as they do not harm or threaten others.

But in practice, Superbiology will eventually produce such a variety of individual genetic types that the question of one group separating itself from the rest will be irrelevant. In the long term, the species as a whole is destined to expand into numerous individual types resulting from the endless possibilities for self-transformation made possible through biodesign.

Our destiny is to diversify, our ethical task to live by the principle of unity-in-diversity.

RISE OF THE ROBOTS

Doom mongers say, "Humankind will be exterminated by superintelligent robots."

According to this argument, man will create evermore complex computers, until machine intelligence far outstrips his own. The resultant supercomputers will themselves design and build intelligent machines so far advanced that they treat us with the same respect and dignity we offer to fleas—that is, none. Machine-life will thus take over and replace us in the evolutionary chain, while *Homo sapiens* will join the dodo in the long list of extinct species.

Such an argument ignores one simple fact—the human instinct for survival.

The human survival drive is such that we are extremely unlikely to allow ourselves to create machines that are capable of destroying us. If such a danger begins to arise, we will use our ingenuity ("creative intelligence") as we always have done, to safeguard our own existence against threats posed against us.

In short, we will devise methods of ensuring that our machine creations are limited to abilities which do not threaten our own survival—for instance, by ensuring that our computer brains remain disembodied and easily unplugged (like HAL, the renegade supercomputer in Stanley Kubrick and Arthur C. Clarke's classic film *2001: A Space Odyssey*)!

Much the same point was made by biologist Julian Huxley, who invented the concept of transhumanism in its modern sense: "Man's distinguishing feature is his intelligence, limited though it may be. If another animal were to show signs of rivaling him, he would certainly notice the fact and take steps to deal with the threat long before it was too late."[1]

In short, humankind will do to its future creations what theists say God has done to *us*—deliberately make us weak and disunited to protect his omnipotence—a phenomenon that might be called the Babel Syndrome. The difference—we will be around to stop the cybermen from taking over, while God is apparently either dead, resting, or absent without leave.

ATTACK OF THE BLUE-EYED BLONDS

Doom mongers say, "Designer babies will lead to a world of identical blond, blue-eyed boys."

If one examines this common bio-Luddite argument logically, one might conclude that parents are most unlikely to rush out en masse to order blue-eyed blond boys, for the simple reason that most parents would generally prefer their genetic offspring to look a little like themselves!

But if some parents are desperate for a blond, blue-eyed baby boy, then who are we to prevent them having their wish—so long as the technology is kept out of the hands of the state! The biolibertarian principle is simple—*leave it to the parents to choose*.

What is far more important than individual aesthetic choice is the prospect of freeing our children from disease, and enhancing our physical and mental health, happiness, and vitality.

In the transhumanist vision of the future, parents will be able to guarantee their children freedom from mental and physical disorders from birth. Having provided them with the best possible start in life (the instinctive goal of every parent), they will grow up in a Self-Enhancement Society which encourages individual self-improvement in all areas of life, whether through exercise for the body, education for the mind—or enhancement of both body and mind, through Superbiology.

COLLAPSE OF THE GENE POOL

Doom mongers say, "Supergenics will reduce genetic diversity, allowing a supervirus to wipe us out."

But what could be better placed to counter a threat to its survival than a species able to alter its genetic constitution at will!

How better able to protect itself from the whims of capricious nature is humankind today, than before the advent of technology? A species less vulnerable to disease, natural catastrophe, or even alien invasion, by virtue of technology is a species better able to survive.

The fear of reduced genetic diversity is ungrounded. Supergenics will extend, not limit, the number of transhuman types by affording ever-increasing variations of biological transformation, so much so that in the future we will begin to speak of the "human species" in the plural, not the singular.

In the DNAge, it will be up to individuals to decide what they make of themselves. So let a million flowers bloom.

OVERCROWDING

Doom mongers say, "Extending the human life span will lead to catastrophic overcrowding."

In the short term, birth rates will continue to sink in the modern world, both through increased use of contraception, and the rise in female careerism. The transition from a male-dominated industrial economy to a female-dominated service economy will continue to reduce population growth by encouraging childbirth later in life, resulting in fewer births.

In the long term, technological advances will allow us to build civilizations under the earth, in the oceans, in outer space, on other planets. We may have a few years to spare before having to worry about overpopulating the universe.

Fear of overpopulation is just another example of **Apocalyptic Scare-mongering Syndrome** (ASS), the inability to prevent the impulse to run around shouting *the end is nigh!*

END OF THE WORLD

Doom mongers say, "All effort is futile as global warming or a passing comet will wipe out life on earth, and the sun will eventually burn itself out."

One notes with wry amusement the enthusiasm with which postmodern scientists are determined to spread their **negamemes** of nihilistic fatalism, by repeatedly warning us about the latest inevitable cosmic disaster ahead. The extent to which they gleefully attempt to negate any belief in human progress makes one wonder about their motives.

First they told us the oil would run out by 1990. Then, after a few cold winters, that we should prepare for a new ice age. Next, after a few hot summers, that the world is heating up because of our selfish prosperity.

And in case that's not enough to demoralize us, we should never be allowed to forget the inevitable extinguishing of the sun and collapse of the universe billions of years hence (this one pops up in the Sunday papers every so often, just in case we're feeling too good about life).

Truly, the end is always nigh for the **negs**. What the voices of pessimism choose to ignore, however, is human ingenuity. Look at the pace at which technological advance has surged ahead, even over the past few years! In tens, hundreds, thousands, hundreds of thousands of years, do these postmodern doom mongers not think that humanity will have developed the capacity to avoid global warming, divert a comet from its path, populate other planets, even create new worlds, new atmospheres, new suns?

The continual advances in human capabilities that characterize the story of humanity are becoming so dramatic that assertions of our inevitable failure to solve future problems threatening our survival must be considered **logically incorrect**.

End-is-nigh doom mongering is an irrational response to the evolutionary progress of the species from Stone Age to Gene Age, ape man to spaceman—human to transhuman.

UTOPIAN DREAM

Doom mongers say, "Designer Evolution is a utopian pipe dream, as much wishful thinking as the theistic belief in heaven and the Marxist belief in world communism."

One notes that the accusation of naïve idealism is most frequently the response of those who have made little attempt to follow developments in biotechnology, instead basing their argument on a "gut feeling" that the possibility of such things is "too far-fetched."

It is not within the scope of this book to list details of the enormous scientific and technological developments currently taking place in the field of biotechnology. This is a book of philosophy, not a review of technology. But as evidence grows supporting the conviction of the inevitable transformation of the species through Superbiology, let the cynic be pointed in the direction of the endless source of information that is now openly available in the public domain—thanks to the greatest wonder of the modern world—the scientific miracle that is the Internet.

In our own time, a revolution is occurring in science. A new world is opening up in front of our eyes. A world of genomics, proteomics, phenomics, neuroscience, neuropharmacology, nanotechnology—a world of Superbiology. Up-to-date developments can now be followed daily on the Internet, where news of current research worldwide is available at the click of a mouse, through published research papers, newsgroups, online newspapers, weblogs, and e-zines which will send you, free of charge, regular updates on breaking developments in the subject of your choice.

Those who have taken the trouble to familiarize themselves with current developments in science and technology would be hard-pressed to deny that mental and physical enhancement through Superbiology is not only possible, but inevitable—given time and, vitally, the survival of the modern world.*

*That such survival is not at all certain is undeniable given our current engagement in a war on two fronts: a conventional war against the premodern forces of totalitarian theo-fascism, and a Meme War against postmodern relativism, both of which seek the destruction of modernity—the belief in human progress through reason.

The information is there, for those who have eyes to see.

The irrational tendency to automatically dismiss a difficult but desirable goal as unattainable might be called impossibilism.

Impossibilism is the refuge of the ignorant—an excuse not to look and learn.

NEGS WILL BE NEGS

A certain tendency toward skepticism is valuable, of course. Caution is an instinctive by-product of our evolutionary history. Survival does not favor the naïve overenthusiast—the Icarus myth tells this story. But there is a world of difference between the logic of rational doubt and the total illogic of the cynic. Cynicism seeks to silence every voice in the wilderness that cries out, *eureka!* For the cynic, the fear of disappointment is too great to permit the joyous excitement of new experiences, new possibilities. Cynicism is a fool's path—a downhill slope to misery.

Transhumanists are neither cynics nor idealists, but **positive realists**. We cannot prove our predictions. We can only say *logic suggests that this will happen*. The eventual eradication of death and disease and the expansion of faculties beyond the limitations of the human condition will eventually be achieved—not through prayer, or the smug superiority of the cynic, but through science. Radical, miraculous, Big Science.

For the first time in human history, the concept of banishing disease from the face of the earth is no longer a pipe dream but a genuine possibility. However far in the future it may be, we now know that even the horror of aging can become a thing of the past—but only if the modern world embraces the emerging Superbiology in the years ahead.

So let us reject the scaremongering of the bio-Luddites, and recognize the potential of biotechnology to improve the quality of our lives.

Let us denounce attempts at state restriction on the rights of individuals to do with their bodies as they wish, providing they harm no others in the process.

Let us call for the freedom of parents to use all available technology in order to offer their children the best possible start in life.

Let us discover a newfound respect for science and for scientists, in particular those engaged in the most important job of all—the quest to defeat disease and death.

Let us look forward to a New Enlightenment—a new spirit of self-confidence, and belief in the development of human potential.

Above all, let us recognize the existence of the Will to Evolve—the instinct to expand our abilities in pursuit of ever-increasing survivability and well-being—that we may one day come to rid ourselves of the tragedy of life by transcending the biological limitations of the human condition.

EUTOPIA

> Wherfore not Utopie, but rather rightely My name is Eutopie, a place of felicitie.
> —Thomas More, *Utopia* (1760)

The Doom mongers say, "Transhumanism is a utopian dream."

Utopia? The word means "nowhere place." So let us speak instead of **eutopia**—the *good* place—where the creative art of eugoics has been developed to such an extent that the misery of human suffering and limitation are no more.

NOTE

1. Julian Huxley, *The Destiny of Man* (London: Hodder, 1959), p. 20.

PART 3
TRANSHUMANISM AS A TOTALIZED PHILOSOPHICAL SYSTEM

Eventually, most philosophical problems may turn out to be biological.
—Robert Ettinger, *The Prospect of Immortality* (1962)

Transhumanism is more than just a defense of Superbiology. Here I present, for the first time, an outline of transhumanism as a totalized philosophical system in the Western tradition.

The bottom line—human beings are biological entities, operating in the world by means of four basic functions: perceiving, feeling, thinking, and doing:

- How should we perceive the world?
- How should we feel?
- What should we think?
- What should we do?

Chapter 7
METAPHILOSOPHY

> The philosopher is nature's pilot, to be in hell is to drift: to be in heaven is to steer.
> —George Bernard Shaw, *Man and Superman* (1903)

Metaphilosophy is the attempt to answer the question, What is philosophy, and what is it for?

SUMMARY OF TRANSHUMANIST METAPHILOSOPHY

Philosophy is a type of computer software program called a **meme map** designed by the self-recognizing, embodied biocomputer we call the human mind, in order to determine the nature of the world and how best to live in it. A meme map may be copied to other brains through communication—or **memetic reproduction**. When spread sufficiently within the **meme pool** to affect the general behavior within it, a meme map becomes a **metameme**. The transhumanist meme map challenges the **dominant memes** in the memetic marketplace of the modern world.

WHAT IS PHILOSOPHY?

The world is information.

Information is a communicable pattern.

A pattern is a stable arrangement of parts.

A metapattern is a pattern of patterns.

Philosophy is the attempt to construct a metapattern of information describing the nature of the world, and how best to live in it.

WHAT IS THE PURPOSE OF PHILOSOPHY?

The purpose of philosophy is threefold: instinct, utility, and pleasure.

Instinct

Curiosity is instinctive. The desire to understand our world is an innate property of the human brain, evolved for the survival value of knowledge.

Utility

The survival value of knowledge may only be refuted if one claims that our mentally less well-endowed ancestors were better able to survive and prosper by virtue of their ignorance, rather than living lives that were "nasty, brutish, and short."

Pleasure

The *joy* of knowledge is a neurochemical pleasure reward for behavior conducive to survival.

PHILOSOPHY AS MEMETICS

A meme is a unit of *cultural* information, as a gene is a unit of *biological* information.

Memes are communicated ideas, beliefs, values, and habits.

A philosophical system is a type of **meme map**—a computer software program uploaded into the memory of the self-recognizing biocomputer that is the human mind, in order to guide behavior.

Philosophy is the attempt to systematically *map out* the fundamental nature of human existence.

The meme map with which we are born is written by our genes to promote behavior that will allow them to survive. Such a meme map could be called **Genethics**—short for "genetic ethics." The three-part genethic software program for human behavior reads, "Survive, reproduce, self-destruct." Genethics condemns every human being to physical decay and death.

When a human brain develops the capacity for self-consciousness—that is, the ability to recognize its own existence—it may begin to reprogram the genethic meme map with its own instructions. Software programs written by the conscious mind to replace those of the "selfish genes" might be called **Nurethics**—short for "neural ethics."

A philosophical system is a nurethic meme map.

METAMEMES

As genes spread through the gene pool, so memes spread through the **meme pool** that is a human culture.

A meme map may be copied to another human brain through communication. A downloaded meme map is saved or deleted by the host brain according to its power. **Dominant memes** are those which spread quickly through the meme pool. **Recessive memes** are those which remain dormant, awaiting a better time in which to prosper.

When a meme map is copied to a sufficient number of brains such that it begins to affect behavior within the culture as a whole, it might be called a **metameme**. A metameme is a culturally influential meme map.

The philosopher concerned with the nature of society cannot afford to ignore the power of the metameme. For a human culture is defined by the dominant belief systems operating within it.

PRIVATE AND PUBLIC PHILOSOPHERS

There are two types of philosopher—private and public:

- The private philosopher seeks knowledge for his own ends alone—his goal is the construction of a personal meme map.
- The public philosopher is equally concerned with the impact of a philosophy on the society or species—his goal is the understanding or dissemination of a metameme.

The best type of philosopher strides both worlds. It is instinctive for the human mind to seek an understanding of its own nature, and that of the world in which it operates. But one's understanding of the world affects both how one behaves, and how that behavior affects others. One cannot morally divorce a private from public philosophy unless one lives in the desert, for what one believes may be passed on to others—by memetic reproduction.

It is the moral duty of the philosopher concerned with society to identify and evaluate the dominant metamemes operating in society, and if they are considered destructive—to challenge them!

The construction of a philosophical system is a moral act.

METAMEMES IN THE MODERN WORLD

In the pluralistic modern world the individual is presented with a number of metamemes from which to choose as possible alternatives to the program of Genethics for survival, reproduction, and self-destruction:

- The theistic software program reads, "Worship God to live forever." Its response to death is to pretend it doesn't exist by inventing a place you go when you die.
- The New Age software program reads, "Abandon the ego and be one with the universe." Its response to death is to pretend it doesn't exist by thinking of it as "returning to the source."
- The ecoist software program reads, "Serve the earth goddess Gaia." Its response to death is to pretend it doesn't exist by thinking of it as going "back to nature."
- The postmodern software program reads, "Revel in the absurdity of life." Its response to death is one of ironic acceptance: "eat, drink, and be cynical, for tomorrow we die."

To the transhumanist, these metamemes are little better than the original genethic meme map condemning us to death, for they all implicitly *accept* the final instruction in our genetic programming for self-destruction.

THE TRANSHUMANIST METAMEME

If all existing alternatives to the genethic software program on offer in society are deemed unacceptable, the intelligent individual finds the urge to construct a new meme map to guide behavior.

Transhumanism may be regarded as the attempt to write new software for the brain, as an alternative both to our genetic programming, and to other programs for behavior offered in the memetic marketplace of the modern world.

Transhumanism is the replacement of Genethics with Nurethics—instructions for behavior written by the "selfish genes," with those written by the self-recognizing, embodied brain we call the human mind, in its *own* interests of ever-increasing survivability and well-being.

The transhumanist software program reads, "Seek out all possible information by which to transcend biological limitations and override the genethic program for self-destruction."

If this is deemed an emotionally "cold" definition of Transhumanism, consider that the impulse to construct such a philosophical system arises in an embodied biocomputer so filled with the wonder of the world that it seeks to devote its processing powers to the discovery of the means by which physical life may be continued—forever.

Chapter 8
NEUROEPISTEMOLOGY

> Of all the faculties of the human mind, it will, I presume, be admitted that reason stands at the summit.
> —Charles Darwin, *The Descent of Man* (1871)

Epistemology is the attempt to answer the question, "What is knowledge?"

By **Neuro-epistemology** I mean the acquisition of knowledge viewed as a process of information retrieval by the evolved, self-recognizing, self-programming, embodied, organic information processor that is the human mind, in its own interests of ever-increasing survivability and well-being.

SUMMARY OF TRANSHUMANIST EPISTEMOLOGY

Knowledge is an accurate representation of the world. Science is the pursuit of knowledge in the interests of increasing survivability and well-being. The construction of a scientific theory is a creative act (**creative science**). Science should not seek to reduce all experience to logic—rather, logic should be used to guide feelings (**warm logic**). There are two sorts of science: **speculative science** produces theories based on unproven correlations between events. **Practical science** produces hard facts which may be of practical use, for instance, in the quest to cure disease. Transhumanists place their faith in the latter. The best attitude to new scientific theories is neither **orthodoxism** (rejection of all but the current scientific orthodoxy) or **credulism** (willingness to believe anything scientists tell you) but the **judicious open-mindedness** of **agnoskepticism**—a sensible middle ground.

The **New Scientism** rejects both theistic absolutism and postmodern epistemological relativism for a renewed respect for the scientific method of observation, hypothesis, and experimentation as the knowledge base responsible for the miraculous technowonderland of the modern world. Science is not the source of our eventual downfall but the greatest hope for our transcendence.

KNOWLEDGE

"Knowledge" is accurate information about the structures, forms, and processes in the world, stored as a meme map in the memory of the evolved biocomputer that is the human brain, to assist survival and well-being.

"Wisdom" is the ability to use knowledge sensibly—that is, by the application of reason.

"Reason" is the ability of the brain to establish accurate causal connections between events in the world. Causal connections are the consistent relationships between bits of information which determine their mutual interactions.

"Logic" is the system of rules governing the art of reasoning. Logic consists of two complementary mental activities, deduction and induction.

"Induction" is the identification of patterns in nature—the attainment of general laws or principles governing nature's operations through consistently verified observations.

"Deduction" is the attainment of specific facts about nature from known general laws or principles governing its operations.

"Nature" is a metapattern of information—a pattern of patterns. Induction is the means of identifying the patterns in nature—the general rules governing her operations. Deduction is the means of identifying specific events within those patterns.

"Learning" is the addition of new information to an existing pattern—the attainment of new information about nature through a process of induction and deduction. Learning is the ability to add new parts to the metapattern of information that is the meme map of human knowledge. The evolution of reason and learning gave rise to forethought.

"Forethought" is the ability to predict future events from deductions based on laws established through induction. Forethought was responsible for the emergence of human civilization. For the ability to make the causal connection between planting seeds today and harvesting crops tomorrow enabled humankind to evolve from hunter-gatherer to farmer, thus attaining the social stability necessary for the development of the complex cultures based on voluntary, mutual cooperation we call "civilization."

Reason (induction, deduction, learning, and forethought) applied to systematic observation is the basic tool of science.

THE NEW SCIENTISM

"Science" is the accumulation of increasingly accurate information about the structures, forms, and operations of nature through the application of reason to systematic observation. The accumulation of scientific knowledge is the construction of a metapattern representing the causal connections between events in nature.

The purpose of science is the enhancement of human well-being through the technological refinement of nature.

In the twentieth century, science was routinely, casually pilloried—while its exponents quietly went about their business making the most extraordinary advances in human knowledge ever seen in the history of the species. The quantum leap in science and technology over the latter decades of the twentieth century, particularly in the twin areas of computer and biological science, is now making itself felt. At the dawn of the twenty-first century we are beginning to see a shifting of attitudes toward science. The modern world is witnessing a revival of confidence in previously unfashionable scientism, after its unwarranted denigration under the yoke of postmodern nihilism.

The extraordinary success of popular science books is just one source of evidence for the rise of a **New Scientism**—a renewal of belief in science and technology as a source of human betterment. The emergence of the New Scientism signifies three things:

- The deep desire of human beings to understand their world;
- The failure of religion to provide an adequate metaphysic for a technological age;
- A newfound respect for the achievements of science and technology.

Twenty-first-century supporters of science might wish to reclaim the term *scientism* from its popular usage as a term of abuse, and begin to employ it proudly in public. Supporters of the New Scientism might be termed **sciphiles**, their attitude, **sciphilia**. Those displaying an irrational fear of science might be called **sciphobes**, suffering from an irrational psychological disorder called **sciphobia**.

Sciphiles should "come out." Science is the new black. After decades of relegation to the planet of the geek, science is fashionable once more.

CREATIVE SCIENCE

The greatest scientists and artists possess brains equally excelling in logical and creative thinking.

"Logical thinking" is the ability of the brain to identify causal connections between events in the metapattern of nature.

"Creative thinking"—or imagination—is the ability to construct *new* patterns in the mind. Creative thinking in science is the ability to identify or construct new patterns in nature.

The best scientists are those most able to combine logic with creativity by identifying existing casual connections between parts in the metapattern of nature, and imagining new connections between those parts.

Creativity is pattern making. The best scientists are creative artists identifying and making new patterns out of the stuff of the world.

In the **creative science** of the future, scientists will be regarded as creative artists engaged in the construction of an accurate metameme—a comprehensive map of the world.

WARM LOGIC

In creative science there is no dichotomy between logic and emotions—thinking and feeling operate together in the retrieval and evaluation of knowledge.

Logic is the ability of the mind to identify the causal connections between parts of the world—emotion is the body's biochemical response to them. The wise individual employs logic to decide whether or not to be guided by the emotions.

The purpose of logic is not to rid ourselves of feelings, but to increase emotional well-being by enabling us to willfully choose which emotions to excite or encourage, and which to inhibit or avoid. The use of logic to enhance emotional well-being might be called **warm logic**.

In the contemporary **technomythology** of *Star Trek*, the collaboration between logic and emotion is symbolized by the relationship between Mr. Spock and Captain Kirk, whose ability to function as the perfect team signifies the ideal synthesis between thinking and feeling:

> Spock: Insufficient data, Captain.
> Kirk: Insufficient data is not sufficient, Mr. Spock. You're the Science Officer, you're supposed to have sufficient data all the time.
>
> —*Star Trek*, "The Immunity Syndrome"

Here, Spock's attitude represents "cold" logic—the simple acceptance of current limitations of knowledge as fact. Kirk's response represents human will—the determination of the mind, equally fired by neurochemical activity, to achieve a desired goal irrespective of present limitations. Action requires the body, which in turn requires the neurochemistry of feeling. Where Spock "understands" through logic, Kirk "wills" through feeling. Together, the will to apply knowledge in action is the ideal combination for the attainment of goals.

In the creative science of the future, the "cold logic" of the old scientism will be replaced by warm logic—reason mediated by feeling in pursuit of ever-increasing survivability and well-being.

EVOLUTIONARY EPISTEMOLOGY

The truth-claims of science are based on their testability. For a scientific theory to be generally accepted one must be able to repeatedly "put it to the test."

The observation that apples fall downward from trees may be consistently confirmed: the assertion that invisible tree goblins are throwing them down may not. The presence of cancer in the breast of a thirty-six-year-old woman can be verified: the presence of fairies at the bottom of the garden cannot. This is the difference between the knowledge claims of science and superstition.

Unlike religious dogma, scientific knowledge does not claim to be absolute. Like everything else in nature, science evolves. Competing theories are put to the test—those that survive are those best able to be repeatedly confirmed through systematic observation. Our ability to examine nature with ever more accuracy inevitably means that knowledge will be revised over time. New theories emerge to challenge and sometimes overthrow old ones. But the fact that science undergoes constant revision in no way negates the legitimacy of scientific knowledge. Imperfection does not infer invalidity. Scientific truth is neither absolute nor relative, but evolutionary—subject to continual, critical reexamination as ever more and increasingly accurate knowledge is attained through the scientific process of reason applied to systematic observation.

The **evolutionary epistemology** of the New Scientism replaces both the absolute truth-claims of premodern religion, and the absolute relativism of postmodern skepticism with the belief in the continual increase in accuracy and quantity of human knowledge over time, otherwise known as *human progress*.

PRACTICAL VERSUS SPECULATIVE SCIENCE

It is important to distinguish between two levels of scientific enterprise—we might call them practical and speculative.

Practical science produces clear evidence of causal connections between events, such as "a mutation in gene X is responsible for disease Y." This is the best kind of science because it offers relatively unequivocal statements of fact which may be practically useful, for instance, in the fight to cure disease. By contrast, **speculative science** merely offers statistical evidence for possible correlations between events. In short, it is far less likely to be accurate.

When the French were found to suffer less heart disease than the British, scientists were quick to make a causal connection with their greater predilection for red wine. Next thing we knew, "red wine prevents heart attacks" was the headline, and out rushed the public to stock up on claret—having received a convenient excuse to get drunk more often. Few bothered to read the original scientific paper. Time and again, the public are fed a media soundbite telling them what to stop putting into their body this month, while closer examination reveals the correlations to be entirely conjectural—any manner of other causal links could be made. Do the French really have fewer heart attacks because of a penchant for red wine—or do French *genes* tend to produce a greater joie de vivre, and hence a more relaxed "type B personality" (in my own neurotypology, a "serotonin neurotype")?

What prevents heart disease, Beaujolais or biology? We don't know, but let us not assume that one or the other is true merely because one or two research papers have suggested a link, and editors smell a sales boost. Scientists are not unaware of the advantages in publishing papers that fire health concerns—truly the stuff that headlines are made of. One is tempted to hypothesize a statistical correlation between publishing a paper suggesting a link between an environmental trigger and a disease, and the likelihood of receiving media attention and new research grants. Or a correlation between the date of scientific papers linking wine to good health, and the release of that year's Beaujolais nouveau.

The basic difference between practical and speculative science lies in the strength of causal correlations involved. If all individuals suffering from disease X are found to lack gene Y, and switching off gene Y in genetically similar animals is found to induce the symptoms of disease X, one may be justified in asserting that a strong causal correlation has been established between X and Y. It is somewhat harder, however, to justify the claim that too many fizzy drinks cause throat cancer (as we have just this second been informed on the news), despite assurances from scientists who have noticed increased rates of both, and decided causality is at work—probably after stuffing some poor mouse with a thousand cans of cola and observing that it appears not at all well.

The essential weakness of speculative science is the failure to recognize explanations other than the one sought—the tendency to find the evidence to prove one's hypothesis by conveniently overlooking other possibilities. In its tendency to assert unlikely connections between events and ignore alternative explanations, speculative science is indistinguishable from magic. If one's Australian cousin phones just as one was thinking of him, it may well be that nonlocal casual connections between brainwaves as yet undetected by science are at work. But it may also be a coincidence. Or an item of mutual significance on the world news may have unconsciously reminded you of one another's existence. The enthusiastic occultist and speculative scientist share a preference for ignoring prosaic explanations they don't like, for more sensational ones they do—just as a child has more fun believing that Santa ate the cookies and milk.

CREDULISM

The sensible attitude toward the pronouncements of speculative science is a middle ground between two extreme attitudes we might call **orthodoxism** (the unwillingness to consider new theories) and **credulism**—the gullible acceptance of every half-baked theory going.

Credulism is credulity as a habit or predisposition—the tendency to accept unlikely propositions without question, adopted as a general attitude to life. **Scientific credulism** is the uncritical public acceptance of dubious scientific pronouncements presented via the media. In a recent case, volunteers were required to carry around a small electronic counter all day, to be pressed whenever they thought about sex. Psychologists concluded from this brilliant experiment that human beings think about sex every two minutes. This discovery has proved most popular with the public. One hears it repeated all the time—especially by husbands to wives. And of course, being told to press a hand-held button in response to carnal desires could not possibly influence the straying of one's mind in any particular direction.

Another example of scientific credulism has appeared this very morning. The headline blazing across the front page of the liberal-leaning, quality UK Sunday newspaper, the *Observer*, reads "Pollutants Cause Huge Rise in Brain Diseases."[1] Note the use of the unequivocal word "cause." In the second sentence of the article, "cause" becomes "has been linked to"—by a scientist at the esteemed Bournemouth University who says, "this has really scared me." Finally, at the end of the article, he offers us something closer to the truth: "There's not one single cause, and most of the time we have no studies on all the multiple interactions of the combinations on the environment. I can only say there have been major changes. It is suggested it's multiple pollution." From "cause," to "linked to," to "suggested," all in one lucrative front page.

If we are wise, then, we should distinguish between reliable and useful practical science of the kind that identifies specific genes responsible for specific diseases, and speculative science, which suggests possible causal correlations between events that may or may not prove to be accurate, but should certainly not be blindly accepted as such, unless one wishes to live in the puritan manner, since few pleasures in life have not been declared carcinogenic by some enthusiastic scientist or other.

ORTHODOXISM

> All our science is just a cookery book, with an orthodox theory of science that nobody's allowed to question, and a list of recipes that mustn't be added to except by special permission from the head cook.
> —Aldous Huxley, *Brave New World* (1932)

At the opposite extreme to credulism, **scientific orthodoxism** is the knee-jerk tendency to reject any theory that questions the current scientific consensus.

It is an unfortunate characteristic of human nature that once an individual has publicly declared his belief in doctrine X, wild horses will not get him to accept that X is anything other than Gospel Truth, even if evidence to the contrary hits him in the face, the reason being one of pride, and fear of losing status through being known as the long-term advocate of a dud theory. Sooner defend a flawed idea to the death than lose face. Orthodoxism—the adherence to an overly inflexible metameme—encourages the stagnation of creative thought, and the stultification of progress. Dogmatic scientism is little better than doctrinaire theism. Sadly, orthodoxism is as prevalent among scientists as among theists.

My point: I am *not* advocating an attitude of naïve subservience to scientists and their multifarious proclamations. It is not speculative science but hard, practical science in which Transhumanists place their hopes for the future—the sort of science that gets results. And this is not wishful thinking—for it is perfectly logical to believe that the continual increase in scientific and technological knowledge that has already led to the control of malaria, yellow fever, bubonic plague, polio, cholera, and measles (not to mention the development of heart transplants and IVF treatment) will eventually result in the elimination of disease and the enhancement of the human condition.

THE JUDICIOUS OPEN-MINDEDNESS OF AGNOSKEPTICISM

The best attitude toward new scientific pronouncements, and any *other* truth-claims, might be called **agnoskepticism**—a midpoint between the professed ignorance of agnosticism on the one hand and the radical doubt of extreme skepticism on the other. Or between gullibility—the belief in anything one is told, and *pyrrhonism*—the denial of the possibility of attaining any dependable knowledge at all.

The agnoskeptic is one who possesses an automatic alarm bell warning against the belief in too many impossible things before breakfast, while being sufficiently open-minded and aware of his own lack of omniscience to carefully examine new theories, however implausible they may seem, instead of dismissing them out of hand. Unlike extreme skepticism, agnoskepticism is rational doubt without irrational dismissal.

The agnoskeptic attitude might be called one of **judicious open-mindedness.** Judiciousness (soundness of judgment, discretion, prudence, good sense) is the ability to logically assess new information rather than accept or reject it uncritically. For the agnoskeptic, new information should neither be incorporated into one's meme map unquestioningly nor hastily dismissed. Truth-claims should be evaluated not positively or negatively, but logically—not by the subjective temperament but the objective, rational mind.

The reason one should be open-minded about new theories should hardly need stating: Where else do new scientific discoveries, breakthroughs, developments come from, if not from those brave and clever enough to challenge the current orthodoxy? Open-mindedness is a prerequisite for creativity, the willingness to look for and construct new patterns of information from new types of associations between events. The creative scientist is open-minded, the uncreative narrow-minded. Scientific discoveries require an expansion of the mind to explore new possibilities, not a contraction to admit only that which has already been confirmed.

At the other extreme, we should avoid the naïveté of scientific credulism, the tendency to uncritically accept scientific speculations handed to us by the media. There is nothing wrong with speculative correlations—this book is full of them! But the intelligent individual would be wise to distinguish between fascinating theories on one hand, and practically useful facts on the other. The free-thinking individual avoids both the naïve "will to believe" and the instant dismissal of the unorthodox.

In creative science, the old scientism of narrow-minded orthodoxism is replaced not by an uncritical credulism, but by the judicious open-mindedness of agnoskepticism.

So let scientists continue with their theories, while we listen with the agnoskeptic's attitude of judicious open-mindedness, and hope that they may develop into useful facts which may assist in the enhancement of the human condition.

* * *

RESPONSE TO THE SCIPHOBES: THEISTS, POSTMODERNISTS, ECOISTS. AND CONSERVATIVES

Acknowledging the imperfection of science is a very different thing from rejecting it entirely as a source of knowledge. Attempts to delegitimize science are based on an irrational fear we might call **sciphobia**. Contemporary sciphobia comes in two popular flavors, **premodern absolutism** and **postmodern relativism**.

Premodern theists claim absolute knowledge based on faith (defined as "hope without evidence"). Science is regarded as a threat to the credibility of their beliefs, and attacked for its supposed "dehumanizing" effects.

Postmodern leftists assert the relativity of all knowledge claims. Science is viewed as merely one way of looking at the world, and attacked as a source of "Western hegemony" (leadership or domination, depending on your spin).

For the transhumanist, knowledge is the gradual accumulation of increasingly accurate information about the structures, forms, and operations of nature.

We should embrace modern science with open arms as the only possible method of defeating disease, enhancing abilities, and extending life.

The transhumanist attitude toward scientific knowledge represents a renewal of modernity—the belief in ongoing human progress through reason, science, and technology.

THEISTIC SCIPHOBIA

Theism is credulism of the superstitious kind—the unquestioning belief in information obtained through faith or indoctrination, as opposed to observation and reason.

Faith is wishful thinking, the belief that something is true because you would *like* it to *be* so. Logic is rational thinking, the identification of the consistent relationships between events in the world as they exist, as opposed to how we might *wish* them to exist, according to our emotional longings. Where theism is a metameme based on faith, transhumanism is a meme map based on reason.

The essence of theism is the belief that human beings were created by an unidentified ghost who intends to reward us with immortality in an undetected, immaterial world if we worship and serve him to his satisfaction.

Religion is a negative philosophy which says, "We are so scared of death and evil that we will invent an afterlife and evil spirits to account for them." Transhumanism is a positive philosophy which says, "Death and evil are biological problems eventually resolvable through technology."

In the transhumanist worldview, death is the absolute end of life, hence the imperative to cure it. The essence of transhumanism is the recognition that human beings are evolved, biological organisms programmed to die. We should thus seek to alter our genetic makeup in the attempt to eradicate disease, extend our life span, and enhance our capabilities.

Transhumanism rejects the irrational metamemes of religious credulism for two reasons:

1. To claim that a world of inevitable, universal suffering was created by an undetectable, omnipotent, benevolent intelligence who wishes to be worshipped for his trouble is logically absurd.
2. The value system resulting from such a belief leads to the ethical rejection of those advances in science and technology which hold out the only prospect of freeing ourselves from the miseries of death, disease, and the limitations of mind and body.

Transhumanism is the belief in science over superstition.

Science as Perilous Knowledge

Sciphobic theists say, "The scientific search for absolute knowledge constitutes a major threat to the future of humanity that is mercifully unrealizable, because of human frailty and transience."[2]

One seems to have heard this somewhere before: "Of the tree of knowledge of good and evil, thou shalt not eat of it: for in the day that thou eatest thereof thou shalt surely die" (Genesis 2:17).

The idea that weakness and transience are our greatest gifts and knowledge our greatest sin is an attitude almost too self-loathing to contemplate. Rather, the instinct to pursue knowledge should be considered the *greatest* of all human attributes, and the biggest *threat* to humanity the propagation of irrational beliefs which surpress the development of the will to evolve by which to free ourselves from the torments of human suffering.

The discouragement of scientific developments by which to enhance the human condition is the product not of compassion, but of cowardice. What humanity needs is not less knowledge, but more—and the will to attain it.

Science as Pseudo-Religion

> Men never do evil so completely and cheerfully as when they do it from religious conviction.
> —Pascal, *Pensées* (1670)

Sciphobic theists say, "The New Scientism will be intolerant of religion."

Intolerance? But it is not *scientists* who display intolerance, every day, all over the world, where war after war is fought, human beings slaughtered in the name of religion.

Sciphobic theists say, "Science is to be worshiped as a new God."

Transhumanists do not "worship" science or scientists but believe in human progress, and recognize technology to be the best means of achieving it—by far the strongest tool in the human armory by which to transcend the limitations of the human condition.

Human transcendence is a goal attainable not through faith and worship, but through reason and technology.

POSTMODERN SCIPHOBIA

> Jacques Lacan, Jacques Derrida, and Michel Foucault . . . share a characteristic philosophical position which is . . . quite radically anti-scientific. They question

> the status of science itself, and the possibility of the objectivity of any language of description or analysis.
>
> —Madan Sarup, *Post Structuralism and Postmodernism* (1988)

Implicit in the construction of a rational philosophy is the assertion that we live in an ordered universe, the relatively stable, if evolving nature of which may be observed. The practice of constructing a rational philosophical system depends upon the premise that there exists an underlying structure to the world—a metapattern which may be identified through the combination of observation, logic, and experimentation (perceiving, thinking, and doing). In short, you have to believe that the world is coherent in order to understand it.

Transhumanism is a wholly rational philosophy that seeks to map out the metapattern of the world by identifying the causal connections between events in time and space. By contrast, postmodernism is a wholly irrational philosophy that sees the world as fundamentally chaotic, and thus unknowable by science.

Transhumanism rejects the irrational metameme of postmodernism for two reasons:

1. The belief that the world cannot be understood by science is false. Science is proceeding wonderfully well in the goal of mapping out the metapattern of the universe.
2. The value system resulting from postmodern skepticism is that of nihilism—the absence of belief in anything at all. The antiphilosophy of postmodern nihilism threatens the modern world with collapse, either by social breakdown through the absence of unifying beliefs, or by attack from outside forces spurred on by the sight of a civilization devoid of self-belief, meaning, and purpose.

Science as Conjecture

Sciphobic postmodernists say, "Science is unreliable as a source of knowledge because today's scientific theories may be overturned tomorrow."

Big deal! Pigs might *fly* tomorrow (perhaps with the help of Superbiology) but airplanes *do* fly, *today*. This is not conjecture, but fact based on well-established scientific rules governing the consistent operations of nature.

Scientific truth is not "the whole truth." We cannot know the whole truth about the nature of existence at this point in our evolution, because we lack the required brainpower. But this is no argument for abandoning the *quest* for truth, any more than a judge would abandon a trial on the grounds of not being actually present when an incident took place. The judge and the scientist-philosopher arrive as near to the whole truth as is practically possible given the necessary limitations of the human condition. Neither decide to abandon the quest for truth because they lack the omniscience of a deity.

Scientists do not claim to offer omniscience, but increasingly accurate and detailed information about the structure, forms, and operation of nature, which may be of benefit in our attempts to improve the human condition by enabling us to alter aspects of the world to our own advantage.

Postmodern sciphobes say, "Science is an unreliable source of knowledge because scientists make mistakes." But the fact that scientists sometimes make mistakes no more invalidates science than the fact that Tiger Woods sometimes loses negates his ability as a golfer.

Science proceeds by trial and error, mistakes are made, hypotheses altered in a process of continual improvement, which—like all self-improvement—requires the ability to endure mistakes and plow ahead.

Science is "evolutionary truth," an ever more accurate representation of the world, as old theories are modified in response to new information.

Scientists sometimes get it wrong, but when they don't the knowledge they accumulate is extraordinarily useful. It is this fact, the sheer utility of science, that makes it by far the most powerful weapon in the human armory—the weapon that will eventually allow us to do away with disease, and transcend the limitations of the human condition.

Science as Cultural Imperialism

> Enlightenment is totalitarian.
> —Theodore Adorno and Max Horkheimer, *Dialectic of Enlightenment* (1944)

Sciphobic postmodernists say, "Scientific truth is just a myth invented to justify Western cultural imperialism."

To understand the absurdity of this postmodern fantasy requires a familiarity with its warped logic.

Essentially, postmodernism is disappointed, post-Soviet Marxism. Here is a typical postmodernist rumination:

> We find ourselves semiotically overloaded and unable to make sense of the past and experience the relation of the past to the future in terms of an eternal disappointment (with socialism, communism, etc.).[3]

In other words, the failure of the Soviet experiment in massive-scale, centralized economic egalitarianism (symbolized by the fall of the Berlin Wall in 1989) left believers in state socioeconomic control with a problem. To solve it, they looked back to an influential book of 1944 by the German philosopher Theodore Adorno, *The Dialectic of Enlightenment,* and decided that the cause of the problem was the "Enlightenment project of modernity" itself—the belief in universal human progress through scientific rationality, originating in the eighteenth-century European Enlightenment, the Age of Reason.

After the Second World War, a new postmodern Left declared that the failure of economic egalitarianism was due to its alignment with an ideology of Western supremacy based on Enlightenment reason. The Left came to see modernity as the ideological basis of the patriarchal, colonialist, imperialist domination of the world that had culminated in two world wars, attempted genocide, the atom bomb, environmental pollution, and the "dehumanization'" of the modern world. The horrors of the gulag and concentration camp were blamed on the arrogant designs of Modern Man. The motto of the postmodern Left, taken from Adorno's book, left no doubt as to its intentions: "Enlightenment is Tyranny."

There was just one small snag. The Enlightenment project of modernity had turned into an incredible success story! The postwar free world was awash with prosperity. A flourishing civilization was increasingly extending civil rights, building a welfare safety net for the least able, and packing the world with technological innovations to make life easier and more fun.

So in order to destroy the popular belief in modernity, an ideology that, despite its flaws and imperfections, had quite clearly "worked," the postmodern Left was forced into an utterly bizarre tactic: the attempt to destroy the validity of holding any totalizing worldview or belief system whatsoever!

According to the postmodern Left, the cause of fascist and communist dictatorship was the ideology of modernity itself—the European-inspired belief in universal human progress through scientific rationality. But such a successful ideology as modernity would clearly prove far too successful to attack head-on. Instead, in order to destroy modernity, its opponents asserted that *any* attempt to propose a general, universal understanding of the world was to be declared invalid. The world was declared unknowable. Scientific reason was not capable of explaining the universe. Science was nothing more than Western patriarchy. The ideology of modernity should be abandoned forthwith.

For the postmodern Left, the only method of eliminating any possibility of the rise of a malevolent ideology was to outlaw ideology itself! And this was, and is, precisely the goal of postmodernism—the elimination of all totalized philosophical systems. The concepts of universal truth, meaning, morality, or progress are condemned as entirely relative and subjective. No culture, lifestyle, or belief is to be considered superior to any another—except, of course that of modernity—for the modern world is to be regarded not as the zenith of human progress, but the nadir of human regression from man to beast.

Thus, in order to prevent any future possibility of totalitarian control, postmodernists announce the abandonment of all universally shared worldviews, and declare the birth of the "Postmodern Condition," in which all strongly held convictions are to be ridiculed as subjective, and seriousness replaced by "playful" irony—in other words, not believing in or taking anything too seriously any more. Postmodernity is thus the condition of a civilization without any shared beliefs or ideals which we may want to realize as a single civilization or species.

In postmodern philosophy, totalized philosophical systems are termed "meta-narratives," to imply that they're nothing but fiction, purely subjective—just "tall stories:"

> It will be a long time before the peoples of Europe will accept any ideology that claims to have a complete, totalised explanation of the world . . . and this is where the novel, created to discuss the fragmentation of truth, comes in.[4]

But if belief systems are outlawed in favor of the "fragmentation of truth," then what is left? Simply the belief in—in what? In *nothing*. Thus postmodernism necessarily results in—has resulted in—nihilism; the condition suffered by the modern world today—an antiphilosophy of absolute relativism which says, "Believe in nothing but the right of others to believe anything they like, for there is no such thing as universal truth, meaning, or values."

The modern world is now characterized by the belief in nothing except the right of others to believe in anything they want. Another name for it is nihilism —postmodern nihilism.

Welcome to the postmodern age.

Let us recapitulate, for an understanding of the true nature of the postmodern metameme is a vital Weapon of Memetic Destruction (WMD) in the Meme War against it.

To the postmodernist, science is not the systematic identification of the structures, forms, and operations of nature for the purpose of improving the human condition, but a metaphysical justification for the domination of the world by Modern Man. In the postmodern equation, scientism equals Western totalitarianism.

Having equated science with the likes of Hitler and Stalin, postmodernists conclude that the only way to prevent the possible emergence of any future totalitarian dictatorship is not merely to rubbish science, but to delegitimize the entire concept of a universal belief system.

In order to prevent the possibility of the world ever again being dominated by a totalitarian regime, postmodernists hit on the idea of promoting the belief in the absolute cultural relativity of truth itself. To achieve the extraordinary task of destroying the very possibility that anyone should ever again hold or promote a totalized worldview, postmodernists seek to convince the world that all claims to truth are relative. Your personal worldview may be true *for you*, but it cannot be considered any more valid than that of anyone else. Thus it may not be legitimately promoted in the name of truth. The dissemination of memes in the meme pool is effectively made taboo. "Thou shalt not spread thy memes" becomes the postmodern dictum, for your memes are no better—no *truer*—than anyone else's.

According to postmodernists, the horrors of fascist and communist totalitarianism were the inevitable result of attempts to impose universal ideologies on the world. Thus in order to prevent any future possibility of the totalitarian imposition of a malevolent ideology, they call for the abandonment of *all* universally shared worldviews and declare the birth of the "Postmodern Condition" in which all strongly held convictions are regarded as absurd and replaced by a sense of irony.

Consequently, in the postmodern diktat, all totalized world pictures are to be ridiculed (or as they prefer to say, "deconstructed") and replaced by the belief in the relativity of knowledge. Postmodernists thus refer to worldviews as "metanarratives," to imply they're nothing but "tall stories." Where before there was truth and falsehood, in the postmodern world, whatever you believe is true *for you*, just as long as you don't try and persuade anyone else to believe it, too.

So according to postmodern epistemology, the claims of the man who declares that the world is made of green cheese are just as valid as those of science.

Sometimes it's hard even for a transhumanist believer in positive thinking not to despair of humanity.

Against Postmodernism

There are four obvious objections to the postmodern metameme.

First, the sheer illogic of postmodernism is clear. Postmodernists seek to replace the belief in any totalizing ideology *with a totalizing ideology of their own!* "Thou shalt have no unifying, totalizing ideology" is *itself* a totalizing ideology!

Second, we are asked to accept the theory that the essential cause of totalitarianism is the ability to hold and promote a totalized worldview, rather than the ability of psychotic, power-hungry individuals to capitalize on the fear and suffering caused by economic meltdown by inciting the primitive, tribal instincts of the masses.

Postmodernists pick on the wrong target. The problem with human beings is not the quest for knowledge but their continual willingness to allow charismatic dictators to lead them into the wilderness of malevolent ideals. The basis of fascist and communist dictatorships is the belief that men are "herd animals," easily controlled en masse. The underlying cause of totalitarianism is the tendency of human beings to allow themselves to be treated like cogs in a machine. The antidote is the opposite: to learn the art of free thought, free speech, *individualism*—and the power of rational discrimination between sense and nonsense, logic and illogic, reason and unreason—good and bad.

Third, we are actually asked to applaud the attempt to relativize truth itself, as though such an act of **memocide** is somehow justifiable; as though there is nothing wrong with destroying the very possibility of unified human beliefs; as though it is perfectly logical to assert that the only way to prevent totalitarianism is to destroy the shared belief in anything at all; as though such an act could have any other consequence than the collapse of shared moral codes and descent into the ironic nihilism of the Postmodern Condition in which all values, ideals, and enthusiasm are to be ridiculed and life deemed nothing but an absurd joke.

The goal of postmodernism is the elimination of all totalized philosophical systems. But if absolute beliefs are outlawed, what is left but the belief in nothing at all. Thus postmodernism necessarily results in the condition suffered by the modern world today—nihilism, an antiphilosophy which says, "Believe in nothing but the right to believe anything you like, for there is no such thing as universal truth, meaning, or values." The inevitable result is the descent into a **Postmodern Anticulture of Nihilism** in which an unmade bed is regarded as great art, atonal cacophony great music, children's books great literature, and in which the willingness to engage in reciprocal altruism diminishes with each generation—for why be good in a meaningless world?

Fourth, and the best argument against the postmodernist assertion that science is just one way of looking at the world, is the simple fact that *science works*. Cell phones ring; iPods play; e-mail arrives (albeit mostly spam). Conversely, those who do a rain dance in a heat wave find that the expected downpour fails to occur.

Contrary to the claims of the postmodern irrationalists, we live in an essentially ordered natural world, governed by regularly occurring causal connections between events. Most of us recognize the truth of the assertion that the sun rises each day in the east and sets in the west. Most of us are capable of understanding the truth of the scientific explanation as to why airplanes stay in the sky—that is, unless they are being flown by theistically meme-washed suicide bombers.

Beware those who seek the abandonment of reason! To abandon reason is to abandon civilization. Where reason is abandoned, barbarism descends.

Let us reject postmodern unreason and renew our belief in the rational mind.

ECO-FUNDAMENTALIST SCIPHOBIA

> I don't see that nature has done such a good job that we can't improve on it. I think it is rather primitive of us to be so fearful of ourselves.
> —Fay Weldon, *Guardian* (February 1997)

By **Eco-fundamentalism**, or **Ecoism** for short, I mean the advocacy of extreme, antimodern environmentalism or Green ideology, in which science and technology are regarded as inherently destructive to the superior force of nature. Ecoism should be distinguished from what we might call **Ecologism**, the perfectly sensible belief in preserving the natural environment on which life depends.

Eco-fundamentalism is essentially contemporary nature worship with an underbelly of misanthropic sciphobia—the belief in Gaia, the benevolent earth goddess, above Man, the evil destroyer of all things natural (like cancer).

The epistemology of Ecoism asserts the primacy of intuition over reason, and holism over reductionism, as tools for the acquisition of knowledge.

Rationality versus Intuition

Sciphobic ecoists say, "Scientific rationality has made us lose our soul—reject it for feminine intuition."

Intuition is the brain's ability to process information automatically.

Intellectual intuition is the presentation of information to the mind without the reasoning by which it was attained, the processing having occurred unconsciously.

Emotional, or *feminine* intuition is the greater capacity of the evolved female brain to recognize subtle body language and emotional cues—a facility originally evolved to assist child rearing by enabling mothers to recognize the needs of children too young to communicate them verbally.

The faculty responsible for civilization was not feminine intuition but reason—the ability of the mind to deliberately and willfully plan and attain desired goals. Without reason, we would very soon return to the primeval slime from whence we came.

Had humankind ever been governed by "feminine intuition" we would still be living in caves.

Some people apparently want us to return there.

Reductionism versus Holism

Sciphobic ecoists say, "Scientific reductionism ignores the organic oneness of nature. Reject it for holism."

Reductionism is the analysis of things through their constituent parts. The recognition that wholes are composed of analyzable parts enables us to understand, repair, alter, and improve the world to our own benefit. We owe to scientific reductionism all the products of the modern world.

Holism is the belief that the whole is greater than the sum of its parts. It is the basis of mysticism, in which the world is regarded as undifferentiated "oneness." But the world is *not* "undifferentiated oneness"—it is billions of organisms eating each other in an effort to stay alive.

Trying to heal the world through holism is like trying to fix a car without looking under the hood.

It's one thing to see the forest as well as the trees, but see only the forest and you'll bump into a tree.

Sciphobic ecoists say, "Reject harsh, reductionistic Western medicine for "holistic health."

"Holistic health" is a euphemism for expensive treatments which sometimes make people feel a bit better because those who sell them are more likely to spend time being sympathetic about minor neurotic complaints than most busy doctors are either able or willing.

Alternative medicine is as much use to the cancer patient as a backrub to a drowning man.

Get real, holists—your mixed-up, mystic gnosis won't cure cystic fibrosis!

Bispectism

To blame scientific reductionism for all human ills is nonsensical. Reductionism and holism are simply the two possible ways of observing a thing—as an arrangement of parts, or a functioning whole. Reductionism is the analysis of parts, holism the observation of the "big picture."

The dichotomy between holism and reductionism could be resolved by embracing the naturalistic metaphysical principle I call **Bispectism**—the recognition that all phenomena can be observed from two equally important aspects: as an arrangement of parts, or a functioning whole. Reductionism is the observation of parts, holism the observation of the whole. A comprehensive understanding of a thing requires equal consideration of its twin aspects.

Bispectism is the principle of "unity in diversity." By uniting reductionism and holism, the principle of Bispectism would see an end to the useless Meme War between reductionists and holists.

CONSERVATIVE SCIPHOBIA

Scientism as Moral Relativism

Sciphobic conservatives say, "Scientific liberalism leads to moral relativism."

Conservatives like to equate science with liberalism, and blame it for the moral decline of the modern world. According to their argument, objective, "value-free" science is the product of a liberal Enlightenment which allowed man to reject the authority of God, and the sanctity of nature. Consequently, a culture dominated by scientism is a culture devoid of moral values.

The root cause of reduced moral consensus in society is not science, but the inability of philosophers to come up with a philosophy *consonant* with science. Philosophers have simply failed to respond to the great surge in scientific knowledge over the past half century that is by far the most sociologically significant phenomenon of our age. It is not *science* that "demoralizes" us, but the failure of philosophers to offer a rational, benevolent system of ethics consistent with our ever-expanding scientific knowledge. Such an ethics would recognize the link between humankind as an evolutionary being and our growing desire and ability to enhance our own condition through biotechnology, and identify the twin sources of benevolent behavior as our emotional instincts and our capacity for reason.

Those who say science offers no support for benevolent ethics are wrong. The ethical implications of evolutionary science are crystal clear, and perfectly in keeping with the conventional morality of reciprocity. Felt and Thought Morality—the two sources of human benevolence—are the product of our evolved nature as biological beings. Practical science allows us to recognize the biological basis of benevolent instincts by identifying the neurochemical correlates of love. And scientific reason allows us to identify the *rationale* behind the human instinct for goodwill, by recognizing that benevolence is simply Sensible Self-Interest, the optimum tactic for mutual survival. In short, we have a better chance of getting on in life by cooperating, whether instinctively through innate feelings of empathy, or rationally through conscious effort.

Science is not some nihilistic bedfellow with postmodernism, but a means of producing increasingly accurate, *useful* information about the structures, forms, and operations of nature, in order to make life more pleasant—for instance, by allowing us to make machines which provide us with light, heat, food, medicine, transport, education, art, and entertainment.

To reject science on the grounds of its moral neutrality is not only logically incorrect, but serves to condemn humanity to the physical horrors of death and disease, and the psychological misery of postmodern nihilism—the *true* amorality, which will rush to fill the vacuum left by the failure of religion if scientific rationality is not allowed to replace it as a source of benevolent ethics for a technological age.

Conservative Stasism

Sciphobic conservatives say, "Scientific progress only makes things worse."

The belief in progress has never been a conservative strongpoint. In the conservative meme map, progress means retrogression—the Beatles song "Let It Be" means "Leave it alone," not "Let it come true." The conservative dictum is "Nothing shall change."

The essence of conservatism is the belief in *conserving* the old, tried, and tested. The essence of transhumanism is the desire to *improve* on the present by embracing the original, the daring, the forward thinking. To be a transhumanist is to embrace the future. In short, Transhumanism is progressivism, and progress for the transhumanist means not only the quest to improve *social* conditions—the basic goal of *liberal* progressivism—but the improvement of the human condition itself through Superbiology.

Conservative sciphobia is based on entelephobia—the rejection of the belief in evolutionary progress through an irrational fear of the new.

CONCLUSION

In Praise of Science

> We are encouraged by a hungry press to get hysterically angry and frightened of remote dangers, blaming science, without appreciating how safe we are off compared to any other generation. . . . We are very bad at celebrating of the great good fortune science has brought us in the past century.
> —Polly Toynbee, *Radio Times*, (November 18–24, 2000)

Sciphobes focus on the very worst of science. To the sciphobic bio-Luddite, "science and technology" means Hiroshima, Chernobyl, and Bhopal. To the Transhumanist, it means the Internet, space travel, and a cure for cancer. Rejecting science because of the bomb is like condemning the body because of cancer. Only a fool throws the baby out with the bathwater by rejecting the good through fear of the bad.

The purpose of science is to understand nature so that we can *improve* it, in order to make life easier, less of a struggle, more pleasant, *better*. Those who refute the claim that science has succeeded in making things better should deny themselves the entire products of technology forthwith—starting, perhaps, with electricity. One suspects few will take up the challenge.

The development of Superbiology promises nothing less than the eventual eradication of disease, aging, and hunger. If such things are achieved, it will not be through the belief in invisible beings no one has ever seen, but through the product of socially concerned endeavors by much-maligned human beings, who work away in quiet corners of the world for the incalculable benefit of all humankind.

Science, like life, is not perfect. Scientists make mistakes. Some are unduly influenced by their financial patrons—like theists by priests—into seeing things that aren't there. Yet no rational person could seriously deny the fact that, overall, science has succeeded brilliantly in formulating the rules governing the operations of nature that have served to dramatically increase the quality of human life—essentially by allowing us to *make* things, in order to make life easier, more pleasant—*better* than in the harsh state of nature untouched by man.

Consilience

To improve upon nature requires an understanding of its methods. A renewed belief in the ability of modern science to produce a comprehensive body of knowledge about the structures, forms, and operations of nature has been expressed by the evolutionary biologist E. O Wilson in his book *Consilience*:[5]

> The belief in the possibility of consilience across the great branches of learning is not yet science. It is a metaphysical worldview shared by only a few scientists and philosophers. [Its] strongest appeal is in the prospect of intellectual adventure—the value of understanding the human condition with a higher degree of certainty.

The belief in such a unification of knowledge is something "in the air" as we enter the twenty-first century. The age of anxiety is giving way to the age of technology. An emerging belief in the power of science for the good of humanity is gradually replacing the gloomy cynicism of the past. The rise of the New Enlightenment will be the latest expression of the Will to Evolve—the instinctive impulse of human beings to expand their abilities in pursuit of ever-increasing survivability and well-being.

Science is the very pinnacle of human creative achievement, the product of a stubborn, stoical persistence and endurance that represents the *best* of our species. Instead of fearing science and technology, let us celebrate the human ingenuity that makes it possible for us to seriously consider the future prospect of a world free from disease, hunger, and suffering.

Science is not the cause of our future downfall, but the greatest prospect for our transcendence.

Let us abandon irrationality and renew our belief in science.

NOTES

1. *The Observer*, August 11, 2004.
2. Don Cupitt, "Face to Faith," *Guardian*, 1987.
3. Matthias Frey, http://www.senseofcinema.com.
4. Salman Rushdie, Herbert Read Lecture, delivered in his absence by Harold Pinter, ICA London, February 6, 1990.
5. E. O. Wilson, *Consilience* (London: Little, Brown, 1998).

Chapter 9
NEUROMOTIVE PSYCHOLOGY

> The first thing that the human species has to do to prepare itself for the cosmic office to which it finds itself appointed is to explore human nature, to find out what are the possibilities open to it (and of course, its limitations).
> —Julian Huxley, "*transhumanism,*" in *New Bottles for New Wine* (1957)

Psychology is essentially the attempt to answer the question, What is the nature of human beings?

Human beings are fundamentally biological entities. We must therefore derive our understanding from the study of the biological basis of brain, mind, and behavior. The word *neuroscience* is currently ahead in the Meme War for umbrella term of choice to describe such a study. A memetic analysis of mutating language indicates the reason why. For unlike other candidates such as biopsychology, psychobiology, or physiological psychology, neuroscience signifies the importance of the *brain and extended nervous system* as the central controller of behavior—the governor, or managing director of the body. This is the meaning of *Homo cyberneticus*—man the *steersman* of his own destiny. *Homo sapiens* will evolve into *Homo cyberneticus* when human behavior is controlled by the conscious mind, as opposed to the selfish genes which demand his self-destruction.

Today we are witnessing a gradual **memetic mutation** in the meme pool—a change in the popularity of terms commonly employed within society from the word *psyche* to *neuro*. To the ancient Greeks, *psyche*—meaning "breath," or "wind"—implied the *soul*, the animating life principle that left the body at death to live forever in a conveniently immaterial realm. So the word *psyche* originally inferred a superstitious view of man. By contrast, the Greek word *neuron*, meaning "sinew," "cord," or "nerve," refers to the brain, spinal cord, and nervous system—in other words, a wholly *physical* phenomenon. Thus the contemporary preference for the prefix *neuro* over *psyche* signifies the belief in the physiological as opposed to metaphysical basis of human behavior—the growing understanding that how we think, feel, and behave are governed not by a ghost inhabiting the body, but by the brain and extended nervous system.

The biopsychological model I shall propose focuses on two main areas: the influence of neurochemicals on feeling and thinking, and the influence of the conscious mind on motivation. Thus I call it neuromotivational or **Neuromotive Psychology**.

The purpose of Neuromotive Psychology is threefold:

1. To encourage recognition of our true situation as evolved biological beings existing in a condition of slavery to a genetic program for self-destruction (**Genethics**).
2. To encourage recognition of the powerful influence of neurochemicals on behavior, leading to the prospect of a future science of neurochemical self-enhancement (**eugoics**).
3. To encourage recognition of the **Will to Evolve**—the instinctive human drive to expand abilities in pursuit of ever-increasing survivability and well-being (**Nurethics**).

The recognition of our biological nature is vital for the continuing evolutionary progress of the species. For as long as we continue to believe that personality is determined by an immortal being living inside of us, or that personality is nothing but the product of our social conditioning, we will forbid ourselves the only opportunity of freeing our minds and bodies from our present state of biological limitation and servitude to a genetic program for self-destruction.

Here, then, is an outline of the fundamental nature of human beings according to Neuromotive Psychology. I shall discuss four topics corresponding to the four basic human faculties, introducing four main new ideas:

- Feeling: Neurochemical Conditioning—Emotion as Genetic Bribery
- Thinking: Neurodualism—The New Cartesianism
- Personality: Neurotypology—The Four Neurohumors
- Doing: Neuromotivation—The Will to Evolve

SUMMARY OF TRANSHUMANIST PSYCHOLOGY

Human beings are evolved, biological entities genetically programmed to survive, reproduce, and self-destruct. Feelings are genetic bribery—a neurochemical incentive to act in ways beneficial to survival and reproductive success. Recognizing the neurochemical basis of feelings in no way undermines their importance in our lives.

The mind is the embodied brain's recognition of its own existence. The evolutionary emergence of consciousness, and the ensuing experiential dichotomy between mind and body, is not the cause of man's downfall from a supposed state of innocent bliss, but a vital catalyst for the further evolution of the human species. The Cartesian duality of mind and body was not a tragic fall from grace, but a necessary catalyst for the development of the Will to Evolve beyond the limitations of the human condition (**Neurodualism**). Personality is based in neurochemistry. Identification of the neurochemical dimensions of personality is a necessary precursor of future attempts to enhance our bodies and minds through Superbiology (**Neurotypology**).

The evidence for the neurological basis of intelligence is incontrovertible, and derives from a combination of neuroscience, genetics, twin studies, and common sense. Raising intelligence through Superbiology will increase the human capacity for benevolence, as "thinking of others" requires the ability to think.

The **New Biology** of the twenty-first century is not the **old biologism** of nineteenth-century Social Darwinism. **Bio-empathy** has replaced **bio-elitism.** We all suffer from biogenetic limitations. But we are more than naked apes ruled by selfish genes. Man is imbued with a **Will to Evolve**—the innate drive of a conscious entity to continually seek ways in which to enhance its condition, in the interests of ever-increasing survivability and well-being.

There are two complementary expressions of the Will to Evolve: the **Prometheus Drive** for Individuation (the Will to Grow) and the **Orpheus Drive** for Integration (the Will to Love). The Will to Grow and the Will to Love form the Will to Evolve. The Freudian drives of Eros and Thanatos correspond to our genetic program for survival/reproduction and self-destruction. Eros and Thanatos are transcended by Prometheus and Orpheus when control by the selfish genes is replaced by the control of the conscious mind. Humankind is limited by its evolutionary heritage. Superbiology may allow us to extend our abilities beyond the limitations of the human condition. The future of psychology lies neither in Freudian fatalism nor humanistic idealism but transhumanist neuromotivation. The main requirement is the Will to Evolve.

NEUROMOTIVE PSYCHOLOGY

A human being is an evolved, biological entity, genetically programmed for survival, reproduction, and self-destruction.

The human brain is a biocomputer—an organic information processor, preserved by natural selection for the survival value afforded by its ability to gather, store, and respond flexibly to information received from the senses.

The human faculties are biological tools evolved by genetic mutation and preserved for their ability to aid survival:

- Sensation is the brain's apprehension of the environment through the senses;
- Perception is the brain's ability to organize sense data into coherent patterns of information, enabling their recognition and evaluation.

Feeling and thinking are the means by which the body and mind evaluate environmental stimuli.

- Feeling is the automatic evaluation of stimuli by the body (Genethics).
- Thinking is the conscious evaluation of stimuli by the mind (Nurethics).

Human beings are not invisible immortal ghosts squatting in mortal bodies, but evolved, organic, embodied information processors, programmed to survive, reproduce, and self-destruct. Once we recognize ourselves as such, we can begin the task of enhancing our biological condition, free from the myths of superstition that serve to prevent our *genuine* transcendence, through Superbiology.

FEELING

Feelings are genetic bribery—an evolved, behavioral conditioning mechanism inducing an attraction or aversion response to stimuli, through a neurochemical reward of pleasure, or punishment of pain, in the interests of survival and reproductive success. We feel that our genes may induce behavior beneficial to their own survival. To have feelings is to be aware of internal, neurochemical activity stimulated by the genes in order to encourage behavior conducive to our survival and reproductive success, so that *they* may survive by "moving house"—leaving the old one to fall into a state of disrepair, dilapidation, and death.

Feelings assist survival and reproduction. We feel because our ancestors were better able to survive and reproduce than those who lacked the ability to feel. Painful feelings condition us to avoid a stimulus. Pleasurable feelings attract us to a stimulus. In a male, the presence of a beautiful woman may stimulate an automatic, neurochemical pleasure reward of testosterone, oxytocin, and dopamine to induce physical, emotional, and psychological attraction in the interests of reproductive success. Conversely, suddenly remembering the trash has yet to be taken out or the dishwasher emptied may induce a neurochemical *punishment* of anxiety, to deter behavior of equally severe *detriment* to survival and reproductive success.

Happiness and Sadness

The most basic feelings of all—happiness and sadness—are the experience of pleasure or pain in response to activity of benefit or detriment to survival and reproductive success.

"Pleasure" is the sensation of **bliss chemicals** such as dopamine and serotonin flooding the brain in response to behavior conducive to survival and reproductive success, in order that such actions are repeated. We automatically experience pleasure in response to behavior that increases our ability to survive and reproduce. Sex, love, friendship, and food are universally experienced as pleasurable because they confer most benefits to survival and reproduction. The continual state of pleasure we call happiness is nature's neurochemical reward for surviving and reproducing in accordance with our genetic programming.

Conversely, such feelings as hunger, loneliness, and sexual frustration are universally associated with *un*happiness because they serve as a neurochemical deterrent against behavior *detrimental* to our survival and reproductive success.

"Unhappiness" is a neurochemical deterrent against behavior that threatens our ability to survive and reproduce. "Sadness" is a genetic punishment for failing to conform to nature's demands. "Grief" is an incentive to protect those who share our genes, or are useful to their survival. Our genes ensure that we care for those whom we need most by threatening us with the pain of loss. Hence, the worst grief is produced by events most detrimental to survival and reproduction—the death of a loved one, the loss of a job. By contrast, news that one's neighbor's best friend's sister-in-law's cousin has died is unlikely to result in a strong feeling of grief (unless one suffers from an anxiety neurosis), because such an individual is of no practical use to one's own survival or reproduction. If, on the other hand, one's spouse, a family member, or a friend should die, one's grief will be considerable—literally in proportion to the potential contribution of the loved one to one's ability to survive and reproduce. The death of a child is the cause of the greatest grief of all, as the survival of our genes depends upon the survival of our children.

We share with certain other animals the instinct to empathize with the suffering of others upon whom our survival depends. Monkeys grieve for their offspring. But feelings of empathy for those upon whom we are not dependent represent an extension of our innate instinct for empathy, facilitated by the ability of the conscious mind to recognize and identify with other people's situations. No ape could empathize with the victims of 9/11. A dog may pine for its owner, but only human beings feel sympathy for strangers. Of this capacity, we should be proud.

Love

Love is . . .

- An instinctive attraction to those useful to the completion of our genetic programming for survival and reproduction.
- An evolved, neurochemical inducement to protect and surround oneself by those of benefit to survival and reproductive success.
- A neurochemical feeling of attraction to those who assist our ability to survive and reproduce.
- A pleasure reward for cooperation and copulation.
- An incentive to pair bond in the interests of the selfish genes.
- Nature's neurochemical bribe to procreate.

To recognize the evolved, neurochemical basis of love is no more to undermine its value in our lives than an understanding of what lies beneath the hood of a car undermines the pleasure of driving.

Lust, Orgasm, Erection, Clitoris

Lust is an inducement to mate—an innate, neurochemical conditioning mechanism encouraging the transfer of genes to a different body, that they may survive, where we perish. Our sexual habits have evolved for the benefit of genetic reproduction. The neurochemical inducements to reproduce are the male and female sex hormones, often referred to collectively as androgen and estrogen.

The pleasure of orgasm is a neurochemical reward for copulation and ejaculation. The contractions accompanying female orgasm probably serve to assist the sperm in its journey to find an egg. The longer period generally required by a female to experience orgasm is probably an evolutionary adaptation (that is, a trait preserved for its survival advantages) encouraging the choice of a healthy male capable of lengthy intercourse, premature ejaculation indicating an overreactive (that is, weak) nervous system.

The independence of erection from conscious control could be seen as beneficial to reproductive success by restricting reproduction to healthy males—impotence being a sign of poor health.

The clitoris was probably preserved by evolution for its benefits to reproductive success. The willingness of a male to stimulate the clitoris in order to please a female—and to delay ejaculation—indicates to her the quality of empathy she requires if her mate is to adequately protect mother and child, rather than desert the family nest. A man unaware of clitoral pleasure is less likely to be a caring—or worldly-wise—husband and father. During sex, there is more unconscious evaluation going on than one may think—or like to think.

Since reproduction is our basic evolved function, it has also evolved to induce our greatest pleasure—or at least, one of them. Thus, much misery is caused by sexual frustration—that is, failure to attain the joy of sex. In the future, Superbiology will allow us to increase our sexual virility by enhancing the genes responsible for the manufacture of sex hormones—if we so choose. One suspects that most people *will* choose.

Feelings as Genetic Instructions

Positive Feelings

Feelings could be regarded as genethic software—programmed instructions to act in ways conducive to genetic survival and reproduction:

"Joy" is the genethic instruction "Continue with this behavior—it is of great benefit to your survival and reproductive success."

"Love" is the genethic instruction "Stay and protect this person—he/she is likely to help you successfully complete your programming for survival and reproduction."

"Romantic love" is the genethic instruction "Protect and mate with this person—he/she seems to be suitable material for reproductive success."

"Affection" is the genethic instruction "Increase intimacy with this person—he/she is likely to engage in reciprocal altruism conducive to your programming."

"Sexual desire" is the genethic instruction "Mate now—this person offers a high probability of reproductive success."

"The pleasure of orgasm" is the genethic instruction "Repeat this behavior regularly for reproductive success."

"Pride" is the genethic instruction "Keep it up, well done!—this behavior is most conducive to your programming."

Negative Feelings

"Sadness" is the genethic instruction "Try another tactic, this behavior is of no use to survival and reproduction."

"Depression" (intense, long-term sadness) is the genethic instruction "Change behavior now! Your lifestyle severely threatens your successful programming!" Hence depression tends to follow failure in relationships or career.

"Grief" is the genethic warning "Protect the well-being of those useful to your programming, or suffer the punishment—the neurochemical pain of loss."

"Guilt" is the genethic instruction "Do not repeat this behavior—it threatens your programming by inducing hostility toward you."

"Shame" is the genethic instruction "Stop this behavior now!—your inadequacies are causing a hostile response from others, which threatens your programming success."

"Fear" is the genethic instruction "Withdraw from this potentially dangerous threat to your programming!" We fear the unexpected because we cannot predict its behavior. As prediction is a property of reason, rational thinking can help dispel fear.

"Surprise" (fear of the unexpected, sometimes followed by joy if the danger fails to materialize) is the genethic instruction "Prepare yourself for action! This new stimulus may threaten your programming."

"Anger" (or hostility) is the genethic instruction "Defend yourself against this threat to programming."

"Jealousy" (hostility + fear) is the genethic instruction "Compete with this rival who threatens your successful programming."

"Disgust" (physical aversion) is the genethic instruction "Avoid!—detrimental to survival and reproduction!"

"Hatred" (emotional aversion) is the genethic instruction "Protect yourself—this person constitutes a serious threat to your programming." "Contempt" is intellectual or cognitive aversion, where hatred and disgust are emotional and physical aversion respectively. Intellectual contempt for malevolence, or evil, is a perfectly logical response to a serious threat to survival. *Irrational* hatred occurs when the conscious mind erroneously evaluates another as a threat, through flawed logic—the failure to accurately deduce another's nature from his or her appearance or behavior.

"Loneliness" is the genethic instruction "Isolation is detrimental to reproductive success—phone a friend."

The emotional acceptance of death experienced by the dying (Thanatos) is the genethic instruction "Abandon the will to live and complete the final instruction in your programming . . ."

Emotional Freedom from the Selfish Genes

What is the value in recognizing our feelings to be an evolved, neurochemical conditioning mechanism in the service of the selfish genes?

The slave unaware of his slavery can never be free. Only by first recognizing that we exist in a state of servitude to our genes can we hope to free ourselves from the limitations of enforced feelings and actions we may not desire.

Only by recognizing that feelings are the products of neurochemistry, not an invisible tenant squatting in the body, can we go on to neurochemically enhance the quality of our feelings in our *own* interests—as opposed to those of the selfish genes which demand our self-destruction—when they've finished with us.

The emotions our genes want us to feel, and those we would *prefer* to feel, do not always coincide. We do not always feel what we wish to feel; we feel what our genes tell us to feel. Human achievements tend to require great tolerance of conditions such as mental or physical strain, which, to our genes, appear to be of no value to the immediate goal of survival and reproduction. Consequently, our genes may inflict upon us negative feelings of depression or anxiety, in order to discourage behavior they see as unproductive.

If you devote your life to science, philosophy, or art, your genes may try to deter you from your task by lowering levels of dopamine or serotonin to induce depression—for to blind nature the strange behavior we call "human achievement" is of no obvious, *immediate* benefit to survival or reproductive success.

Similarly, if you seek to overcome your fears by sailing around the world single-handedly, running into a burning building to save lives, or going back to college as an adult, your genes may try to discourage such apparently useless behavior through the neurochemical infliction of anxiety by raising adrenaline or lowering serotonin.

Why should we restrict our behavior to basic instincts, or condemn ourselves to distressing mood states imposed by our genes? Why not liberate ourselves, by learning how to produce positive feelings of well-being at will? Positive thinking is a good start and may help to raise levels of bliss chemicals. But our genes are powerful taskmasters—we cannot overcome their hold on us through wishful thinking alone. Genuine liberation from negative emotions requires the assistance of the technology opening up in front of our eyes—Superbiology.

There can be nothing bad in wishing to feel good. So let us allow ourselves to feel how we wish to feel, not how our genes tell us to feel. And if we are honest with ourselves, how we truly wish to feel is *better than well*!

EMOTIONAL FREEDOM FROM MISERY

Some say we should enjoy all the emotions our genes induce in us, good and bad, positive and negative, pleasurable and painful, as all a part of life's rich tapestry.

Fine—if you consider such feelings as misery, depression, and anxiety to be beneficial, go ahead. Experience them. No one will stop you. But please don't try telling clinical depressives, anxious neurotics, or generally unhappy people whose lives are crippled by the neurochemistry of misery to do the same.

Experientialists fool themselves into believing that anyone can be happy through positive thinking alone. Numerous individuals have made their fortune selling such a message. Unfortunately, it is quite untrue.

Unpalatable though it may be, the fact is that some people are born with a genetic susceptibility to negative emotions. An individual dominated by the personality trait I call the Opioid Neurotype, characterized by an oversensitive sympathetic nervous system, is much more likely than others to suffer from the condition of extreme emotionality that used to be called neuroticism. Neurotics are, more often than not, unhappy.

Neuroticism is essentially a neurogenetic susceptibility to excessive emotional anxiety. Most neurotics would rather not feel that way. The recognition that feelings are simply the result of neurochemical activity programmed by our genes is the first step toward helping them achieve emotional freedom. The second step is to learn how to enhance the neurochemistry of feelings.

Who would seek to deny the opportunity for anxious, depressive, Opioid Neurotypes to transform themselves into the cheerful, relaxed, Dopamine and Serotonin neurotypes? Those who regard such a goal as an unacceptable assault on freedom and individuality should ask themselves in what way the chronic neurotic is really free. Free for what? Free to be miserable? What of the freedom to improve one's emotional well-being?

Let us embrace the technology by which to enhance the way we feel, that we may allow ourselves to feel *better than well.*

* * *

THINKING

> A thing of no avail he was, until a living mind I wrought within him, and new mastery of thought . . .
> —Aeschylus, *Prometheus Bound*

Here I summarize my position on the nature of the brain, mind, consciousness, self, and intelligence. I introduce a new mind-brain theory—**Neurodualism**—defining consciousness as recognition; self-consciousness as self-recognition, and intelligence as the general information-processing power of the self-recognizing biocomputer that is the human mind.

Summary of Thinking

The brain is an evolved, organic computer. The mind is not an illusion but the brain's recognition of its own embodied existence (**Neurodualism**). The emergence of mind was a pivotal development in human evolution, for it signaled the awareness of mortality, and the consequent emergence of the cognitive drive to defeat death and disease by enhancing abilities. Neurophilosophers who deny the existence of mind or self serve only to encourage the regression of man from a rational, freethinking individual to unthinking zombie, controlled by the forces of state, media, or commerce (**No-Selfism**). Intelligence is the general power of the organic information processor we call the brain. The identification of the genes responsible for intelligence is a necessary precursor to the enhancement of brainpower through Superbiology. Ideologues who seek to deter the investigation of intelligence on supposed moral grounds—the threat of discrimination against the "intellectually challenged"—serve only to hold back the development of the means by which to increase the (brain) power of the least advantaged.

Brain

The brain is a biocomputer, an organic information processor evolved to regulate the functions of the body and guide behavior conducive to survival. The brain serves to scan the environment through the senses (perception); store data as recallable patterns (memory); add new information to memory (learning); identify causal connections between events (logic); plan behavior accordingly (forethought); and manipulate nature to improve survival (creativity/design).

- "Thinking" is information processing—the ability of the organic computer we call the human brain to translate the objects of perception into symbols, store them in memory as causally connected patterns, and utilize the information to assist survival.
- "Reason" is the ability of the brain to recognize the causal connections between events in order to assist behavioral decision making,
- "Logic" is the system of rules governing the practice of reasoning. Logic makes possible learning, forethought, design, and creativity.
- "Learning" is the ability to add new parts to existing information patterns stored in the memory, representing new information about the nature of the world.
- "Forethought" is the ability to predict future sequences of events, based on causal connections between existing patterns in nature.
- "Creativity" is the ability to construct *new* patterns of information, first, in the mind through imagination, then symbolically through design, then practically through technology (when of practical use) or art (when not).
- "Language" is the ability to communicate information by mean of symbols.
- "Writing" is the ability to transfer the symbols of oral language to the physical symbols of text.
- "Thought" is the ability to transfer oral and physical symbols to cognitive symbols, existing only in the brain.

The development of thought and language way beyond that of our ape ancestors was the product of an evolved complexification of the brain. A genetic mutation has been identified that scientists believe reduced the jaw size of our early ancestors, thus placing less stress on the skull—and removing the restraints on its growth. The result was the gradual complexification of the cerebral cortex—the latterly evolved, convoluted outer layer of the human brain (or "gray matter")—and with it, the development of complex language, thought, logic, learning, and creativity—all faculties made possible by the emergence of the human mind.

Mind

The nature of the mind and its relationship with the body is the oldest of philosophical problems. The dominant strain of Western thought on the subject, brain-mind dualism, is perhaps most associated with the philosophers Plato and Descartes.

Metaphysical Dualism

Platonic Dualism

Plato shared the belief of Socrates that the mind is an immortal entity, which returns to an eternal, immaterial realm of pure "Ideas" on the death of the body.[1] Plato's ideal world is the classic philosopher's notion of heaven as a sort of great library in the sky, containing the blueprints for the material world. Interestingly, as we shall see later, a similar concept returns, "scientized," in the emerging view of the world as a sort of cosmic computer, processing bits of information called events or "things." Unfortunately, however, there is no evidence to indicate that the human mind survives the death of the body to "merge" with the cosmic computer. As far as we know, when the body dies, the mind dies with it. The self-recognizing electrochemical activity we call the mind requires the physical body in order to function—at least at this point in our evolution.

Cartesian Dualism

Platonic dualism is repeated by the French Enlightenment philosopher René Descartes—often considered the founder of modern scientific rationality. Descartes conceived of the mind as an immaterial entity living in the brain (in the pineal gland, he believed), which operated entirely separate from the body. Cartesian dualism is expressed in Descartes' famed pronouncement, "*Cogito ergo sum*"—"I think therefore I am," which distinguishes the mind from the body, and associates selfhood with the former.[2]

Cartesian dualism is known today, disparagingly, as the doctrine of the "Ghost in the Machine." The phrase derives from the Greek word "mechane," describing the mechanical crane used in ancient Greek theatre to hoist up an actor portraying God, who would conveniently appear at the end of a play to resolve all problems. In contemporary antidualism, then, the concept of the mind as an entity distinct from the body is ridiculed as metaphysical wishful thinking—simply a means of attaining immortality. Metaphysical dualism is an example of **The Grand Delusion**—the belief in the immortality of mind, sometimes called "spirit" or "soul"—invented to alleviate the fear of death.

Platonic and Cartesian dualism are essentially philosophical versions of theistic metaphysics—the belief in an immortal soul.

Neural Monism

The opposite of metaphysical dualism—sometimes called neural monism—is the currently **Fashionably Orthodox Position Among Scientists (FO PAS)** on the mind-brain question.

Neural Monism holds that the brain and mind are two aspects of the same, material thing—the mind is simply "the brain at work":

> The idea of mind as distinct from the brain, composed not of ordinary matter but of some other, special kind of stuff, is dualism, and it is deservedly in disrepute today. The prevailing wisdom, variously expressed and argued for, is materialism: there is only one sort of stuff, namely matter the physical stuff of physics, chemistry, and physiology, and the mind is a physical phenomenon. In short, the mind is the brain.[3]

One would not argue with the view that the brain and mind are the same in the sense of both being "material stuff"—cells, chemicals, and electrochemical activity. However, *from the perspective of the actual experience of a conscious human being, brain-mind dualism is no illusion, but an experiential fact of life!* For *experientially*, the mind is quite different from the body. Thinking is a fundamentally different and more complex mode of response to the world, which cannot reasonably be equated with the organic hardware in which it occurs. There are two elements present, not one: biological and cognitive. To try to reduce the two to one is to eradicate the vital cognitive recognition of our own biological condition—in short, the fact that *we are self-recognizing brains in decaying bodies subject to a genetic program for self-destruction.*

The neural monists who tell us that the mind is an illusion are living in cloud cuckoo land—a very dangerous place to live. Next I shall explain why.

No-Selfism

> Making a decision turns out to be not a matter of self-control and will-power, but allowing the false self to get out of the way.
> —Kenan Malik, *Man, Beast and Zombie* (2000),
> on *The Meme Machine* by Susan Blackmore

Recent attempts to expound a scientifically based antidualist position have resulted in a bizarre phenomenon—an unholy alliance between New Agers and neurophilosophers, both of whom declare that there is no such thing as either the mind or the self. The New Agers' rejection of brain-mind dualism might be described thus:

The mind is "the false ego"—a delusion of consciousness from which we must free ourselves. In reality, nature is "spiritual" (or, if one has been reading the Tao of Physics, "quantum energy"). The "separate self" (thinking mind) is an illusion disguising the reality of "cosmic oneness." The conceptual separation of mind from body heralded by Descartes signaled the dawn of a Modern Age in which man (not woman) lost touch with his body and emotions, and came to regard nature merely as material to be manipulated by the cold, unfeeling, rational, male mind. The result has been Man's inhumanity to Man, caused by the desire to defend the "grasping ego"—which does not in fact exist at all.

With disturbing similarity to this hogwash, an antidualist argument has been proposed by former parapsychologist turned neurophilosopher, Susan Blackmore, in her book, *The Meme Machine*.[4] The essence of her argument might be described thus:

The mind is not a thing, but a description of "brain activity." Descartes was wrong to distinguish mind from brain, because there is no such thing as a mind. Mind is synonymous with self, therefore, neither is there such thing as a self. We therefore have no mind, and no self—there is no willful, self-governing, independent "I" in control of behavior, just brain activity which trundles along entirely of its own accord, independent of human volition. Since there is no self, human beings therefore have no Free Will, because there is no self to do the willing. Humans therefore have no mind, no self, and no will. Instead, behavior is controlled by "entities" called memes, which infest brains in their own survival interests. We should learn to love our condition of having no mind, self, or will; allow memes to make our decisions for us, and abandon ourselves to the moment, living in the perpetual present without goal or purpose.

Just as the Buddha thought, sitting under his tree—doing nothing.

It comes as no surprise to find that Dr. Blackmore is an active (or should that be passive?) practitioner of Zen Buddhism. Though careful not to mention the B word in her book, she gives the game away when introducing the phrase "false self"—the standard New Age neo-Buddhist description of the "ego"—the conscious, rational, thinking mind. On her Web site (susanblackmore.com), though, a rather clearer picture emerges:

> There is no reason to believe we would all behave much worse if we accepted the illusory nature of the self. Indeed many people argue that the self is the root of all human suffering, and that practices like meditation, which aim to undermine it, lead people to behave better, not worse. . . . This is where the moral objections of the critics come to the fore. They argue that without a sense of self—with free will and personal responsibility—we could not have effective

> legal systems or expect people to behave morally. I disagree. . . . We begin to see the world around us as the result of selfish memes competing to use us for their own propagation, rather than seeing us as rational beings in charge of our world.

"Rather than seeing us as rational beings in charge of our world."

Really.

Dr. Blackmore also lists her activities and published works. They are most illuminating—or perhaps one should say, enlightening. I confess to not yet having got around to reading her 1990 paper, "No Self in Buddhism and Psychology."

Perhaps my selfish memes don't want me to.

The foreword to what he calls "Susan Blackmore's admirable book" was written by one Richard Dawkins, Professor for the Public Understanding of Science.

It is somewhat disturbing that neurophilosophers have managed to get themselves into the position of agreeing with the beliefs of their most irrational of opponents. One wonders which view is more absurd, the No-Selfism of the New Agers or the New Age neurophilosophers.

Against No-Selfism

Selfish Gene theory holds that genes replicate for their own benefit at the expense of the human bodies they inhabit. To be strictly analogous to a gene, then, a meme must be a unit of cultural information that replicates independently of the will of the members of the culture it inhabits. Hence, a religion or a catchy popular tune replicate within minds independently of human will. But this is not the case. Rather, *memes replicate because human beings have deliberately spread them in society—for power or financial gain.* Pop songs make money—their composers deliberately try to make them as catchy as possible, in order to make as much money as possible (whether they admit it or not). Reverse baseball caps are a lucrative commercial product. Political ideas are immensely strong power tools. *For he who holds the power, is he who controls the meme.*

Memes are words, thoughts, ideas, habits. These things are deliberately placed in the minds of others for the basic purpose of control—both in the worlds of politics, media, and advertising and in personal relationships, when we influence others to adopt our favored habits (as when, for instance, a husband is persuaded to go shopping, and a wife to have sex). If memes truly operated of their own accord, we would enter Dr. Blackmore's neo-Buddhist world of No-Selfism in which we have no will of our own, but act according to the dictates of "entities" which control our thoughts and behavior in their own survival interests—an idea that must be music to the ears of potential authoritarians, whether in the world of politics, big business, or the media.

In 2003 the public was memewashed to accept a particular political policy —the invasion of Iraq—through the continual, daily repetition of three memes

—the idea that Iraq was ruled by an "*evil dictator*" who "*gassed his own people*" and possessed "*Weapons of Mass Destruction.*" As a result of daily indoctrination on both sides of the Atlantic with these three simple memes (over and over and over again), we began to hear them echoed by the public, parrot fashion—as the opinion polls slid away from the former majority in favor of peace, to a new consensus for war.

Memes do not act independently at all—they are deliberately placed into minds by others, for a specific purpose.

That purpose is power.

No-Selfism and Schizophrenia

The suggestion that human beings are powerless pawns of entities called memes, which infest their minds for the purpose of their own self-replication—that we are merely mindless zombies ready to be brainwashed into repeating whatever ideas and behavior to which we happen to be exposed—is both false and extremely dangerous.

When Richard Dawkins concluded his book *The Selfish Gene* with the line, "We, alone on earth, can rebel against the tyranny of the selfish replicators," he meant, specifically, that the conscious mind may override the dictates of both genes *and* memes—both the biological determinism of selfish instincts and the social indoctrination of undesired cultural conditioning.

But Dr. Blackmore reverses the evaluation of selfish memes. She deems them good (or, perhaps, God?). We should get the rational mind or self out of the way and let memes rule our lives. Here is Dr. Blackmore writing on her Web site about technology, which she sees not as the product of willful, rational, autonomous human beings, but of entities living in our heads.

> But have we really created the technology? Look at this from the meme's eye view. They are the replicators, and they don't care about our welfare any more than genes care about the welfare of the creatures they create. So could all these computers, net servers and web sites be the creation of the selfish memes? . . . Perhaps I am getting carried away here, but this is the power of the meme meme. Once you understand it the world looks different. Indeed "I" look different too, for "I" am only a co-adapted meme-complex—or, as philosopher Daniel Dennett puts it, an ape infested with memes.

So according to Dr. Blackmore, willful human beings don't create technology of their own accord—creatures living in our brains do.

Schizophrenics, too, believe their brains are inhabited by entities that rule their lives. The late postmodern philosopher–psychologist Gilles Deleuze believed schizophrenia to be a privileged position characteristic of the Postmodern

Condition, in which a stable sense of self is replaced by an endless series of masks. Thus the schizophrenic is "free to be mad." He called his psychology Schizo-analysis.[5] It sounds much like that espoused by Dr. Blackmore, who also thinks we should abandon our sense of self—to the Meme Spirit, presumably.

New Age neo-Buddhists, too, think of schizophrenia as a privileged position. They like to say that the only difference between the schizophrenic and the mystic is that "one sinks, where the other swims," and refer to a mental collapse as "not a breakdown but a breakthrough."

Beware those who preach the dissolution of self. Where selfhood perishes, madness descends.

No-Selfism as Non-Sense

Camille Paglia has expressed the truth about selfhood and sanity with characteristic clarity:

> Man is not merely the sum of his masks. Behind the shifting face of personality is a hard nugget of self, a genetic gift. The self is malleable but elastic, snapping back into its original shape like a rubber band. Mental illness is no myth, as some have claimed. It is a disturbance in our sense of possession of a stable inner self that survives its personae.[6]

She's right. The self is no illusion—except for the insane. When memes act independently of the self—that is, of the human will—we call it brainwashing, mind control, or social conditioning, In the naïveté of her neo-Buddhist No-Selfism, Dr. Blackmore is implicitly advocating our regression into mindless zombies, controlled by the dictates of church, state, business, or media.

The adoption of ideas and habits does not have to be the product of unconscious social conditioning. The rational human mind may *choose* to adopt one idea or habit rather than another. When we choose to think whatever we like, we call it "free will" or just "freedom." It is, one might suggest, something worth protecting.

Neo-Buddhist New Age neurophilosophers are quite entitled to go around "thinking" they have no self, but one wishes they would keep their delusions to their non-selves.

The neo-Buddhist doctrine of No-Selfism is nonsense. When left to New Agers, the belief that the autonomous, rational, thinking, willing self is an illusion may be fairly harmless. But when espoused by influential philosophers, scientists, academics, it becomes positively dangerous. For the promotion of No-Selfism denigrates the importance of the autonomous mind—and by extension, the notion that an individual is master of his fate, captain of his "soul." And there are those who could wish for nothing better than the descent of society into an unthinking mass of docile robots conditioned to believe what the scientists and philosophers have told them—that they have no autonomous, decisionmaking, independent, willful self—that they are, literally, zombies.

No-Selfism as Zombification

The dangers of No-Selfism have been expressed in a book by the British writer Kenan Malik, *Man, Beast and Zombie*,[7] which presents an excellent overview of our changing understanding of human nature.

The "beast" of Malik's title is the view of man being inadvertently promoted by neo-Darwinist evolutionary psychologists: a slave to his biological instincts, or selfish genes. The "zombie" is the human being "unselfed" by the New Age neurophilosophers, and consequently dominated by *social* forces outside of his control. Malik rejects the view of both neo-Darwinists and neurophilosophers—"man as beast and zombie"—in favor of rational, autonomous selfhood. One can but agree. No-selfism encourages the kind of mindless passivity desired and required by all totalitarian dictatorships, whether from politics, business, media—or a deadly collaboration between all three.

Picture for a moment a terrible future, an image of things to come if we should allow ourselves to be "unselfed." It is the mid-twenty-first century. Big Brother is broadcasting a message directly into the cybernetically implanted brains of the populace:

Your self and your mind are illusions—you have no will of your own. Watch cruelty TV to safely release your aggression. Pay attention to the hourly, automatic net broadcasts, for news of our glorious leader's latest victories against the Evil Dictator whose Weapons of Mass Destruction threaten us all. Don't think—chill out with Somadope. And remember—you are not alone—techno-surveillance is absolute. Big Brother **is** *watching you.*

The philosopher Hannah Arendt expressed a similar concern for the future. It was, she said, as if:

> Individual life had actually been submerged in the over-all life process of the species, and the only active decision still required of the individual were to let go, so to speak, to abandon its individuality, the still individually sensed pain and trouble of living, and acquiesce in a dazed, "tranquilized," functional type of behavior. It is quite conceivable that the modern age—which began with such an unprecedented and promising outburst of human activity—may end in the deadliest, most sterile passivity history has ever known.[8]

It is not, however, (as Arendt believed) through technology that such a Brave New World could come about, but through our willingness to abandon the power of the rational mind and the autonomous self.

NEURODUALISM

> The brain may be regarded as a kind of parasite of the organism who dwells within the body.
> —Arthur Schopenhauer, *The World as Will and Idea* (1818)

Those who deny the existence of the mind or self are wrong. The mind is a self-recognizing information processor aware of its own embodied existence. The brain collects information about the world through the senses, and stores it in memory as electrochemical symbols (exactly how and where is one of the remaining mysteries of neuroscience). At a certain point of complexity, the brain becomes able to include a symbol of *itself* in memory. It thus becomes aware of its own existence—that is, it becomes self-conscious. *Self-consciousness is self-recognition*—the ability of the brain to recognize that it exists. The self is an embodied computer that recognizes its own existence as a functioning entity, through the ability to *feel* that it exists, by means of the sense organs, and the ability to translate this sensation of self into a symbol among symbols stored in its internal memory, through the capacity for internalized language—that is, thought. "I" is the brain's symbolic representation of the physical sensation of being alive. When a brain can say "I," it has become self-conscious—it is no longer merely a brain, but a brain *and* a mind.

Mind and brain are two distinct concepts, signifying two distinct parts of human experience. The mind is a self-recognizing (or conscious) brain aware of its existence distinct from the body in which it is "housed." When I look at my hand, a self-recognizing brain is perceiving a part of the body with which it exists in a symbiotic relationship, but from which it distinguishes itself ontologically. "I" am not my hands precisely because I am able to distinguish between my body and my self. The self is the mind aware of its existence as a self-recognizing information processor, and of the body in which it exists. The self resides *within* the body because thinking occurs *within* the brain. If someone were to cut off my hands—perhaps as a punishment for unacceptable beliefs—I would not have lost my self. "I" am a thinking entity—I only "lose my self" when I am unaware that I exist, that is, when I am unconscious. I, my self, am the conscious (self-recognizing) aspect of the embodied biocomputer that is a human being.

The self-recognizing brain is aware of the distinction between itself, as a conscious information processor, and the rest of the body, which is less complex and therefore unconscious—unaware of its existence. The fact that we instinctively say "my body" demonstrates that we draw an experiential distinction between body and mind. The essential aspect of a human being—unlike, say, a cow—is not his body as a whole, but the self-conscious brain he calls his mind. Man is ultimately his mind, not his body—this is the real meaning of *Homo sapiens*—the thinking animal. It is the mind—the brain endowed with the capacity for self-recognition that distinguishes humanity from the less-evolved animals, with simpler brains.

The emergence of the human mind allowed man to inhibit instincts induced by the selfish genes, in order to increase his own survivability and well-being. It was the evolved frontal lobes of the cerebral cortex—the convoluted outer layer of "gray matter" beneath the skull constituting nine-tenths of the brain—that was responsible for the emergence of the mind.

Self-consciousness—the ability of an entity to recognize its own existence—allows for the *voluntary* inhibition of the emotional impulses produced by the older, more primitive (that is, simpler) parts of the brain in its own interests.

It is the evolved, autonomous human mind, with its capacity for forethought through reason that drives human beings to forego the pleasures of the moment for the rationally deduced prospect of future rewards. The brain's capacity for reason permits the inhibition of immediate gratification for the promise of future gain. It was the ability of the conscious brain to defer gratification by inhibiting genetic instincts that made possible the emergence of human civilization.

The ability to define oneself as mind, not body, is a vital development in the evolution of the species. It is instinctive of a self-recognizing brain, aware of its own mortality, to seek to override the body's genetic program for self-destruction. I am not my body but my mind. I know it, because I recognize that I want to live and my body is programmed to die!

We are self-recognizing biocomputers encased in bodies that are programmed to self-destruct, and doomed to do so unless our minds act autonomously of the bodies in which they are housed, by employing reason to override their programming for self-destruction.

Mind-body dualism is no illusion—mind and body are two. The mind wants to live, but the body is programmed to die. We therefore have a choice: to remain in a condition of slavery to the selfish genes that demand our self-destruction, or to develop the technology by which to rewrite our genetic programming in our own interests—ever-increasing survivability and well-being.

IN DEFENSE OF DESCARTES

Plato and Descartes may have been mistaken in the belief that the mind is an immortal entity that lives forever after the body's demise. But the no-self New Age neurophilosophers who seek to reduce mind and self down to the level of nothingness are equally self-deluded—or at least, they would be, if they believed they *had* a self.

Those who regard the Cartesian "dichotomy between mind and body" as the source of all human ills are sorely mistaken. *Cartesian dualism is not the cause of Man's problems, but the beginning of the solution.* Descartes's "Cogito ergo sum"—I think therefore I am—signifies the brain's recognition of its own existence distinct from the body. The evolution of the human mind allowed man to wake up to the horror of his condition of slavery to a genetic program for self-destruction. When Descartes defined existence in terms of his thinking mind, he did not usher in an age of collapsed human values, but the start of the brightest era in human history—the quest for freedom from slavery to the selfish genes which demand our self-destruction.

Descartes is not the villain but the hero of the piece. Descartes's cogito marks the beginning of human evolution from *Homo sapiens* to *Homo cyberneticus*—man the steersman of his own destiny.

The New Cartesianism

The neuro-Cartesian meme appears to be spreading. In a recent article on the mind-body problem in literature, the English novelist A. S. Byatt tells a significant story. After reading the fashionable popular expositions of anti-Cartesian neural monism—the current scientific orthodoxy—she unexpectedly came across a different view: the Canadian philosopher Ian Hacking,[9] lecturing on "Body and Soul in the twenty-first century":

> He has come to think, he says, that we are in a new Cartesian phase—we feel our minds to be separate from our bodies because of what we can do with things like body parts, genetic material, pace makers and cyborgs. . . . I was quite shocked by this view of things, as I had been reading Antonio Damasio . . . on Descartes' Error—which was to suppose that there was anything that could be called the mind which differed from the message system of the nerves and neurons. . . . I am naturally sympathetic to Damasio, . . . but I have to say that Hacking is saying something true and important—the more we control and mess with the wet stuff, the more difficult it is to feel the controlling intelligence as simply part of flesh, blood and wet stuff (and silicone, possibly).[10]

Precisely. While the chattering classes lap up the antidualism they have been told is the scientific orthodoxy, those capable of thinking for them*selves* recognize the absolute necessity of **neuro-Cartesianism**—for to deny the reality of dualism is to deny the existence of the mind, the self, the personality, and consequently human freedom—the ability to determine one's own destiny.

Unable to face death, antidualists seek to abolish the self that recognizes its fate. By denying the reality of mind, self, and will, the no-selfers doom themselves to a genetic program for self-destruction.

Don't listen to those who tell you your self is an illusion—they are playing with words, and they are wrong. You have a self—a self called you. Your self is your brain's recognition that it exists as a physical body—flesh, blood, and neurotransmitters in the real world of time and space.

Don't let anyone take it away from you.

Mind Uploading

Descartes believed that the mind could exist without the body, and the body without the mind. The theory of neurodualism opens up the possibility that, contrary to popular belief, he may have been right.

The mind is the body's recognition of itself, a computer program stored in the organic brain, which includes memory of its own embodied existence as a symbol among symbols.

It seems likely that the initial emergence of mind required the presence of sensory apparatus. The ability of the brain to recognize itself as an embodied entity may have been made possible only by means of sensation. The brain may have come to recognize it exists because it could *feel* that it exists. Sensation requires a physical body. Selfhood may require sensation. Thus, it seems likely that artificial intelligence could achieve consciousness only by means of sensory apparatus. (Interestingly, such an "e-skin" is now being developed by scientists.)

However, although the emergence of mind may *initially* require the presence of sense organs, once the recognition of embodied existence we call self-consciousness has been stored in brain memory, it is theoretically possible that the continued existence of the mind may no longer depend upon the continued existence of the body—for physical sensation can be removed, and the mind remain capable of self-awareness. Numb the body, and the mind still thinks.

The mind is a software program uploaded into the hard drive we call the brain, which includes recognition of its own existence. Software programs can, of course, be copied. The software program that is the mind could therefore potentially be copied and downloaded again into another body—whether organic or artificial.

If the self is a computer program—software rather than hardware, "mentality" rather than the brain in which thinking takes place—then it may potentially be possible to "upload" the mind-self to another computer, thus preserving it even after the death of the original body. A brand-new body could then be "grown" through Superbiology, and the self "uploaded" into the new brain. "Mind Uploading" thus becomes a speculative possibility as a future method of attaining cognitive immortality through technology.[11]

Flesh, Blood, and Neurotransmitters

The concept of "mind uploading" raises a difficult question about identity, for if there are several existing copies of the software program that is your self, but no original body, then which copy is "the real you"? The answer, of course, is that they are *all* you, because the essence of selfhood is a software program which can be copied indefinitely. Thus one must consider the complications of a scenario in which several versions of "you" could be uploaded into different physical or mechanical bodies, creating the phenomenon of "cybertwins."

But here we shall leave the subject. The possibility of cognitive immortality through mind uploading is just one of many futuristic ideas contemplated by transhumanists. In the present context, it may be used as a metaphor to help explain my concept of neurodualism, by distinguishing between the mind and brain, and identifying selfhood with the former.

In the twenty-first century, the human body will increasingly be capable of cybernetic enhancement—that is, the insertion of computer implants to improve functioning. Artificial pacemakers, of course, are already routine. Implants enabling the control of electronic equipment are now being developed. Soon it will be possible for the mind to control a computer, the TV, or the Internet by thought alone. In the twenty-first century, telekinesis—the dream of the parapsychologists—will be realized through technology.

But there is no necessary requirement to leap from the support of cybernetic enhancement to the belief in the complete replacement of the human body with non-organic material, as some transhumanists advocate. Many of us prefer to think more in terms of strengthening and enhancing the bodies we have at present, rather than entertaining contemporary versions of 1950s sci-fi "brain in a jar" scenarios likely to play into the hands of bio-Luddites keen to dismiss transhumanism as puritanical, neo-Platonic dualism born of body-loathing—a "fear of the flesh" we do not share. For we rather like being embodied, organic entities, made of flesh, blood, and neurotransmitters.

In fact, we like it so much, we'd like to go on doing it forever.

INTELLIGENCE

> It's almost impossible to study the genetics of intelligence either in the US or the UK because it is socially contentious. You can say that the net effect is that it helps to perpetuate a system where people are dumb.
>
> —James Watson, *The Independent* (February 3, 2003)

There is much argument as to the nature of intelligence—most of it bunk.

Intelligence may be defined as the measure of the brain's ability to process information; the general power of the self-recognizing organic information processor that is the human brain, the most important aspects of which include:

- Speed: the capacity for quick thinking, analogous to the speed of a computer's internal processor

- Logic: the ability to determine regular causal connections between events
- Learning: the ability to add new bits of information about the world to existing causally connected patterns stored in memory
- Creativity: the ability to construct entirely new patterns of information
- Inhibition: the ability to withstand a stimulus—that is, to prevent the body from acting on emotional impulse

The capacity for memory, though vitally important, cannot properly be considered an aspect of intelligence as, contrary to popular belief, the ability to remember facts is independent of the ability to connect them together logically. Memory may help you win a TV quiz show, but it was intelligence that enabled John Logie Baird to invent the TV.

The Genetics of Intelligence

> The tendencies to deny all genetic inequality in mental ability are rooted in cloud cuckooland, however laudable their political or moral motivation.
> —Richard Dawkins, foreword, *The Genetic Revolution and Human Rights* (1999)

The seat of intelligence lies in the frontal lobes of the cerebral cortex, the latterly evolved "gray matter" built up around the more primitive parts of the brain we share with our prehuman ancestors. The evidence for the neurogenetic basis of intelligence derives from a combination of neuroscience, genetics, twin studies, and common sense. For instance:

- A 2001 study of twins at the University of California identified a link between high IQ scores and density of neurons in the frontal lobes.[12]
- Scientists examining records of Einstein's brain found that the area of the frontal lobes associated with cognitive skills was significantly larger than average.[13]
- A mutated gene has been identified with a form of mental retardation (GRP56). Researchers believe a fully functioning version of the gene may have contributed to the evolution of human intelligence.[14]

In addition, genetically identical twins brought up apart, in different homes, with different families, have consistently been found to score very similar levels on IQ tests for reasoning ability (logic), while genetically dissimilar children living *together* in the *same* home show *no* correlations on their scores—clearly indicating that genes, not environment, is the main determinant at work in determining intelligence.

And of course, plain common sense tells us that intelligence is innate. All parents of more than one child know that some children are born cleverer than others, just as other aspects of their personalities differ widely, such as introversion or emotionality, artistic or musical ability. Few people outside of New Age therapy groups believe we could all be rocket scientists or nuclear physicists if we learned to love ourselves more. All but politically motivated ideologues know what is meant by the difference between Einstein and an idiot. If there really were no neurobiological differences in intelligence, it follows that we could *all* become Einsteins, and the dunce we remember from our schooldays must actually have possessed the same abilities as the Harvard high flyer—if only his parents had been able to afford daycare, perhaps. In short, unless one considers intelligence to be the product of a disembodied spirit that squats in a fertilized egg at the moment of conception, to regard intelligence as the product of genes is the only—*intelligent*—view.

Bioegalitarianism versus Civil Egalitarianism

Unfortunately intelligence is not *always* to be found at the forefront of scientific enterprise—that Janus-faced trickster, ideology, also has a nasty habit of rearing its duplicitous head. Research into the genetic basis of intelligence has been held back by those who believe we should not seek to define or examine, let alone enhance, intelligence in case it leads to discrimination against the "intellectually challenged." This is like wanting to ban skyscrapers so as not to discriminate against people with vertigo.

Those who oppose the investigation into the neurogenetic basis of intelligence regard any recognition of innate differences in brainpower as socially unacceptable, because such knowledge could potentially be used to justify prejudice and discrimination on the grounds of "un-braininess." They thus claim that humans are born with identical potential abilities, such that all social inequality must be the result of cultural as opposed to biological disadvantage. We might call this position **bioegalitarianism**, to be contrasted with the wholly sensible and compassionate position we might call **civil egalitarianism**, defined as "the belief in equality of respect and equality under the law, irrespective of abilities."

Transhumanists are completely disinterested in the question of who might be cleverer than whom. One-upmanship is a dismal parlor game—what matters is the freedom of individuals to seek to enhance their intelligence, and that of their children, through Superbiology. Such a goal requires a comprehensive understanding of the neurogenetic basis of intelligence.

Of course, providing the optimum environment and decent education for every child is a must. The emerging science of epigenetics will increasingly emphasize the vital importance of avoiding maternal stress, depression and anxiety, drug and alcohol abuse, poor diet, and exposure to pollution, all of which may detrimentally effect the proper functioning of genes in the newborn child, and reduce the capacity for cognitive functioning. But optimizing the pre- and postnatal environment can only ensure the optimal functioning of genes that are already present. Superbiology offers the possibility of *enhancing* genes to increase the *potential* intelligence which may be brought out by an excellent nurturing environment.

The Value of Raising Intelligence

The prospect of increasing brainpower through Superbiology is one of the most exciting of all. What might be the benefits of raising intelligence? One is shocked that the question is so often asked! Is it truly possible that our culture is now so dumbed down by the postmodern metameme of cultural relativism that the public is incapable of recognizing any value in intelligence at all?

Some benefits of raising intelligence include:

- The social value of innovation, ingenuity, creativity, invention, discovery.
- The value of being able to get a better job.
- The joy of knowledge; the simple pleasure in using one's brain to the maximum capacity.
- The value of decreased antisocial behavior. Reason is the ability to identify causal connections between events. Intelligent people don't throw litter on the ground, because they are capable of reasoning that the end result will be filthy streets for all. Stupid people are less able to make the causal connections between present actions and likely future consequences. They act without forethought—without thinking. How many times have we heard, or used the excuse "I'm sorry, I didn't think!" A society in which everyone thought before they acted would enjoy the benefits of dramatically reduced antisocial behavior.

Will we really learn how to raise intelligence by altering genes? Yes, of course, just as we continually increase the processing power of the brains we call "computers." But to improve something, you have to understand the nature of what it is you're trying to improve. The necessary precursor of the attempt to improve brainpower through Superbiology is the identification of the neurogenetic basis of intelligence. The purpose is not to discriminate against the unbrainy, but to learn how to improve brainpower. Many if not most parents would welcome the opportunity to raise their children's level of intelligence. What parents would not wish their children to have as strong a head start in life as possible? And what parents would deny that high intelligence provides such a head start?

* * *

PERSONALITY

NEUROTYPOLOGY: THE NEUROCHEMISTRY OF PERSONALITY

> I suspect that most temperaments will be traceable to physiological differences derived from neurochemistry.
> —Jerome Kagan, psychologist, Harvard University, *Galen's Prophecy* (1994)

Summary of Personality

Personality may be defined as the habitually observable characteristics of individual behavior. Here I seek to identify the neurochemical basis of personality as a precursor to a future science of eugoics—the neurochemical enhancement of the self, in the perfectly natural desire to feel *better than well.* I propose a theory as to the evolved function of two of the neurochemicals most likely to be the target of neuroenhancement, dopamine and serotonin, and outline my own neurochemical personality theory, the **Four Neurohumors**. Finally I respond to the experientialist bio-Luddites who deny and oppose investigation into the biological basis of personality through fear of categorization, prejudice, and discrimination. The **New Biology** is not the **old biologism**. **Bio-empathy** has replaced **bio-elitism**. We are *all* victims of biogenetic limitations. Eugoics will be of greatest benefit to those who suffer most from the disadvantages of "genetic injustice"—an unlucky roll of the genetic dice, causing impediments of body or mind.

The Neurochemistry of Personality

"Who am I?"

The question lies at the very heart of both philosophy and psychology. But what exactly makes one personality differ from another?

Theists say, "Personality is the product of an invisible ghost that enters an egg at the moment of fertilization by a sperm." Sorry, can't go for that one.

Liberal egalitarians say, "We're born as identical blank slates—personality is conditioned by society." In the image of the state, presumably. Good news for all dictators.

The model of psychology I propose, **Neuromotive Psychology**, says, "Personality is rooted in biology. Subtle neurobiological differences produce unique temperaments, or **neurotypes**, which are further refined by life experience."

Neither invisible spirits nor social conditioning, but neurobiology lies at the heart of personality.

Central to the biology of personality is the role of neurochemicals. There is now a growing recognition of the strong influence of neurochemistry on all aspects of personality, including the tendency toward extroversion or introversion; emotion stability or instability; cheerfulness or gloominess; gregariousness; conscientiousness; logic, or abstract reasoning ability; artistic creativity; even personal attitudes, such as political and ethical beliefs, and aesthetic preferences.

Major advances have been made in our knowledge of neurochemistry during the past half-century or so. Most significant has been the discovery of neurotransmitters—brain chemicals that function rather like glue, emerging from cells temporarily to connect them together, in order to transmit electrical impulses between them. Neurotransmitters play a vital role both in maintaining health and influencing personality traits.

When the next report appears in the newspaper informing you that scientists have linked a particular personality or behavioral trait to a specific gene, look a little more closely and you will see somewhere (usually toward the end of the report) that the gene in question "codes for" a specific neurotransmitter such as serotonin, dopamine, noradrenaline, or GABA. In other words, it isn't genes themselves that directly affect behavior, but the type and level of neurochemical they produce.

In short, neurochemistry is the basis of personality.

Neurohumors

> Today, in the era of the neurohumors, temperament has made a comeback.
> —Peter D. Kramer, *Listening to Prozac* (1994)

On their discovery in the late twentieth century, neurotransmitters were initially referred to as **neurohumors**, signifying the link with the first attempt to correlate biochemicals with behavioral traits—the ancient Greek personality theory of the four humors—originating some twenty-three hundred years ago!

The power of the four humors is demonstrated by its longevity. The theory retained its popularity through the Middle Ages, into the Renaissance and beyond, and is with us today still, through our continued use of the terms *sanguine*, *phlegmatic*, *melancholic*, and *choleric*, to describe the basic personality traits we might call cheerful (or friendly), relaxed (or "chilled"), gloomy (or depressive), and touchy (or neurotic). Which of these traits are preferred by the vast majority of us can be demonstrated by a quick glance at a list of advertisements for flat mates in your local paper. The two most common adjectives describing the ideal living companion are "friendly" and "chilled" (that is, easygoing or laid-back). These two characteristics correspond to the sanguine and phlegmatic temperaments respectively.

The psychiatrist Peter D. Kramer revived the use of the term *neurohumors* in his seminal book, *Listening to Prozac*,[15] which signals the beginning of the transition from twentieth-century psychotherapy to twenty-first-century neuropsychology. Here I shall continue the use of the term *neurohumors*, for four reasons, none of which are of any great importance:

- First, *neurohumors* may be used as a convenient umbrella term to describe all classes of behaviorally significant endogenous chemicals, such as hormones, neurotransmitters, neuromodulators, and neuropeptides.
- Second, as with the continued use of the term *Hippocratic Oath*, the word *neurohumors* emphasizes the link between ancient Greek humanism and *trans*humanism—the belief in the ongoing quest for knowledge (for the transhumanist agrees with Socrates: "an unexamined life is not worth living").
- Third, a link with the theory of the four humors is appropriate because it turns out to have been on the right lines—it does indeed seem possible to correlate four basic dimensions of personality with four types of biochemical or "humor"—not those of the original theory (black bile, yellow bile, blood, and phlegm), but *neuro*humors such as serotonin, dopamine, adrenaline, and opioids.
- Finally, *neurohumors* just sounds nicer. Why shouldn't science be aesthetically pleasing? It certainly wasn't an aesthete who invented a biological term like *phylogeny*, for instance.

The Function of the Neurohumors

At the point of writing, there is no absolute scientific consensus on the behavioral function of the neurohumors. One problem is the fact that they don't operate in "splendid isolation" from one another. Thus, it is as "biologically incorrect" to speak of "*the* neurohumor for happiness" as to speak of "*the* gene for criminality." The matter is further complicated by the fact that neurohumors may serve more than one function. However, so much is now known about the behavioral effects associated with the best-known neurohumors that one can certainly speculate about their main, original, evolved function. Here, then, is a list of some of the main neurohumors and the types of behavior they may have made possible for our Stone Age ancestors:

Acetylcholine	Waking and remembering (deficient in Alzheimer's disease)
Glutamate	Thinking (the most widespread "excitatory" neurohumor in the brain)
Dopamine	Planning (deliberate/intentional/volitional activity or "will")
Serotonin	Feeling confident (by reducing sensitivity to stimuli)
Adrenaline	Hunting (by inducing physical activity)
Opioids	Reducing pain (by flooding the body with endorphins)
GABA	Relaxing (by reducing mental and physical arousal)

Testosterone	Mating (by inducing sexual activity)
Estrogen	Gestating (at its peak in pregnancy)
Oxytocin	Nurturing (the "maternal love chemical")
Adenosine	Feeling tired (accumulates during the day)
Melatonin	Sleeping (released in darkness from the pineal gland)

As our understanding of neurohumor function has risen, the first serious attempts have been made to identify the neurochemical basis of personality traits. Perhaps unsurprisingly, such attempts have come not from neuroscientists, but from the world of psychiatry—that is, from those who actually have to *prescribe* drugs which change mood and behavior by altering neurochemistry.

Once the causal correlations between specific neurohumors and dimensions of personality have been firmly established, we will see the development of a comprehensive theory of the neurochemical basis of personality, and the emergence of a new science we might call **Neurotypology**.

As an example of the sort of neurochemical personality theory that will increasingly emerge in the coming years, here I shall offer a theory of my own, focusing on the possible evolved function of the two neurohumors perhaps most likely of all to be targeted by a future science of neuroenhancement—dopamine and serotonin.

The Function of Dopamine and Serotonin

My theory in a nutshell: Scientists refer to neurochemicals either as "excitatory" or "inhibitory"—that is, they either increase or decrease activity within the nervous system. Living organisms maintain their existence through a process of continual self-adjustment to changing environmental conditions—a process known as "homeostasis." Neurochemicals function as a homeostatic regulation mechanism, helping to maintain the body in a state of dynamic balance conducive to survival, through continual self-adjustment to changes in external conditions. I suggest that the basic original evolved function of dopamine and serotonin was the regulation of mental excitation and inhibition—or activity and rest.

Dopamine is an excitatory neurohumor that excites the central nervous system to facilitate conscious, willful control of the body. High levels of dopamine correlate with a brain that is "fully turned on," hence the psychological effect of stimulants such as caffeine and cocaine. Low dopamine correlates with a brain "turned down low," hence the inability of Parkinson's disease sufferers to *willfully* instigate movement. Genes which lead to the production of high quantities of dopamine produce an extrovert, goal-seeking, driven dimension of personality—the **Dopamine Neurotype**.

Serotonin is an inhibitory neurohumor that reduces the sensitivity of the brain to external stimuli, thus reducing the reactivity of the sympathetic nervous system, resulting in mental calmness, relaxation, or "peace of mind." The pleasurable, stress-reducing effect of serotonin is nature's reward for attaining the ideal state of homeostasis—that of complete relaxation. High serotonin correlates with a sympathetic nervous system that is "turned down low," reducing its sensitivity to stimuli—hence the desensitizing effect of eugoic drugs such as Prozac, now used to treat anxiety in preference to benzodiazepines such as Valium. Low serotonin correlates with a sympathetic nervous system that is "fully turned on," hence the emotional hypersensitivity of low serotonin-induced, anxiety-and-depressive disorders. Genes that lead to the production of high quantities of serotonin create a calm, relaxed, self-assured dimension of personality—the **Serotonin Neurotype**.

In short, dopamine and serotonin are mood-raising pleasure rewards encouraging mental activity and calmness respectively. Dopamine makes you sanguine (mentally lively). Serotonin makes you phlegmatic (physically calm). Together, they represent the ideal state of *mens sano in corpore sana*—a sound mind in a sound body. Happy and relaxed—or "friendly and chilled."

Next I shall present evidence for my theory of dopamine and serotonin function in a little more detail.

Dopamine and a Sound Mind: The Increased Intentionality Hypothesis

> Altogether it can be suggested that a correlation exists between dopamine innervation and expression of cognitive capacities.
> —A. Nieoullon, *Progress in Neurobiology* (2002)

At the time of writing, the **Fashionably Orthodox Position Among Scientists** holds that dopamine is the chemical correlate of pleasure, or as we might put it more romantically, dopamine is a bliss chemical. Undoubtedly this is so. But a chemical that is required to instigate the movement of the body such that the inability of the brain to produce sufficient levels results in life-crippling Parkinson's disease must be considered far more than just a source of pleasure. I suggest that dopamine is a pleasure reward for the effort of *will*—the ability of the mind to rule the body, and consequently, the neurohumor most responsible for the development of the human mind. In short, dopamine is a "consciousness chemical."

Dopamine and the Evolution of Consciousness

A scientific paper published in 2002 proposed that dopamine plays an important role in "regulating cognitive functions." In other words, dopamine facilitates *thinking*! André Nieoullon, a French neuroscientist, hypothesized that brain levels of dopamine increased in the latter stages of mammalian evolution, producing "progressively more developed cognitive capacities related to increased processing of cortical information."[16] In other words, the evolution of man from ape was the result of genetic mutations which increased brainpower by raising levels of dopamine. *Dopamine was the neurochemical responsible for the evolution of man from ape.*

Dopamine and Inhibition

Dopamine functions by allowing the frontal lobes of the neocortex—the most recently evolved outer layer of the brain—to prevent or *inhibit* instinctive emotional responses governed by the older, underlying areas we share with our ape ancestors. The mental capacity for inhibition—the ability to rein in physical and emotional impulses (rather than responding automatically to the dictates of our genetic programming)—is the faculty that makes us truly human. Language, reason, learning, creativity, and volition are all products of inhibition. Dopamine appears to be the neurohumor above all others responsible for the faculty of inhibition, and thus the chemical which allowed *Homo sapiens* to develop brainpower way beyond the ability of his ape ancestors—and to dominate planet Earth.

Dopamine and Will

The paper goes on: "Dopamine has thus to be considered as a key neuroregulator which contributes to anticipatory processes necessary for preparing voluntary action consequent upon intention."[17] Translating this *academowaffle* into plain English, dopamine is the neurotransmitter which allows the mind to control the body—a facility sometimes referred to by psychologists as "intentionality," but described perfectly adequately by the everyday word *will*. As fertility expert Professor Robert Winston writes in the book accompanying his BBC TV series *The Human Mind*, "Will is a brain faculty like any other, one that can be strengthened with use, or lost with damage."[18]

Dopamine may be the chemical correlate of will, the ability of the mind to control the body.

Low Dopamine and Depression

Low dopamine is associated with depression—a state in which *thinking* processes are reduced and the mind grinds to a halt, ceasing to control behavior; hence the typical physical lethargy associated with depressed mood. Essentially, depression is a collapse of will—that is, a collapse of the mind's ability to control behavior—a facility governed by dopamine. The mental correlate of physical lethargy is apathy—the inability to instigate behavior. Lethargy and apathy are traits characteristic of depression that are often confused with (or condemned as) laziness, thus helping to reduce even further the collapsed confidence of the depressive. Hence, dopamine-raising eugoics such as the antidepressant and smoking-cessation drug Wellbutrin (trade name of bupropion) have been found to reduce apathy.[19]

Low Dopamine and ADHD

The childhood behavioral condition ADHD (Attention Deficit Hyperactivity Disorder) is characterized by the inability to sit still or concentrate. Treatment with the amphetamine-based dopamine-raising drug Ritalin allows the *mind* to take over control of the body, thus inhibiting the physical impulsivity triggered by the "emotional brain"—the more primitive limbic system inherited from our ancestors. Most of the brain's dopamine is found in the part of the forebrain associated with ADHD. In short, ADHD is a genetic disorder in which the brain fails to increase dopamine production as it develops in childhood, resulting in the inability of the *mind* to control the body.

Low Dopamine and Parkinson's Disease

Abnormally early death of dopamine-producing cells is the cause of Parkinson's disease, characterized by the inability of the *mind* to direct the behavior of the body. The sufferer will typically say such things as, "*I* want to move, but my body doesn't." Without the drug L-Dopa—synthetic dopamine—the Parkinson's sufferer "seizes up," becoming unable to move—a state often accompanied, understandably, by depression. Again, dopamine is implicated as the neurochemical correlate of will—the ability of the mind to govern the behavior of the body.

High Dopamine and Happiness

Stimulants such as cocaine and amphetamines increase mental arousal, agility, and mood by raising levels of dopamine. A high sensitivity to dopamine results in the continual quest for mental stimulation—the buzz of mental excitement. Complex thought is extremely pleasurable—that's why philosophers and scientists do it. The "Eureka Moment" of discovery at which a new, important piece of the jigsaw in the metapattern of nature suddenly presents itself to consciousness is the ultimate burst of dopaminic bliss—an intellectual orgasm. The raised mood associated with high levels of dopamine is nature's reward for the survival value of mental agility and strength of will—the ability of the mind to rule the body. Willing and thinking are pleasurable because they help us to survive (though you would be forgiven for doubting this were so, in our present dumbed-down society).

High Dopamine and Confidence

Similarly, the surge in confidence experienced by cocaine users is caused by a sudden rise in dopamine levels. Confidence, or its fashionable contemporary version, "self-esteem," is a state of "assurance, boldness, fearlessness." Dopamine raises confidence as a pleasure reward for social dominance resulting from lack of fear. Hence the lavatorial habits of secretly shy pop stars (music is a mating call, and thus a common method of self-advertisement by the otherwise timid). How does dopamine reduce fear? By increasing the size of your ego! Raising dopamine increases thought processes—and hence one's sense of self. Suddenly and dramatically raising dopamine by introducing it straight to your brain through your nose dramatically increases your sense of self. Notoriously, all the cocaine user can talk about is "Me, me, me."

Excess Dopamine and Schizophrenia

In short, dopamine raises mental agility, concentration, mood, and confidence. Why, then, have not more antidepressants been developed which focus on raising dopamine levels? Perhaps because of something the establishment calls "abuse potential"—in other words, they don't want people to be *too* high, *too* confident, or too *mad*, for dopamine levels have also been found to be higher than normal in schizophrenics, who typically exhibit overcomplex, confused, "manic" thought processes, while the "neuroleptic" drugs used to treat schizophrenia achieve their effects by *lowering* dopamine levels in the brain. Thinking *too much* is detrimental to survival because it creates confusion. The cocaine user is almost indistinguishable from the schizophrenic in the inability to control the surge of thoughts flooding the brain. The ideal level of dopamine, then, is enough to optimize the capacity for thinking and feeling without ascending into the madness of an "unnatural high."

Serotonin and a Sound Body: The Reduced Sensitivity Hypothesis

> Serotonin is important for adequate coping with stress.
> —*European Journal of Neuroscience* (December 2002)

If dopamine is nature's reward for mental agility and strength of will, serotonin is its reward for the complementary state of calmness, serenity, or emotional stability.

Serotonin is needed to make melatonin, the chemical that induces sleep, secreted from the pineal gland in response to darkness. Serotonin may have similarly evolved to help us sleep, through a reduction in sensory stimulation, achieved by reducing the sensitivity of the brain to external stimuli.

Serotonin appears to function as a neurochemical behavioral regulation mechanism analogous to a TV remote control, "turning down" the level of an overaroused brain and body. The consequent reduction in oversensitivity serves to alleviate anxiety and depression, thus increasing social ease and confidence and, in turn, raising mood.

Sufferers of anxiety disorders characterized by hypersensitivity to stimuli are the victims of serotonin levels that are naturally far too low, producing brains and bodies that wildly overreact to events. Genetic evidence supporting such a theory has now been produced, as Steven Pinker writes in his book on the biology of human nature, *The Blank Slate*: "If you have a shorter than average version of the stretch of DNA that inhibits the serotonin transporter gene on chromosome 17, you are more likely to be neurotic and anxious, the type of person who can barely function at social gatherings for fear of offending someone or acting like a fool."[20]

Serotonin and Prozac

Eugoic drugs such as Prozac alleviate depression by raising serotonin levels in the brain. I suggest they achieve their effect by reducing the brain's sensitivity to external stimuli. The brain operates by scanning the environment and sending a signal to the sympathetic nervous system in response to any potential danger, thus triggering the automatic defensive response of the body. The "autonomic nervous system" (ANS) springs into action, pumping out adrenaline in preparation for a response to a potential threat—either attack or retreat, "fight or flight." The feeling experienced when the body is flooded with adrenaline is the negative emotion we call "anxiety"—otherwise known as fear. But if the brain's reaction to external stimuli is *dampened* by serotonin, a sort of "neurochemical blindfold and ear plugs," it will cease to send out signals triggering the sympathetic nervous system. The automatic flight-fight response will *not* spring into action; adrenaline will *not* flood the body; no anxiety will be experienced, and the individual will remain calm. If one no longer feels fear, by definition one feels braver—and happier, since a state of fear is not a happy one. Thus, I suggest that serotonin reduces anxiety by decreasing sensory stimulation, which in turn naturally results in raised mood and confidence.

A predisposition to shyness, nervous anxiety, and depression is symptomatic of a particular dimension of personality or temperamental trait I call the Opioid Neurotype. When personality is dominated by the Opioid Neurotype, it produces an emotional introvert whose hypersensitive nervous system habitually overreacts to stimuli, causing the body to be flooded by endogenous tranquilizers, or opioids. The introverted behavior of the Opioid Neurotype results from the instinctive attempt of an overaroused brain and body to reduce environmental stimuli in the attempt to alleviate the discomfort of sensory overarousal. Introversion—the predilection for peace and quiet above partying—is the attempt to reduce a habitually hyperaroused brain and body. Hence the antidepressant effect of serotonin-raising drugs such as Prozac may be the cognitive consequence of reducing anxiety by "turning down" the brain's oversensitivity to external stimuli. Such a reduction in sensitivity naturally results in confident extroversion (stimuli-seeking, nonfearful behavior), and generally raised mood, because one is not afraid of something to which one's nervous system does not react. Fear is simply a state of physiological arousal in preparation for attack or retreat—fighting or running away. If the autonomic nervous system is not triggered by the brain to respond to the environment with an adrenaline rush of fear, then the environment is not experienced as fearful.

The fact that serotonin-raising drugs such as Prozac have completely replaced the benzodiazepines such as Valium as the medical treatment of choice for anxiety disorders supports my theory. For this would come as no surprise if the *primary* effect of such drugs is to relieve anxiety rather than depression.

The **Reduced Sensitivity Hypothesis** (RSH) also explains why not all depression is alleviated by serotonin-raising drugs. Such drugs may be effective primarily for the kind of depression which results from "anxiety disorders"—what used to be called "neurosis." The "other sort of depression," I suggest, results not from serotonin depletion, but from abnormally low levels of *dopamine*, leading to a condition one might call "boredom depression"—in other words, lack of mental stimulation.

RSH also explains the unsolved mystery as to why serotonin-raising antidepressants generally take two weeks or so to have a significant effect on depression. If serotonin primarily decreases physiological anxiety—to which depression is a natural cognitive response—it obviously takes time for the mind to adjust to the fact that the sensitivity of the nervous system is no longer "turned up" way too high—that the body is becoming calm, relaxed, insensitive to stimuli, relatively fearless. In short, a gradual realization occurs that there is no longer any need to feel afraid.

In *Listening to Prozac*,[21] a seminal book on the personality-enhancing effects of eugoic drugs, psychiatrist Peter D. Kramer relates how a colleague referred to serotonin as "the police," to describe the feeling of psychological security produced by serotonin-raising eugoics. Just as one is less fearful of criminals when the police are around, so the body is less fearful of the environment when serotonin is around—the feeling of "psychological security" it produces is the basis of emotional contentment, bravery, and confidence.

Serotonin and Confidence

Decreasing the brain's reactivity to stimuli naturally results in greater confidence by decreasing fear of the environment. A knight protected by shining armor is likely to feel more confident than one without. Drugs that act on the brain to reduce sensitivity of response to the environment are like a neurochemical suit of armor.

Serotonin and Dominance

Confidence is synonymous with bravery, extroversion—and dominance. Psychologists tell us that dominant apes have higher serotonin levels than submissives, and winning sportsmen higher levels than losers. Being dominant, winning, requires the absence of fear. Clearly, wearing a neurochemical suit of armor has its survival advantages. Reducing reactivity to stimuli decreases general fear of situations, resulting in greater feelings of confidence, bravery, extroversion, and dominance. The archetypal dominant "alpha male," symbolized by the figure of James Bond, always remains calm, unruffled—the epitome of "cool" in response to even the greatest dangers, because high levels of serotonin ensure that the brain and body do not cause fear by overresponding to external stimuli. The most dominant members of society are those with the highest serotonin levels because high serotonin provides them with a head start—neurochemically induced confidence.

Serotonin and Happiness

A serotonin pleasure reward is automatically released in response to dominant social behavior because winning naturally increases the prospect of reproductive success (that is, "winning a mate"). This is the reason why "winners" are happier than "losers." Happiness is nature's neurochemical reward for survival behavior. Hence serotonin also raises mood as a pleasure reward for increased genetic survivability.

Serotonin and Empathy

The recreational drug ecstasy is thought to achieve its effect of inducing feelings of benevolence by raising serotonin levels in the brain. Why should serotonin increase emotional empathy? Because of the principle "I like you because I do not fear you." Excessive fear is simply the body's overresponsiveness to stimuli. By reducing the "fight or flight" response of the sympathetic nervous system, serotonin decreases fear of others, thus increasing the capacity for benevolence.

Serotonin and Decreased Aggression

Recent studies have also correlated low serotonin levels with aggression. This again supports the Reduced Sensitivity Hypothesis. Aggression is a biochemically mediated impulse, induced by the autonomic nervous system in self-defense against a potential threat. Dampening down the ANS response thus has the effect of reducing aggressive impulses. This suggests that serotonin-raising eugoics may be of use in treating the personality trait I call the Adrenaline Neurotype, characterized by impulsive, or "choleric," behavior, which results in a greater likelihood of "getting into trouble" through the inability to control aggressive impulses. The Adrenaline Neurotype is characteristic of the ADHD child. One wonders, then, if serotonin-raising eugoics may be of use not only in the treatment of ADHD, but also in the attempt to tackle the problem of antisocial behavior generally. But I shall leave the subject of the potential for eugoic drugs in criminal neuropathology to another day, having no doubt ruffled quite enough feathers for one book.

Dopamine and Serotonin: A Sound Mind in a Sound Body

The ideal state of well-being has long been considered that of a "sound mind in a sound body." Dopamine and serotonin may be the neurochemical correlates of such a state.

The ideal "sound mind" is lively and active, while the ideal "sound body" is relaxed or serene. A sound mind in a sound body means a high level of mental arousal and physical relaxation. A high, balanced level of both represents an optimum state of well-being—*mens sano in corpore sana* (a sound mind in a sound body). Thus, the optimum state of mental health may require high, balanced levels of both dopamine and serotonin, providing clear thought and raised mood in a calm, relaxed body.

The future enhancement of both dopamine and serotonin levels through Superbiology may contribute toward an ideal state of neuropsychological well-being. The next step in eugoics (personality enhancing, or "mood-brightening" designer drugs) may well be the development of a drug which raises both dopamine and serotonin levels, producing an active mind in a calm body, and hence a continual state of "mild bliss"—a plateau experience conducive to daily functioning in society, rather than the brief high experienced through street drugs such as dopamine-raising cocaine and serotonin-raising ecstasy.

(Update: In mid-2004, the US National Institute of Mental Health announced the first clinical trial of combination serotonin- and dopamine-raising mood-brighteners for the treatment of depression, saying "data suggests that treatments which affect the serotonin and dopamine systems will be more effective than agents which use a single mechanism.")[22]

And genetic evidence supporting my Reduced Sensitivity Hypothesis for the function of serotonin appeared in the May 8, 2005, edition of *Nature Neuroscience*. Dr. Daniel R. Weinberger, director of the Genes, Cognition and Psychosis Program at the National Institute of Mental Health, reported the identification of a shortened version of a gene resulting in low serotonin production, which

> contributes to general anxiety and the risk of depression following major life stresses. . . . Those with at least one short copy of the gene had less effective circuitry in the part of the brain that controls responses to fear. . . . Researchers linked the effectiveness of the circuit to the subjects' vulnerability to depression and anxiety. That makes sense, Weinberger said, because in affected people "the problem isn't that you're fearful, it's that you can't stop being fearful, you can't turn it off." . . . The potential link between gene variations and brain changes "underscores that tiny changes in the DNA code have the potential to cause far-reaching changes in the person."[23]

THE FOUR NEUROHUMORS: A THEORY OF THE NEUROCHEMICAL DIMENSIONS OF PERSONALITY

Having proposed a theory as to the evolved function of two of the main targets for a future science of neuroenhancement, it is possible to go further still and propose a theory as to the neurochemical dimensions of human personality based on variations in habitual levels of behaviorally significant neurohumors.

A comprehensive understanding of the neurochemical basis of personality is a necessary precursor to the development of a future science of neuroenhancement. Just as today someone unhappy with the shape of their nose or breasts may choose to enhance their body through aesthetic surgery, so tomorrow those unhappy with aspects of their temperament may choose to enhance their *personality* through *gene* surgery.

Of course, this is not to everyone's taste—some people think we should stick with the personality handed to us at birth. But equally, many others would embrace the opportunity to alleviate a life-crippling tendency to shyness, gloominess, lethargy, or impulsivity, or to enhance their capacities to think and feel *beyond* the level of "normal" health, in the perfectly reasonable desire to feel *better than well.*

Here, then, I briefly outline my own neurochemical personality theory, as an example of the sort that will increasingly emerge in the coming years as our knowledge of neurochemical function increases. I shall begin by briefly mentioning the main bio-personality theories occurring between that of the four humors and my own theory—the **Four Neurohumors**.

Jung and Sheldon

Jung's Four Functions

In the early twentieth century, the psychologist Carl Jung updated the theory of the four humors by correlating them with four basic human functions he called sensing, feeling, thinking, and intuition. I have incorporated the idea of four basic functions into my own theory. However, Jung's book on psychological types[24] is so complex—running to a mammoth six hundred pages—that one ends up far more confused about human personality than when one started. Jung was notoriously interested in the paranormal, and while he shows many insights into human personality, one cannot escape the feeling, when trying to work one's way through the complex descriptions of his proposed eight personality types, that one is reading a glorified astrology column in a Sunday supplement.

Sheldon's Somatology

The English psychologist William Sheldon proposed a simpler model in his book *The Varieties of Human Physique*,[25] by correlating personality differences with three different body types he called ectomorph, endomorph, and mesomorph. In other words, he noted that thin, fat, and muscular people tend to have different personalities. Everyday observation tends to support this theory (as does the recent discovery of the neurohumor MSH, which suggests an association between weight gain and raised mood, and vice versa). But clearly, body shape is not the only factor affecting personality. And today's PC world may not appreciate a personality theory that tends to reinforce the stereotype of fat people as cheerful, thin as taciturn, and muscular as oafish.

Eysenck's Dimensions of Personality

By far the most important physiological theory of personality is that of the late-twentieth-century behavioral psychologist, Hans Eysenck.[26] Like Jung, Eysenck also looked to the Four Humors, but combined them with the utterly modern physiological personality theory of the behaviorist psychologist Pavlov. The result was a simple but comprehensive theory, which initially narrowed personality traits down to two basic dimensions, which he correlated with physiological states:

> **Introversion versus extroversion:** governed by the central nervous system, specifically the "basal cortical arousal state" (BCAS), the habitual arousal level of the cerebral cortex—the level of the brain's reactivity to stimuli.
>
> **Emotional stability versus instability:** governed by the sensitivity of the peripheral, autonomic nervous system (ANS)—the level of the *body's* reactivity to stimuli.

Eysenck then correlated all possible combinations of these two dimensions with the theory of the Four Humors, to produce a table such as this:

STABLE EXTROVERT (Sanguine) Unreactive ANS Low Cortical Arousal	STABLE INTROVERT (Phlegmatic) Unreactive ANS High Cortical Arousal
UNSTABLE EXTROVERT (Choleric) Overreactive ANS Low Cortical Arousal	UNSTABLE INTROVERT (Melancholic) Overreactive ANS High Cortical Arousal

The brilliance of Eysenck's theory lies in its simplicity. Essentially, the basic dimensions of personality are explained by variations in the arousal level of mind and body.

Because Eysenck's theories largely preceded today's knowledge of neurochemistry, he did not go on to correlate personality types with neurotransmitters. But today we are in a better position to extend his theory.

Cloninger's Model

An early attempt to correlate neurochemicals with personality traits was made by psychiatrist Robert Cloninger in the 1980s.[27] Cloninger linked three neurotransmitters with three traits:

- dopamine = Novelty seeking
- serotonin = Harm avoidance
- noradrenaline = Reward dependence

Cloninger's model seems like a good start. However, using rather obscure terms like "harm avoidance," "reward dependence," and "novelty seeking" which mean little in terms of everyday experience seems rather unnecessary. "Novelty seeking," for instance (others have called it stimuli seeking), sounds suspiciously like the more familiar trait "extroversion," while "harm avoidance" seems like an obscure way of describing introversion—an aversion to excessively arousing stimuli.

An alternative tactic for correlating neurochemicals with specific behavioral traits is to start at the beginning by asking "What are the most basic human functions?" Man evolved from the simple to the complex. As complexity originated in simplicity, it makes sense to go back and start with the simple in order to unravel the complex. Thus, to identify which neurohumor influences which type of behavior, we might begin with the simplest types of behavior of all.

The bottom line: a human being consists of a brain and a body, both of which can be either active or at rest. The most basic types of behavior of all, then, are "mental activity," "mental rest," "physical activity," and "physical rest." By correlating these four basic types of behavior with the four neurohumors most associated with them, it is possible to produce an updated version of the theory of the four humors—the Four Neurohumors.

The Four Neurotypes

We start, then, by asking which neurohumors are most clearly associated with the basic functions necessary for survival—mental and physical activity and rest—and conclude as follows:

- MENTAL ACTIVITY
  Arousal, agility, or alertness is most clearly associated with dopamine. High quantities of dopamine produce a goal-seeking, driven, highly motivated, extrovert personality trait—the **Dopamine Neurotype**.

- MENTAL REST
  Serenity, tranquility, or undisturbed peace of mind is most clearly associated with serotonin. High quantities of serotonin produce a calm, relaxed, self-assured personality trait—the **Serotonin Neurotype**.

- PHYSICAL ACTIVITY
  Vitality, vivacity, or vigor is most clearly associated with the hormone adrenaline. High quantities of adrenaline produce a physically impulsive personality trait—the **Adrenaline Neurotype**.

- PHYSICAL REST
  The attempt to dampen down physical arousal is governed by the opioid system. Lack of genes producing serotonin results in a body which floods with opioids, producing a nervous, anxious-depressive, or "neurotic" personality trait—the **Opioid Neurotype**.

Thus, I add a fourth group of neurohumors to the three neurohumors in Cloninger's model. The word *opioid* means "having the form of an opiate." The "opioid peptides" are a group of neurohumors including endorphins and enkephalins, which serve as tranquilizers, dampening down the sensitivity to pain and reactivity to stress. The endogenous opioids are mimicked by the opiate drugs: opium, morphine, codeine, heroin, and methadone, collectively known as analgesics (Greek for "pain-removers"). Opium is produced from the seeds of the opium poppy and contains morphine, from which codeine and the faster-acting heroin may be extracted. Methadone is a synthetic heroin substitute dished out daily to addicts as a poor substitute for treatment.

I shall use the term *Opioids* as an umbrella term to describe all those inhibitory neurochemicals that reduce physical activity or sensation. Other vital inhibitory neurohumors include GABA—currently believed to be responsible for the tranquilizing effect of alcohol and cannabis—and MAO—which reduces levels of the stimulatory neurotransmitters dopamine and noradrenaline.

Similarly, the neurotransmitter noradrenaline is the brain's equivalent of the hormone adrenaline, but here, for the sake of simplicity, I shall use the term *adrenaline* to cover all neurohumors which stimulate physical activity. The male sex hormone testosterone, for instance, is another.

A table of the neurochemical dimensions of personality, then, looks something like this:

THE FOUR NEUROHUMORS AND NEUROTYPES

Neurochemical Dimensions of Personality

	EXCITATORY **ACTIVE**	**INHIBITORY** **PASSIVE/RECEPTIVE**
	NURETHIC NEUROTYPES	
Neurohumor:	**DOPAMINE**	**SEROTONIN**
Activity:	MENTAL AGILITY (activity)	MENTAL SERENITY (rest)
Faculty:	THINKING	PERCEIVING
Drive:	PROMETHEUS	ORPHEUS
	GENETHIC NEUROTYPES	
Neurohumor:	**ADRENALINE**	**OPIOIDS**
Activity:	PHYSICAL IMPULSIVITY	EMOTIONAL SENSITIVITY
Faculty:	DOING	FEELING
Drive:	EROS	THANATOS

The Four Neurohumors are four neurochemicals associated with the four most basic types of behavioral trait, and the four basic faculties by which human beings operate in the world—perceiving, doing, feeling, and thinking. For instance, a body dominated by the dimension of personality I call the Opioid Neurotype is typically overreactive to stimuli, resulting in excessive emotional sensitivity. The subsequent "hunger" for opioids is the attempt to reduce the overarousal of a hypersensitive nervous system. Therein lies the biopsychology of heroin addiction, "tragic" romanticism, and masochism—all consequences of the desire of an oversensitive Opioid Neurotype to reduce the emotional pain of a supersensitive body.

The Serotonin Neurotype may be associated with the faculty of perception because high levels of serotonin allow the mind to calmly observe the world objectively, with a decreased sense of self, resulting from decreased physical sensitivity to stimuli. The Dopamine Neurotype also allows the mind to observe the world, but here by means of analysis, not synthesis—identifying the trees, not experiencing the forest. If the Dopamine Neurotype is a scientist, the Serotonin Neurotype is an artist.

I emphasize, however, that no one is *all* Dopamine, Serotonin, Opioid, or Adrenaline Neurotype. The "General Neurotypes" represent personality *traits*, not *types*. Just as in the ancient theory of the Four Humors an individual personality consisted of a unique mixture of each of the four humors, so in the theory of the Four Neurohumors *each individual personality consists of a unique balance between each of the four neurohumors*. A personality *type* (or Personal Neurotype) consists of a unique mixture of different *traits* (or General Neurotypes). The relative dominance of each trait (General Neurotype) determines overall personality type (Personal Neurotype). Thus, neurotype in the above context refers to General not Personal Neurotype—a neurochemically influenced personality *trait*, not *type*.

Equally, it must be emphasized that these four neurohumors are not the *only* ones governing human behavioral traits. Inevitably, more than one chemical is involved. For instance, Acetylcholine—the neurohumor raised by nicotine and lowered in Alzheimer's disease—is vital to brain arousal, and may well turn out to be of even greater benefit to neuroenhancement than dopamine. Similarly, glutamate, the most common chemical in the brain, is a vital stimulatory neurohumor, while other important inhibitory, stress-reducing neurochemicals include GABA, MAO, cortisol, and cholesterol.

Thus, here I am merely identifying traits by the name of neurohumors that current evidence suggests are most clearly, significantly involved in their production. It is also, for present purposes, simpler and more practical to speak of a Dopamine or Opioid Neurotype, as opposed to, say, a Dopamine-acetycholine-glutamate or Opioid-GABA-cortisol-estrogen Neurotype.

The point of my model is not to define a one-to-one causal correlation between neurohumor and function, but to convey the point that models of neuropersonality are fast becoming possible, will inevitably be developed in the coming years, and demonstrate the importance of neurohumors to personality and behavior.

The Netropolis: From Genethic to Nurethic Neurotypes

The four motivational drives of Eros, Thanatos, Prometheus, and Orpheus that I associate here with the Four Neurohumors, I shall explore in detail in my theory of human motivation. In a nutshell, the Prometheus and Orpheus drives are cognitive or "nurethic" drives for Individuation and Integration, which may transcend Eros and Thanatos, the "genethic" Freudian-neo-Darwinian (or psycho-Darwinist) drives for survival/reproduction and self-destruction. Genethic neurotypes correspond mainly to physical activity (controlled by the selfish genes, through instincts), while nurethic neurotypes correspond to mainly mental activity (controlled by the conscious mind, through will). The Adrenaline and Opioid Neurotypes are neurochemically induced personality traits that help facilitate the genethic drives of Eros and Thanatos for survival, reproduction, and self-destruction. The Dopamine and Serotonin Neurotypes are neurochemically induced personality traits that help facilitate the nurethic Prometheus and Orpheus drives for Individuation and Integration. In other words, dopamine helps you think positively, rationally, and willfully, while serotonin helps you feel at peace with the world.

Culturally, if it can escape the yoke of postmodern retrogression, the modern world is likely to evolve from a genethic "survival society" dominated by Adrenaline and Opioid Neurotypes to a nurethic Self-Enhancement Society dominated by Dopamine and Serotonin neurotypes. Such a transition will signify a process of psychosocial evolution from genetic to cognitive drives—from control by the genes, to self-rule by the conscious mind. Why should such an evolution of consciousness occur? Because human beings cannot be held back forever from acting on their innate Will to Evolve—the instinct of a conscious entity to expand abilities in pursuit of ever-increasing survivability and well-being.

The modern world is increasingly dominated by a phenomenon social theorists have called the **ideopolis**—culturally influential, often university-based cities, which attract a large percentage of young "brain workers," especially in the field of IT. Because of the increasing influence of the Internet in modern technoculture, the ideopolis might equally be called a **netropolis**. The rise of the netropolis in the modern world is evidence of the inexorable rise of the Dopamine and Serotonin neurotypes.

In the twenty-first century, the transition from a Genethic to a Nurethic culture will be hastened by our ability to enhance the Dopamine and Serotonin neurotypes through the Superbiology of eugoics, Supergenics, Cybernetics, and Nanotechnology (neuropharmacology, gene therapy, computer implants, and molecular medicine).

As we begin to replace Darwinian with Designer Evolution, the modern world will increasingly move from a genethic Survival Society dominated by Adrenaline and Opioid neurotypes to a nurethic Self-Enhancement Society dominated by Dopamine and Serotonin neurotypes. Left behind will be the old, genethic culture of brute physical survival instincts on one hand, and sentimental, emotional hypersensitivity on the other. Both cultural and biological evolution are leading us in the same direction—from unconscious control by the selfish genes to conscious control by the rational mind.

The evolution of consciousness is leading us from the will to survive to the Will to Evolve.

Whole Brain Functioning

As is well known to New Agers and neurophilosophers alike, the human brain is divided into two sections, left and right, which specialize in different functions. The left hemisphere specializes in language—the symbolic representation of objects and experience—and is thus responsible for both the human sense of self, and the capacity for abstract logic (both of which, of course, require language). The right hemisphere governs perception—the visual scanning of the environment, the ability to see the overall Big Picture, or the "pattern-making" facility often associated with the word "intuition" (actually "unconsciously processed information") that is the basis of artistic creativity. As has often been said, the left brain is a scientist, the right brain an artist.

Further research needs to be done to ascertain the extent of neurotransmitter hemispheric specialization, but we might tentatively associate the Dopamine and Serotonin neurotypes with left and right hemispheric activity respectively. If so, the future social dominance of the Dopamine and Serotonin neurotypes will signify the evolution of human consciousness through the development of "whole brain functioning"—high-level, logical, and creative thinking combined.

Such a development may be reflected in a New Renaissance synthesis between the arts and sciences in the twenty-first century, as passion and reason combine in a new ethos of **technoromanticism**—the fusion of the romantic impulse for human transcendence with a renewed faith in the power of technology to attain it.

* * *

But here I shall leave my speculations. The Four Neurohumors is just one example of the type of neurobiologically based personality theories made possible by advances in our knowledge of the neurochemical influence on attitudes and behavior. More will follow in the coming years. When they do, one hopes their authors will acknowledge the important contribution made in the twentieth century by the behavioral geneticist Hans Eysenck.

Eysenck's death in 1997 was largely ignored by the media, except for the obligatory, largely unenthusiastic obituaries in some—not all—of the broadsheet newspapers—a fact one might contrast with the recent press coverage in England of the death of multiple child killer Myra Hindley, which resulted in page after page after page devoted to her thrilling life story.

RESPONSE TO THE EXPERIENTIALISTS: BEYOND GENETIC INJUSTICE—IN DEFENSE OF BIOPSYCHOLOGY

> We're afraid to try to affect the future, whereas some people are born with a lousy future.
>
> —James Watson, Skeggs Lecture, Youngstown State University (November 2003)

Eysenck's groundbreaking work on the physiological basis of personality was widely overlooked—and sometimes aggressively vilified—in a postwar climate that effectively made the study of the biological influence on behavior a taboo subject for half a century.

Opponents of biopsychology assert that life experience, not biological makeup, determines personality and behavior. I shall call such a view **experientialism**, in preference to the alternatives, "nurturism," which sounds ugly to my ears, and "environmentalism," which is too easily confused with the unconnected attitude I call ecologism—the entirely sensible belief in the importance of preserving the natural environment necessary to sustain life. (In turn, we should distinguish ecologism from the hard-line ideology I call ecofundamentalism, or ecoism—others have called it "deep ecology"—the extreme version of ecologism, which places the well-being of nature above that of human beings.)

Experientialists insist that biological makeup has a minimal effect on human behavior compared to life experience—so we should limit ourselves to seeking sociopolitical (rather than biotechnological) solutions to life problems. The target of experientialist bio-Luddites is the belief they call genetic or biological determinism—the view (actually held by no scientists at all) that all aspects of human behavior are entirely preprogrammed by the genes—that nothing can alter the way we are genetically destined to behave.

Genetic determinism is essentially a "geneticized" version of nineteenth-century Social Darwinism—the combination of a psychology which held that social inequality reflects innate biological differences, and an ethics which held that the principle of "survival of the fittest" requires an absence of state support for the socially disadvantaged. Such a belief, which sought to entirely ignore the effect of social factors such as poverty, class, race, or sex discrimination on life failure, might be called the **old biologism**. Its accompanying ethic, the belief in prejudging others according to innate biological behavioral characteristics determined by their sex, race, or class, could be called biological elitism—or **bio-elitism**.

From Bio-Elitism to Bio-Empathy

It cannot seriously be doubted that the nineteenth-century ideology of biologism and bio-elitism led causally in the twentieth century to the rise of eugenics, fascism, and attempted genocide by encouraging the belief in racial supremacism—the innate biological superiority of one race over all others.

Therein lies the real motivation behind Experientialism, as the late evolutionary theorist and Marxist John Maynard-Smith explained in an interview:

> I suppose I'm really showing another aspect of my upbringing. I was a young man when Hitler was in power. I was in Berlin in 1938 leading up to the Munich settlement, and the whole of my thinking about the world has been much influenced by belonging to that generation. For me, the application of biology to human beings means Rosenberg and the race theories, so I'm obviously a bit reluctant to get involved in biological applications to human behavior.[28]

In other words, the fear that biopsychology might once again lead to prejudice and discrimination on grounds of biological differences led to a powerful will to believe that personality and behavior are not affected by biology, but are solely the result of life experience—that is, a combination of factors such as birth order, upbringing, and, in particular, educational and social opportunities.

But the New Biology is not the old biologism. Today, all reasonable people agree that bio-elitism was not only morally reprehensible but based on a false premise. The obvious success of different ethnic groups and the extent of racial cooperation and mixing in the shrinking Global Village have made belief in the absolute superiority of any one racial group ludicrously untenable. Anyone talking about a master race today—whichever one they might prefer—would be laughed out of court. And, of course, the horrific practical consequences of the attempt to enforce such a belief have been demonstrated only too clearly, as war after war continues to rage in the name of preferred race or accompanying religion. An escalation to global conflict in a nuclear age would threaten the survival of the entire species.

Bio-elitism should be replaced by an attitude we might call **bio-empathy**—the recognition that *everyone* is subject to biological limitations. We are *all* handicapped by nature. Death, disease, and the fragility of body and mind make (sometimes strange) bedfellows of us all. Bio-empathy is the *emotional* basis of the transhumanist desire to eradicate disease, enhance abilities, and defeat death, as Nurethics is its *rational* basis.

Nature + Nurture = Personality

The New Biology recognizes the influence of both biology *and* environment on behavior.

There are two causative factors determining personality: temperament—an individual's unique biological makeup, and character—habits formed by life experience. Nature determines temperament, nurture determines character.

Together, nature and nurture, temperament and character, produce the observable, habitual, characteristic traits of an individual we call "personality."

Temperamental differences derive from biogenetic differences originating from birth. We each inherit a nervous system predisposing us to certain ways of behaving. Temperamental predispositions, or tendencies to act in certain ways rather than others, we call "personality traits." They include the tendency to be introverted or extroverted, calm, anxious, aggressive, cheerful, moody, or glum, as well as general levels of intelligence, and creative or sporting talents. The *likelihood* of individuals displaying such personality traits depends upon their particular biogenetic makeup. Their *actual* behavior depends upon the way these innate dispositions are influenced by experience—upbringing, learning, and education—that is, the effect of the environment on a given temperament, or neurotype.

Both the prenatal environment of the womb and the postnatal environment of the world affect behavioral predispositions. Prenatally, maternal health and nutrition during pregnancy can be of significant influence on a child's subsequent behavioral temperament. Factors such as maternal depression or anxiety, dietary deficiency, smoking, drug or alcohol abuse can all alter habitual hormonal balance, damage fetal growth, and in turn affect neurotype—the neuropsychological aspect of personality.

Postnatally, the development of character is influenced by all aspects of life experience, including parental care, affection and discipline, exposure to trauma, birth order, social and sexual relationships, economic circumstances, educational opportunities, and simply good or bad fortune.

Personality thus consists of two complementary aspects—the biological and the social—temperament and character. Temperament is the biological aspect of personality—the predisposition to act, think, and feel in certain ways rather than in others because of one's particular, unique biological makeup. Character is the socially learned aspect of personality—the effect of a particular temperament (or neurotype) on a given environment dictated by an individual's unique life experience; the influence of parents, family, teachers, schoolmates, colleagues, friends, media, advertising, society, and life in general.

"Nurture" is the development of character through experience. "Nature" is the expression of innate biological temperament. Personality is the combination of nature and nurture.

In Defense of Neurotypology

All scientists and psychologists agree, then, that personality is determined by a combination of biological and experiential factors. The only area of disagreement is the question of the relative significance of the role played by each. I consider this a dull question. Transhumanists are not interested in the silly game of trying to establish the precise proportion of behavioral traits determined by genes and environment. The important question is rather, What are the possibilities for self-enhancement of mind and body? We already know the possibilities for *experiential* enhancement—for instance, improving diet, exercise, relationships, work and educational opportunities. Now we want to know the possibilities for *biological* self-enhancement, through the emerging techniques of Superbiology. A necessary precursor to the biological enhancement of mind and body is an understanding of the biological basis of behavior.

Human beings are fundamentally biological entities—to improve our mental and physical health, we must first understand our biological makeup. We are *all* subject to the biological limitations imposed upon us by nature. We *all* possess weak bodies and minds that are subject to death, disease, and decay. Both the psychological and biological aspects of personality (character and temperament) are malleable, particularly in childhood before habits have formed, and this is certainly reason to ensure that we try to instill children with a positive feeling of self-confidence in a stable, "nurturing" environment, and to try to create a society free from poverty and prejudice.

However, transhumanists are not naïve wishful thinkers but positive realists, who recognize the biological deficiencies that reduce our ability to thrive and prosper, and seek to overcome them, through Superbiology. The brute fact remains that no amount of environmental improvement can fundamentally alter a basic temperamental predisposition, and not all neurotypical traits confer the same level of social advantages. A neurotypical imbalance may result in the misery of life failure by inducing a tendency toward socially inadequate behavior.

The personality type most likely to succeed in life is the emotionally balanced, calm, confident, happy, sociable, extrovert associated with a combination of Dopamine and Serotonin neurotypes. By contrast, traits such as acute shyness, nervousness, moodiness, glumness, or uncontrollable impulsivity, characteristic of Adrenaline and Opioid neurotypes, do not bode well for social success. Positive reinforcement through love or encouragement alone cannot transform an Adrenaline or Opioid Neurotype into an emotionally stable, socially confident Dopamine or Serotonin Neurotype. To accomplish such a feat would require the enhancement of individual temperament through Superbiology.

Most sensible people would agree that some people are naturally more relaxed or cheeful than others, and that such traits assist them to get on in life; just as if we lack intelligence, we are not "free" to be neurosurgeons, no matter how many courses in positive thinking we may attend. Success in life is far easier if one naturally possesses qualities such as emotional stability, gregariousness, or high intelligence. For an emotionally unstable introvert of low intelligence, the pathways to social success and happiness are few and far between. Providing individuals with the opportunity to enhance their neurotype through Superbiology may have a strongly beneficial effect on their well-being. Temperamental attributes such as emotional stability, geniality, high spirits or cheerfulness, and mental attributes such as the capacity for self-control, logic, and learning, will all be targets for a future science of biogenetic enhancement.

A comprehensive understanding of the neurotypology of behavior is a necessary precursor to any future science of neuroenhancement. Once the correlations between neurochemistry and temperament have been confirmed, scientists can begin the real work of finding ever more advanced ways to enhance the neurochemistry of temperament through the development of biotherapies—the Superbiology of eugoics, Supergenics, Nanomedicine, and beyond.

Beyond Experientialism

The neurochemical influence on personality is now public knowledge. Words such as endorphins, dopamine, and serotonin are now as commonplace as the ego, unconscious, and Freudian slip for a previous age. The unwillingness of experientialists to recognize the influence of biological makeup on personality will become increasingly untenable in the age of the decoded genome, when every day sees a new research paper demonstrating the correlation between genes and behavior. Public recognition of the biological basis of variations in individual temperaments will continue to increase over the coming years, as the New Biology advances. If experientialists continue to ignore the biological basis of individual differences for the sake of ideology—the erroneous belief that "differences" is synonymous with "discrimination"—they will not only do a disservice to their profession, to society, and to the species, but also be left with intellectual egg on their faces as more and more proof of their error emerges in the years ahead.

Truth must come before ideology. One can fully understand the sentiment behind the view that any kind of connection between biology and behavior should remain forever taboo, on the grounds that eugenicists made the same connection some eighty years ago, and the terrible end result was the ideology we might call "Hitlerianism"—the attempt to impose the belief in racial supremacy through war, genocide, and totalitarianism. Fear of biopsychology is a perfectly understandable reaction to the horrors of the twentieth century.

But the New Biology is not the old biologism. Supergenics is not eugenics. And it's not just a semantic difference—a change of language. New words describe new phenomena, new ideas, reflecting a new world picture, a new metameme for a new century, a new millennium—a new age.

Bioegalitarian Experientialists seek to ignore the biology of behavior through the fear that knowledge of biological variations will lead to prejudice and discrimination. But they fail to realize that Superbiology will be of *most* benefit to the very *least* able—those suffering from what James Watson has called "genetic injustice"—a bad "roll of the genetic dice," resulting in mental or physical impairments. If you promote the false belief that biology has no effect on behavior, you will ignore the root causes of suffering and deny the opportunities provided by Superbiology for those who need it most. You cannot help the biologically underprivileged by denying that any deprivation exists. As James Watson has said: "There is among human beings a great spread of abilities, and to deny these differences is also to deny we can change them."[29]

We do not have to pretend that no biological differences exist between people to justify moral behavior. The recognition of biological variations is perfectly compatible with the belief in equal rights and dignity for all human beings, regardless of sex, race, personality, or abilities. Psychologist Jerome Kagan put it thus: "Temperamental variation does not pose a threat to democratic and egalitarian ideals. . . . Temperamental differences do not imply differential access to freedom or power; acceptance of inherent variation is perfectly compatible with democracy and egalitarianism."[30]

We all have different abilities, we have all survived the evolutionary journey together, and we must learn to get on with each other. This moral message has the virtue of being both true and sensible. In the words of Hans and Michael Eysenck: "Recognition that we are incorrigibly different from one another is the beginning of all wisdom."[31]

So much antagonism results from the simple inability to recognize and accept the fact of our innate individual differences. The Roman poet and thinker Horace, ever the keen psychologist, put it thus: "Sad people dislike the happy, the happy the sad; the quick thinking the sedate, the careless the industrious."[32]

Differences need not mean division. We should recognize that diversity has evolved because of its value to mutual survival—different types have different talents. No man is *truly* a one-man band. Even the greatest composers need an audience to hear their music. Neurotypology may help us appreciate and respect the different beliefs, values and behavior of other people, through the recognition of a simple, but rarely spoken truth—we are all different! Our minds and bodies are not the same, and the behavior that flows from them is a reflection of our unique nature as individual human beings.

Contrary to the views of the bio-Luddites, for whom any recognition of biological differences automatically means social divisiveness, prejudice, and discrimination, neurotypology can help to *increase* mutual respect through an understanding of the simple fact that we are, each of us, what we are—uniquely individual human beings. I am not you, and unless we wish to live in a state of perpetual war, we both have to accept the fact that everyone is not going to share our own habits, beliefs, and values, and learn to live together in peace, goodwill, and mutual tolerance of our inevitable differences.

When the American founding fathers declared that "all men are born equal," they did not mean that everyone is born with identical biogenetic constitutions, and hence abilities (**bioegalitarianism**), but that all are equally worthy of respect and treatment under the law (**civil egalitarianism**). There is no dichotomy between the recognition of innate differences in ability and the ethical belief in equal respect of all, *irrespective* of abilities.

Biopsychology is not a tool for oppression, discrimination, and prejudice, but a necessary precursor to the development of the Superbiology by which we may enhance our abilities to ever-increasing levels of well-being. So let us embrace the possibility of transcending our biological limitations with open arms.

* * *

DOING: NEUROMOTIVE PSYCHOLOGY

> The idea of "transhumanism" put forward by Sir Julian Huxley would have struck Freud as some kind of sublimation.
> —Colin Wilson, *Beyond the Outsider* (1965)

The study of motivation in psychology asks, What drives human beings to behave as they do?

Clearly a vital question in both psychology and philosophy!

Here I reject both the pessimism of Freudian psychoreductionism, and the overidealism of Maslovian humanism, in favor of a new model of motivation for the twenty-first century I call neuromotivational or **Neuromotive Psychology**.

Summary of Neuromotive Psychology

Human beings are not merely naked apes acting on primitive instincts in the service of selfish genes, but hard-wired with an evolutionary instinct. The evolved, conscious mind instinctively seeks to extend knowledge and abilities in its own interests of increasing survivability and well-being. The **psycho-Darwinist** drives of Eros and Thanatos may be transcended by the **Prometheus and Orpheus Drives** for Individuation and Integration. **The Will to Evolve** is not a drive for appropriation (Nietzsche's Will to Power) but for self-development. However, the complete "self-realization" envisaged by humanistic psychology is impossible given our current state of biogenetic limitation. Genuine transcendence of human limitations requires the assistance of Superbiology. Free will—the capacity for motivational choice by the conscious, rational mind—will be increased in the DNAge by eugoics, the science of neuroenhancement. **Neuromotive Psychology** represents a positive alternative to both the pessimistic fatalism of psychoreductionism and the naïve idealism of Maslovian humanism and its contemporary expression in the Culture of Therapism—the belief that complete human satisfaction is possible in our present biological condition.

Psycho-Darwinism

The dominant view of human motivation today is a combination of Freudian psychoanalysis and neo-Darwinist reductionism. Basic instincts, the ego and the unconscious on one hand, selfish genes and naked apes on the other, form the accepted popular human understanding of human psychology. I shall argue that the Freudian and neo-Darwinist models of motivation constitute a single ideology of psychoreductionism, which ignores higher (more complex), cognitive drives, to the detriment of both individual and social well-being.

Psychoanalysis is today generally regarded as "unscientific," but Freud himself, trained in medicine, graduating in neurobiology, regarded psychoanalysis as a physical science. Two quotes illustrate the point. This one from Freud's first work, *Project for a Scientific Psychology*, was written in 1895: "The intention is to furnish a psychology that shall be a natural science."[33] And from his 1929 work, *Civilization and Its Discontents*, on the subject of religious belief: "One is justified in attempting to discover a psycho-analytic—that is, a genetic—explanation of such a feeling."[34] Freud's use of the word "genetic," here, may surprise many. Take away the Freudian term *psycho-analytic*, and this could have been written by a contemporary neo-Darwinist, neuroscientist, or behavioral geneticist.

Freud began, then, by viewing human beings as biological organisms with biologically based drives and psychopathologies. "Anxiety neurosis" (or *angstneurose*: his own term) he regarded as a disease of the nervous system. Only gradually did Freud begin introducing *experiential* theories of psychosexual conflict resulting from problems in childhood development—the notions of oral, anal, and phallic periods, representing a supposed universal psychological response to the experience of breast feeding, toilet training, and the recognition of having or not having a penis.

But if one ignores the speculative psychosexual developmental theories (such as the strange idea that all little boys wish to murder their fathers and copulate with their mothers), it is evident that Freud's original biological model of human motivation is perfectly in accordance with the neo-Darwinist doctrine of universal behavioral traits inherited from our evolutionary past.

Freud's proposed motivational drives—the life and death instincts of Eros and Thanatos—clearly correspond to the neo-Darwinian view of human behavior driven by genetic predispositions to act in the service of the selfish genes. Eros is the genetic program for survival and reproduction. Thanatos is the command to self-destruct when programming is completed. With the drives of the unconscious id on the one hand, and the predispositions of the selfish genes on the other, the Freudian and neo-Darwinist models of human motivation constitute a single **psycho-Darwinist** psychology of motivation.[35]

Freudian Psychoreductionism: Sublimation

In the Freudian model of human motivation, the entire products of human culture are effectively dismissed as an unconscious channeling of the energy required for survival and reproduction into socially acceptable pursuits. The word chosen by Freud to describe the channeling of basic instincts into cultural pursuits was "sublimation."

The word "sublime" originally meant to raise aloft, (from the Latin *sub*— "up to" or "toward" + *limen*—"limit," and later, the lintel above the door). In the Middle Ages the term was used by the "mystic chemists," the alchemists, both as noun and verb, to describe the process of turning a solid to a vapor by heating. Sublimation thus meant improvement by refinement or purification. Here is the twelfth-century alchemist Artephius (who claimed to be 1,026 years young), writing on the subject in *The Art of Prolonging Human Life*: "The spirit ascends on high to the superior part, which is the perfection of the stone and is called sublimation. . . . Nature dissolves and joins itself, sublimes and lift itself up, and grows white, being separated from the faeces by such a sublimation, conjunction, and raising up, the whole, both body and spirit are made white."[36]

In the eighteenth century, the philosophers Edmund Burke[37] and Immanuel Kant[38] both wrote influential books on the nature of the sublime—regarded as the human perception of awe and wonder in the presence of objects of vast or infinite size, such as an ocean or a starry night. Once again, then, the word was associated with transcendence of limitations—an advancement to something higher or better than before.

Nietzschean Psychoreductionism

By contrast, their successor, the German philosopher Friedrich Nietzsche, used the word "sublimation" in precisely the opposite sense—to deflate the idea that the products of human culture were anything more than a mask of self-deception, disguising a universal drive for primacy he called "the Will to Power." Thus, in the opening aphorism of *Human, All Too Human* on the subject of human morality, he writes: "There exists, strictly speaking, neither an unegoistic action nor completely disinterested contemplation; both are only *sublimations*, in which the basic element seems almost to have dispersed, and reveals itself only under the most painstaking observations."[39] And in his collected notes, *The Will to Power*, Nietzsche deflates the supposedly higher motivations of the artist, scientist, and theist alike to pure self-aggrandizement, by according to each in turn the phrase "*What drives they sublimate*!"[40]

It was from Nietzsche that Freud "borrowed" the reversal in meaning of sublimation, and duly followed him in reducing all cultural activity to the level of brute animal instinct for survival—the entire products of human civilization debased to the level of diverted drives for sex and aggression.

Sublimation as a Meme Attack

The semantic alteration of the word "sublimation" is a perfect example of a **meme attack**—the deliberate modification of a word, phrase, or idea in order to alter the attitude conveyed by its original meaning. In a classic meme attack, the word "sublimation" was captured, its inference reversed, from implying a "raising up" synonymous with progress or improvement, to convey precisely the opposite—a pathological self-delusion disguising lower drives—culture as nothing but the release of pent-up sexual energy caused by the social requirement to inhibit the sex drive. By reversing the meaning of the word "sublimation" from its original sense of transcendence, the psychoreductionism of Nietzsche and Freud effectively eradicated the belief in the possibility of any kind of transcendence of the human condition beyond that of our ape ancestors.

Certainly, the sublimation of "basic instincts" does occur. Cheering on our team or nation in sport is a safe and exciting method of releasing an evolved instinct for tribal aggression. Sport is war without tears. But to regard *all* cultural behavior as *nothing but* a diversion of basic instincts is to imply that a crowd cheering on a boxer as he beats his opponent into a pulp is no different than Shakespeare writing a sonnet, Bach composing a fugue, or Einstein devising the theory of relativity. In other words, utter nonsense. Human beings clearly have far greater motivational drives than those dictated to them by their genetic programming. We are more than naked apes acting on primitive instincts! To restrict one's view of human motivation to those of a goat is not only absurd, but serves to debase the value of humanity, with dangerous implications for society.

Neo-Darwinist Psychoreductionism: The Dawkins Doctrine

And today's neo-Darwinists continue the trend of discouraging human transcendence. For the secular gospel of neo-Darwinism—the Doctrine of the Selfish Gene—implicitly echoes the reductionistic view of human motivation governed by unconscious instincts. In the postmodern age, motivational control by Nietzsche's "Will to Power" or Freud's "unconscious id" has mutated into control by the selfish genes. Here is Richard Dawkins writing about our "biological nature" in *The Selfish Gene*:

> If you wish, as I do, to build a society in which individuals co-operate generously and unselfishly towards a common good, you can expect little help from biological nature. Let us try to teach generosity and altruism, because we are born selfish.[41]

"We are born selfish" says Professor Dawkins. Rather a clear message for the public.

Ah yes, he says, but "*we, alone, on earth can rebel against the tyranny of the selfish replicators.*"[42] In other words, the mind can rule the body. Just as Freud declared, "Where id was, let ego be." Fine in theory. Yet in practice, both paint a picture of human beings as slaves of forces beyond their control. For just as Freud's ego can only control the id at the expense of neurosis (hence we are effectively ruled by the id), so Professor Dawkins tells us we can "expect little help from biological nature" if we wish to be good, because "we are born selfish." In other words, we may "***try*** to teach generosity and altruism" (and how significant is his inclusion of that little word!), but in an age of biopsychology in which the previous generation's naïve belief in experientialism and hence the power of cultural conditioning over neurotype has been crushed, what's the point in trying to be good, one might reasonably ask, when our "biological nature" is selfishness?

The Dawkins Doctrine has served to encourage the demoralization of man.

Are we really "born selfish" as Professor Dawkins declares? What about the innate neurochemistry of benevolence? Oxytocin? Serotonin? (I call it "Felt Morality.") Ah, *but the professor was writing before the rise of biopsychology*, at a time when experientialism—the belief that experience not biology determines human nature—dominated culture. So it didn't matter if we were "born selfish"—society could condition us any way it chose. *But that was thirty years ago!* Now that we are permitted by society to acknowledge the strong neurochemical influence on behavior, the assertion that "we are born selfish" should be recognized for what it is—both false and socially irresponsible. For while claiming to stand for Enlightenment Reason, in reality, both the Freudian and Dawkins Doctrine paint a gloomy pseudoreligious picture of a species chained by Original Sin, subject to the Fall of Man, for whom the spirit may be willing—but the flesh is weak:

> Whereas 19th-century Darwinists saw evolution as the story of the ascent of humanity from its brutish origins, today's Darwinists want rather to tell the story of the Fall of Humanity back into beastliness.[43]

Bio-Fatalism: The Psychoreductionist Triumvirate

So there we have it—Nietzsche, Freud, and Dawkins—a psychoreductionistic triumvirate who, wittingly or not, have served to encourage the belief that higher, cognitive drives should be "deconstructed" into the more fundamental instincts that underlie them.

Nietzsche's Will to Power, Freud's Unconscious Id, and, today, Dawkins's Selfish Genes—the Nietzschean, Freudian, and neo-Darwinist models of motivation constitute a single ideology of psychoreductionism, which downgrades our higher (that is, more complex), specifically human, cognitive drives for human transcendence to the level of "sublimation." All imply that our behavior is the result of biologically based forces largely beyond our control. All effectively downgrade cerebral-cultural activity to the level of a mask of self-delusion disguising our "real" drives—sex and aggression in the service of survival and reproduction, whether controlled by the Will to Power, Unconscious Id, or Selfish Genes.

By downgrading the likelihood of transcending primitive instincts through the conscious mind, psycho-Darwinism serves to reinforce the belief in **biological fatalism**—the idea that man is the slave of his basest instincts, incapable of escaping his lowly condition. By viewing human beings as essentially slaves of unconscious drives or selfish genes, psycho-Darwinism encourages the fatalistic view that man is an unimprovable "fallen" angel, doomed to act out his genetic programming without regard for the moral consequences.

Psycho-Darwinism reinforces the belief in Original Sin. Just as the belief in Original Sin delayed the rise of humanism by its acceptance of human suffering and limitations, so the pseudoreligious belief in biological fatalism fostered by psycho-Darwinism holds back the rise of *trans*humanism, by encouraging doubt as to the possibility of human transcendence from genetic to conscious control of behavior.

The denial of any greater human motivational drives than those of our animal ancestors is not only false, but dangerous to the psychological health of individual and civilization alike. By encouraging the belief in biological fatalism, psycho-Darwinism overrides human faith in evolutionary progress.

The result has been predictable, and depressing to behold. The combined intellectual product of a great triumvirate of nineteenth-century thinkers—Nietzsche, Freud, and Darwin—has all but destroyed the belief in the existence of any higher human motivational drives beyond those of apes.

Psychoreductionism has served to debunk the Enlightenment Project of Modernity—the belief in human progress through reason. The Enlightenment vision of continual advancement has been overthrown by the cultural imposition of psychoreductive psychology. The propagation of psychoreductionism without thought for its ethical implications is speeding the descent of society into postmodern nihilism.

It is Nietzsche and his secret disciple, Freud, who are the true founders of the postmodern condition, and the New Darwinists its unwitting heirs.

Psychoreductionism is not only an erroneous view of human nature, but also ethically dangerous. How people behave is influenced by how people expect them to behave. Tell a child for long enough that he is naughty, and he will begin to act according to the expectations placed upon him. Tell the public that their fundamental, biological motivating drives are no different than those of apes, and don't be surprised if they end up behaving like them.

I therefore find it necessary to refute the claims of the psychoreductionists and challenge the validity of the psycho-Darwinist model of human motivation.

In the next chapter, I shall discuss the meme of the selfish gene. So here, let us look a little closer at the psychological profile of those two "masters of suspicion," Nietzsche and Freud.

Debunking Nietzsche and Freud

Here is Freud's view on the nature of human motivational drives:

> Human beings are not gentle creatures who want only to be loved, who simply defend themselves if they are attacked; on the contrary, a powerful measure of desire for aggression has to be reckoned as part of their instinctual endowment. As a result, their neighbour is to them not only a potential helper or sexual object, but also someone who tempts them to satisfy their aggressiveness on him, to exploit his capacity for work without compensation, to use him sexually without his consent, to seize his possessions, to humiliate him, to cause him pain, to torture and to kill him. . . . As a rule this cruel aggressiveness waits for some provocation or puts itself at the service of some other purpose, whose goal might also have been reached by milder measures, [but] in circumstances that are favorable to it, when the . . . counter-forces which ordinarily inhibit it are out of action, it also manifests itself spontaneously and reveals man as a savage beast to whom consideration towards his own kind is something alien.[44]

One might ask, do we really wish to base our understanding of human motivation on the views of one who considers the motivational drives of his fellow human beings to be no different than those of a serial killer?

Similarly, Nietzsche regarded the basic drive of all living things as the will to dominate others. He thus regarded the drives for war and evil not merely as innate (like Freud) but as positive virtues (like a psychopath). He is generally believed to have written the majority of his work under the influence of progressive mental degeneration due to insanity-inducing syphilis. Likewise, one cannot help but wonder about the extent of Freud's notorious cocaine addiction. Too *much* dopamine can move a man in mysterious ways.

Neither man is believed to have had much in the way of sexual experience. Freud was apparently "never interested in his genitals."[45] The hermit Nietzsche's probable only nonmasturbatory experience was a single encounter with a prostitute from whom he contracted the syphilis which destroyed his mind. In other words, the grand theories of Nietzsche and Freud, based on the view that all human activity can be reduced to the level of displaced drives for sex and aggression (Freud) or dominance (Nietzsche), were largely written under conditions of prolonged sexual abstinence and frustration.

It is somewhat revealing that men who devoted their entire lives to conceptual thought at the expense of sexual experience should feel the psychological need to downgrade intellectual activity and behavior in favor of the sex drive. The Nietzsche-Freud view of human culture as the product of repressed sex and aggression sounds suspiciously like a universal projection of their own sexual frustration. The madness of Nietzschean-Freudian psychoreductionism is probably the product of sexual frustration and neurochemically induced psychological derangement resulting respectively from syphilis and cocaine.

Beyond Psychoreductionism

According to Freud, all cultural activity is a sort of self-deception, a distortion of reality—a defense mechanism against anxiety—an unconscious channeling of sexual energy into cultural activities. According to the neo-Darwinian evolutionary psychologists, behavior is the product of evolved, genetic predispositions for survival and reproduction. Thus, the basic psycho-Darwinist motivation behind, say, writing a book of philosophy is the impulse for sex and aggression in the service of genetic survival and reproduction, channeled into a socially acceptable pursuit.

But an unconscious drive is not coercively driving the philosopher to write a book against his will! On the contrary, the conscious mind is deliberately utilizing energy in its own, cognitive self-interest. The philosopher is consciously choosing to direct energy in the service of increasing knowledge, for the purpose of increasing his own survivability and well-being. The motivational drive is not merely the diversion of energy from its "proper" function, but the conscious, cognitive will of the mind to extend its capacities in its own interests.

The mind is the brain evolved to a level of complexity at which it attains the ability to recognize its own existence. A brain aware of its existence is automatically aware of its fate—impending death. It is instinctive of the conscious brain to question the body's drive or genetic program for self-destruction, thence to replace it with a drive of its own—the Nurethic drive of a conscious entity to act in its own cognitive interests rather than those of the unconscious id or selfish genes (take your pick). As man learns to consciously control his brain, he becomes increasingly capable of inhibiting and overriding his instinctive, Genethic drives—the basic instincts of Eros and Thanatos—with his own cognitive or Nurethics drives, in his *own* interests of ever-increasing survivability and well-being—the goal of the evolved conscious mind as opposed to the genetically programmed body which has us down for self-destruction.

The evolution of the neocortex in *Homo sapiens* has resulted in the emergence of the self-recognizing (that is, conscious) brain we call the mind. As a result, even when the basic drives for survival and sex have been met, man is not satiated. He begins to want more out of life. And to get more, he begins to explore the possibilities provided by the greatest tool at his disposal—his brain—in the words of Arthur C. Clarke, "the most complicated structure in the known universe."

Man wants more out of life in proportion to the complexification of his mind, for the greater the brainpower, the greater the range of possible behaviors conceivable by the mind, and the potential increase in survivability and well-being. The products of the rational mind are not some sort of a distortion of lower impulses, but a raising up from the lower to the higher—that is, from the simple to the more complex. Complexification increases freedom because the complexity of intelligent thought increases the possible range of flexible responses to the environment. Man is free in proportion to his ability to *think for himself.*

From Sublimation to Transcendence

Nietzsche's Will to Power and Freud's Eros and Thanatos simply describe the Darwinian drives for survival/reproduction and self-destruction. But merely to recognize one's slavery is not to free oneself from subjugation. To acknowledge our subjection to unconscious impulses is not to transcend such a condition.

Freud's Eros and Thanatos imply that the lower (less complex) drives are somehow more "real" than the higher—that the higher can never dominate human motivation, but only exist as a sort of temporary diversion from the primary, basic instincts. Such a view denigrates the ability of the conscious brain to override genetic control of behavior—the facility known as "will."

Once a human being can say "I think, therefore I am"—once he is capable of making plans and acting them out successfully—it is simply no longer tenable to suggest that his motivational drives are controlled by the unconscious. Once man becomes capable not merely of inhibiting the impulses produced by the selfish genes, but of altering their very nature in his own interests, he becomes, in effect, something other than man, because the species called *Homo sapiens* is defined not only by its ability to think, but by its condition of slavery to the programming of the selfish genes.

While the unconscious animal has no other motivating drive than its genetic program to survive and reproduce, the conscious animal extends its motivation to the expansion abilities, knowledge and experience—a drive entirely unknown to the less-evolved animals incapable of reason, learning, creativity, or self-consciousness.

Once the brain is sufficiently complexified to allow for complex thought, we see the emergence of higher-level, cognitive drives—the specifically human faculties responsible for the evolution of human civilization. Once we have developed our minds to the level at which they demand more than the mere gratification of animal instincts shared by our evolutionary ancestors, we may begin to transcend our genetic programming by means of more complex, cerebral drives. We may do so in two complementary ways—by continually expanding our knowledge and abilities, and by developing a conscious feeling of integration or "at-homeness-in-the-world"—conditions unknown to both ape and alienated man.

The self-recognizing mind may replace the unconscious Freudian drives of Eros and Thanatos with conscious drives for Individuation and Integration. I call them the **Prometheus and Orpheus Drives**.

The Prometheus Drive

> Prometheus is . . . a personification of the growth process.
> —Arnold Toynbee, *A Study of History* (1934–61)

When the conscious mind takes over behavioral control from the selfish genes, Eros is transcended by Prometheus.

Prometheus, the Titan punished for stealing fire from the gods to give to man, signifies the human quest for ever-greater knowledge, abilities, and experience. The Prometheus Drive is the instinct of the evolved, embodied, self-recognizing brain to continually seek to enhance its abilities in pursuit of ever-increasing survivability and well-being.

The Prometheus Drive is the human drive for cerebral complexification—literally to increase the information content of the conscious biocomputer we call the human mind. Increasing knowledge extends survivability by increasing flexibility of response. A computer only able to process a few bits of information at a time is more limited than one able to assimilate vast amounts. The human brain is far better able than that of the ape to employ logic—the system of rules governing the act of reasoning—to determine the causal connections between events in the world. The drive to accumulate knowledge is an ability preserved by natural selection for its enormous value to survival. Knowledge = power = survivability. The joy of knowledge is the pleasure awarded by nature for increasing our ability to survive.

The Prometheus Drive is thus primarily, although not solely, a matter of cognitive development—an increase in brainpower, reason, intellect, knowledge, and the willingness to employ such power tools in our own interests of ever-increasing survivability and well-being.

The Prometheus Drive is the impulse that makes the child, the scientist, and the philosopher each ask questions about the world, in pursuit of the knowledge by which to improve their condition. The Prometheus Drive is the instinct for curiosity which causes the child to endlessly ask questions of its frustrated parents: "What's this for, Mommy?" "What does that do, Daddy?"

Education is part of the Prometheus Drive, for a high level of knowledge provides a head start in the struggle for life. Higher education tends to get you a better-paid job and a lifestyle allowing you to live longer and more comfortably. Statistically, poverty and life span are inversely proportional.

And the Prometheus Drive is the impulse that drives men to seek power, fame, and fortune—all goals representing an extension of abilities in the interests of increasing survivability and well-being.

Prometheus versus Eros

The difference between Prometheus and Eros lies in the concept of "growth." The sole purpose of Eros is survival and reproduction. The purpose of the Prometheus Drive is personal growth or self-development.

To Freud, all cultural activity was essentially a pathological perversion—not the product of growth and development, but sexual displacement activity. But the cultural activity made possible by the evolution of the rational mind can in no reasonable way be considered a pathological side effect of a repressed sex drive! The conscious, rational mind "takes on a life of its own." It "thinks for itself." It begins to seek knowledge purely for the pleasure of knowing. And it begins to want more—more knowledge, more life—an expansion of possibilities. Becoming as opposed to being. From survival to growth. None of this is present in Freud's Eros—in which culture is regarded as a sort of mistake made by animals thinking too much.

The drive to "want more out of life" is a conscious extension of the basic instinct to survive. In a conscious organism programmed to age and die, the drive for "more"—for self-development or "personal growth"; for evolution in its original sense of the word—the unfolding of potential—is a rational, inevitable response to the knowledge of impending death, the brevity of life, and the options available for action.

The Prometheus Drive is the Will to Grow—to be all that one can be. It may be expressed and strengthened through activities such as exercise, learning, and education. However, it must be recognized that mere effort of will alone cannot override our biogenetic limitations. To transcend the human condition itself requires the appliance of science. The science of Superbiology will enable us to dramatically enhance our ability to act on the Prometheus Drive. For instance, increasing habitual levels of dopamine may serve to increase the faculty of will—the ability of the mind to control the body—assisting in the instinctive attempt to be the *best* we can be.

The Orpheus Drive

> The dull ears that are deaf to the unearthly music of Orpheus's lyre are well-attuned to the drill sergeant's raucous word of command.
> —Arnold Toynbee, *A Study of History* (1934–61)

Complementary to the Prometheus Drive for Individuation, the Orpheus Drive is the drive for psychological Integration—the natural desire of a conscious entity to feel *at home* in the world.

Orpheus, the legendary musician of Greek mythology, symbolizes here the state of psychological unity with the world generally called "spirituality." The sublime melodies played on his lyre reflected the harmonious Music of the Spheres, signifying the underlying unity of the cosmos. The power of music lies in its ability to "take us out of ourselves," that is, to temporarily free us from the bonds of subjectivity—the preoccupation with our own, petty, personal problems, and allow us to experience a sense of unity with and empathy for the whole of creation. Beethoven's setting of Schiller's poem, "Ode to Joy," its lyrics espousing the brotherhood of man, is perhaps the archetypal expression of the Orpheus Drive for Integration.

Orpheus may thus be used to symbolize the instinctive human drive for cognitive Integration—the yearning of a conscious brain to experience itself as one with others and with nature, as opposed to an isolated aspect of the world, Self separated from Other by the "curse of consciousness." The Orpheus Drive is the instinct of a self-conscious brain to identify itself with the greater world in which it exists.

Essentially, Orphic Integration is attained through *creativity*—the ability of the mind to form harmonious patterns out of the stuff of the world—to make into cohesive wholes the disparate objects of perception. Where Prometheus (creator of man, giver of fire) is a scientist, Orpheus is an artist. The Orpheus Drive is an impulse for *creative thought*—the drive to make wholes from parts, to contemplate the forest rather than count the trees—to consider the lilies—to see the world as a unity-in-diversity, and oneself as an integral aspect of the totality of things.

The goal of the Orpheus Drive is to experience a sense of unity with the evolving cosmos of which we are a conscious part. Satisfaction of the Orpheus Drive is the feeling of being at one with universe—an aspect of the evolutionary process of nature. Thus, the Orpheus Drive is sublimation in the *original* sense of the term—the experience of being raised up and out of one's self—whether through music, mathematics, or mystic rapture.

But Orpheus is not Buddha. Unlike in the world of neo-Buddhist No-Selfism, satisfaction of the Orpheus Drive does not require the abandonment of the complementary Prometheus Drive for deliberate, determined, willful action. The Orpheus Drive is not a death wish, a desire for the nothingness of nirvana. A sense of conscious identification with the universe need not come at the expense of selfhood.

Equally, the feeling induced by satisfaction of the Orpheus Drive is precisely the opposite of the kind of "existentialist angst" associated with Jean-Paul Sartre and his followers in gloom. The willingness to act upon the Orpheus Drive represents a positive **New Existentialism**—the will to *overcome* the feeling of alienation, and feel oneself *at home* in the world.

Orpheus versus Thanatos

Where Prometheus transcends Eros, Orpheus transcends Thanatos.

Freud is widely believed to have invented his concept of Thanatos (named after the Greek god of death) as a pessimistic response to both the death of his young daughter and the mass destruction of life in the First World War, concluding that such barbarism could only be explained as an unconscious death wish. Thanatos is not an anomalous mistake, as even some Freudians believe, but rather, a recognition of the third and final instruction in our genetic program—the command to self-destruct. Thanatos is the biological process known as *apoptosis*—programmed cell death. We see Thanatos in the peace which descends on the dying, once Eros, the survival instinct, has been overridden with the Final Instruction, and the body floods with soothing opioid tranquilizers as death approaches.

However, while a "will to die" may certainly manifest itself in certain circumstances, to suggest that *all* our behavior is ruled by a death wish is clearly nonsense. The more common impulse behind Thanatos is not the will to die, but rather, the desire to transcend the feeling of alienation which may result from the experiential dualism of mind and body—the feeling of separation between the conscious mind and the external world—Self and Other. The nineteenth-century American transcendentalist Ralph Waldo Emerson expressed this feeling applied to human relationships, with painful poignancy: "Two human beings are like globes, which can only touch in a point."[46]

When not a neurochemically aided readiness to accept death, Thanatos is the yearning desire to overcome the feeling of alienation resulting from the state of consciousness. Such alienation may be transcended by the Orpheus Drive—not the drive to self-destruct—"the will to die"—but the impulse of a conscious entity to feel a sense of psychological Integration with the world—the instinct to feel oneself a part of things, to feel that one belongs not merely to one's family, community, nation, race, or even species but to the universe, nature, or creation as a whole. Satisfaction of the Orpheus Drive is thus a form of naturalistic or **neurotranscendence**—the uplifting of the mind to a sense of psychological unity with the world.

The ability to satisfy the Orpheus Drive can be increased both by improving relationships (reuniting self and other) and by finding activities by which one can feel integrated with the world, whether it be fishing, rock climbing, hang gliding, surfing the Net, science, art, music, philosophy, or the contemplation of nature.

Superbiology offers the prospect of dramatically increasing our power to satisfy the Orpheus Drive. By raising levels of neurohumors such as serotonin and oxytocin, we may enhance the neurochemical correlates of empathy, or love, thus increasing the feeling of Integration between self and other, man and nature, microcosm and macrocosm.

Prometheus, Orpheus, and the Bicameral Brain

While Eros and Thanatos are physical, or Genethic drives, Prometheus and Orpheus are cognitive, or Nurethic drives.

Where Eros and Thanatos correspond to the two basic faculties of the body—doing and feeling, Prometheus and Orpheus correspond to the two fundamental faculties of the mind—logic and creativity. Logical thinking is the *analysis* of the objects of perception by breaking them down into constituent parts, then examining how they connect together sequentially—corresponding to a straight line. Creative thinking is the *synthesis* of the world into patterns of wholeness—corresponding to a circle. Logical thinking relates to the dimension of *time* through the analysis of *causal* connections. Creative thinking relates to the dimension of *space* through the ability to perceive a number of parts of information at once, as a whole.

Logic and creativity are the two fundamental types of human brain activity, corresponding to the differences in functioning between the left and right hemispheres. The left hemisphere specializes in logical analysis through language, the right brain in perceptual analysis through visual symbols—or pictures. Logic analyzes, creativity synthesizes. Logic is the examination of the trees, creativity the perception of the forest. The left brain is a scientist, the right brain an artist. The ideal philosopher is a scientist-artist, committed to the satisfaction of both Prometheus and Orpheus Drives for Individuation and Integration, through logic and creativity—diversity-in-unity—growth and love.

Prometheus, Orpheus, and Bispectism

Prometheus and Orpheus are Eros and Thanatos transcended through cerebralization—conscious control of behavior in the mind's own interest of ever-increasing survivability and well-being.

The Prometheus and Orpheus Drives represent the two aspects of human existence—the autonomy of the self-conscious, thinking individual, and the dependence of the individual on the society and natural environment for survival. They thus represent a psychological application of the universal principle of nature I call **Bispectism**—the recognition that all things have two aspects by which they may be observed—whole and parts, equal consideration of both being required for a full understanding and appreciation of their nature.

The two aspects of the human psyche are Individuation and Integration, corresponding to parts and the whole. Individuation is the individual's sense of Self as distinct from Other. Integration is the sense of wholeness or unity between Self and Other. Prometheus is Individuation; Orpheus is Integration. Where Individuation is the Will to Grow, Integration is the Will to Love (for what is love but the feeling of unity between self and other?). Prometheus and Orpheus together represent a state of psychological satisfaction through growth and love.

The Prometheus and Orpheus Drives offer the possibility of neurotranscendence through the increased complexity and survivability afforded by conscious control of behavior by the self-recognizing brain we call the human mind. They are not mere "sublimations" of lower-level drives, but qualitatively different, higher drives evolved from the lower by means of conscious control.

My use of the word "higher" to describe the Prometheus and Orpheus Drives is not a case of wooly-minded, romantic idealism. A "higher" drive means a "more complex" drive. Evolution is complexification. Complexification of the body and mind assists survival by increasing flexibility of response. The thinking, conscious brain is an evolutionary complexion of the body. The development from the human drives of Eros and Thanatos to those of Prometheus and Orpheus represents the evolution of the human brain from control by the selfish genes to self-rule by the conscious brain—from Genethics to Nurethics. By acting on the Prometheus and Orpheus Drives for Individuation and Integration, we may transcend our basic Genethic drives of Eros and Thanatos. Together, the Prometheus and Orpheus Drives constitute an evolutionary instinct—the **Will to Evolve**.

The Will to Evolve

The Will to Evolve encapsulates the essence of my philosophical system, for it is simultaneously a metaphysical, psychological, and ethical concept.

Metaphysics

As a metaphysical, or cosmological concept, the Will to Evolve is the impulse for complexification in nature—the observable directionality in the process of evolution toward a continual increase in the overall information content of the cosmos. The Will to Evolve is "what nature does." The unfolding of new forms, structures, and operations is the most basic property of the natural process we call evolution. The word "evolution" means to unroll, like a carpet or a scroll. The Will to Evolve is the impulse in all living things to extend themselves to their full height—to realize their potential—to grow. It is equivalent to Aristotle's term *entelechy*, meaning "containing its future direction within." Entelechy is the innate directionality from simple to complex, potentiality to actuality, that turns an acorn into an oak tree. The Will to Evolve implies no preconceived end point—but rather a process of perpetual growth.

Psychology

As a psychological concept, the Will to Evolve is the mental determination to *steer* the natural impulse for complexification to one's own ends. This is the meaning of *Homo cyberneticus*—man the steersman of his own destiny. By consciously taking control of our own behavior, we can direct our natural instinct for growth in our *own* best interests. Thus, the Will to Evolve is the mind's ability to act *in accordance* with the evolutionary process of complexification, by steering the expansion of capabilities in the direction which offers the greatest benefit to the fundamental, innate goal of ever-increasing survivability and well-being. To evolve is to complexify. To complexify is to increase abilities—to become a more complex animal. Complexification increases survivability, for the more we know, the better able we are to survive. And the more complex our behavior, the happier we are—for nature rewards increased knowledge and abilities with pleasure as an incentive to grow in the interests of survival. The ape knows nothing of the joy of knowledge, but for man the pleasure of personal growth is its own reward. Every time we learn a new habit, we engage our Will to Evolve—the drive to be more than we are—to be the *best* we can be.

Ethics

As an ethical concept, the Will to Evolve is the belief that the best way to live is to deliberately, consciously cultivate one's innate evolutionary instinct for complexification, by continually seeking to expand abilities in pursuit of ever-increasing survivability and well-being. Today, evolutionary complexification occurs in two ways—Darwinian evolution through genetic mutation, recombination, and selection; and *cultural* evolution—through an increase in knowledge and abilities. In the future, evolutionary complexification will proceed in a *third* way—through Designer Evolution, as we make use of our ever-increasing knowledge to enhance our biogenetic makeup and free ourselves from the constraints of the human-all-too-human condition, which condemns us to death, disease, and the biological limitations of body and mind.

Schopenhauer's World as Will and Idea

By using the expression "Will to Evolve," I automatically create a meme-link to the meme maps of the great nineteenth-century German philosophers, Arthur Schopenhauer and Friedrich Nietzsche. I do so deliberately, both to make clear that I consider transhumanism to represent a continuing link in the ongoing chain of modern philosophy, and to clarify the *distinction* between transhumanism and the philosophies of the past. For philosophy, like everything else, evolves. The Will to Evolve is not the same as Schopenhauer's concept of the World as Will and Idea, or Nietzsche's Will to Power.

Schopenhauer regarded nature as an all-embracing, cosmic "Will to Live." "Will-power" in this sense meant something other than that inferred by its contemporary usage—not the ability of mind to guide behavior, but precisely the opposite—genetic instinct—the innate, blind impulse in all living things to survive. This idea may be illustrated by a Schopenhauerian metaphor: *Will power is to the mind like a strong blind man who carries on his shoulders a lame man who can see.*[47] In other words, for Schopenhauer, the mind is merely the passive observer of a more basic motivational drive—the will to live.

Schopenhauer shared the opinion of the seventeenth-century English philosopher Thomas Hobbes, that nature had contrived to produce a world in which life was "nasty, brutish, and short." He therefore regarded the "cosmic will" as intrinsically wicked, and advocated a neo-Buddhist response to it—the renunciation of the will to live in favor of what he called the "will as idea"—a state of detached, philosophical, or aesthetic contemplation of the world, rather than any attempt to assert oneself upon it. For Schopenhauer, an admirer of Platonic idealism and Buddhist nihilism, it would be better not to have been born at all. "Will as idea" was, in effect, the "will to die" (that is, in *theory*—in practice he apparently showed quite the opposite inclinations—the spirit being willing . . .).

The concept of "world as idea" is a classic case of neo-Buddhist passivity—the renunciation of the will to live—aestheticism replacing action. Nirvana as the snuffing out of the self. Orpheus *without* Prometheus. There is nothing wrong with philosophical and aesthetic contemplation (this book could not have been written without it), but the maximum operation of the faculties that lends itself to the optimum state of well-being requires that contemplation be balanced by action—Orpheus by Prometheus. Unless death is to be regarded as preferable to life, self-affirmation must be regarded as superior to self-renunciation. Better to make a mark on the world than to tread lightly on the earth through fear of hurting an ant.

Nietzsche's World as Will to Power

Nietzsche began his philosophical quest as a youthful admirer of Schopenhauer, thus accepting the notion of the cosmic will and its fundamentally wicked nature. However, being his neurotypical opposite, Nietzsche naturally found himself rejecting Schopenhauer's phlegmatic attitude of Platonic, neo-Buddhist renunciation of the world. Instead, Nietzsche decided to *love* nature precisely *for* her cruelty. Thus began his fascination with the Dionysian aesthetic—the belief in sensual instinct over reason, and his self-identification with Dionysus, the Greek god of nature, wine, women, and song. Unfortunately, however, Nietzsche rejected wine, while women rejected him.

Schopenhauer regarded the cosmic will as a universal survival drive in nature. Nietzsche regarded Schopenhauer's attitude as neo-Buddhist nerdiness, and retorted that the basic drive of the cosmic will was not mere survival but the drive to *dominate*—to be more powerful than others; king of the jungle; leader of the pack—an utterly amoral impulse he called the Will to Power. He thus regarded evil as a positive virtue. The Will to Power is *not* the desire for "empowerment" but *appropriation*—the taking possession of something for oneself. In *Beyond Good and Evil*, Nietzsche makes his position crystal clear: "Life itself is essentially appropriation, injury, overpowering of the strange and weaker, suppression, severity."[48] Such an attitude must clearly be considered pathological, probably born of a combination of syphilis, solitude, and sexual frustration. The Will to Evolve is *not* the Will to (over)Power but the determination to be the *best* one can be.

However, to reject Nietzsche's philosophy in its entirety because of its obvious excesses would be wholly illogical. One doesn't throw away the fruit because of the pips. Today's culture has placed an unwritten taboo on Nietzsche because of his appropriation by the Nazis as their "house philosopher," enthused by his hysterical condemnation of the Christian values of pity and meekness. Such a response is symptomatic of an unfortunate tendency one might call **dichotomal thinking**—the "all or nothing" attitude by which a philosophy or idea must either be accepted wholesale or rejected as worthless—as opposed to the sensible attitude of judicious open-mindedness—which in this case involves taking from each philosopher those ideas one considers to be of value and rejecting the rest. Indeed, Nietzsche himself advocated such a tactic:

> The philosopher believes that the value of his philosophy lies in the whole, in the building: posterity discovers it in the bricks with which he built and which are then often used again for better building: in the fact, that is, to say, that the building can be destroyed and *nonetheless* possess value as material.[49]

In such a spirit, the sensible transhumanist might find himself in agreement with many of Nietzsche's attitudes, such as the belief in:

- Facing up to reality as it exists instead of deluding oneself with theistic wishful thinking;
- Living life to the full, without repressing one's natural instinct to "be what one is";
- Regarding man as a halfway house on an evolutionary journey from ape to superman (or transhuman);
- Opposing cultural decline through leveling (or dumbing) down of standards in the name of egalitarianism.

While at the same time rejecting other aspects of the Nietzschean meme map, such as:

- The belief that the ideal state is one of automatism—instinct without thought (Genethics above Nurethics);
- The belief that the evolutionary progress of the species is an illusion;

- The belief that the basic human impulse is a drive for domination over others.

For transhumanists believe in the will to increase their own abilities, not to overpower others.

The Great Pessimists

Let us sum up the positions of Schopenhauer, Nietzsche, Freud, and the neo-Darwinists.

Schopenhauer said the world was evil. Nietzsche agreed and said we should celebrate the fact. Freud agreed with Schopenhauer. The Freudian death wish, invented to explain war and death in the absence of religious belief, is essentially the declaration that the world is wicked and death the true purpose of life. This is pure Schopenhauerian pessimism. Freud actually makes his affinity with Schopenhauer explicit: "We have unwittingly steered our course into the harbor of Schopenhauer's philosophy. For him death is the 'true result and to that extent the purpose of life' while the sexual instinct is 'the embodiment of the will to live.'"[50] Nietzsche was no less pessimistic, referring to his philosophy variously as "Dionysian pessimism" and "*amor fati*"—essentially the belief in biological fatalism—that one should revel in the tragedy of life, and joyfully accept one's fate.

Thus, the **Great Pessimists** Schopenhauer, Nietzsche, and Freud reduce all human activity down to brute survival and reproductive instincts—whether controlled by the Cosmic Will, the Will to Power, or the Unconscious Id.

And the biological fatalism of the Great Pessimists is repeated today by the neo-Darwinist evolutionary psychologists, who speak of a "universal evolved psychology"—eumeme for "basic instincts"—thus replacing behavioral control by the Will or the Id, with control by the Selfish Genes.

The Great Pessimists represent the collapse of humanism and a return to the belief in Original Sin—subjugation to God becomes subjugation to the Cosmic Will, Unconscious Id, or Selfish Genes.

To oppose the Great Pessimists is to fight a Meme War in defense of humanism against those who seek to destroy it.

Beyond the Great Pessimists: The Will to Evolve

Where Nietzsche said, "This world is nothing but the will to power," and Freud said, "The goal of all life is death," transhumanism says, "The purpose of life is to evolve."

For the transhumanist, man is not ruled by animal instinct but guided by the Will to Evolve.

"Will" for Schopenhauer meant nature as a blind drive for survival operating through man. "Will" for Nietzsche meant nature as a blind drive to dominate others operating through man. "Will" for the transhumanist means the ability of the self-recognizing brain we call the human mind to steer the behavior of the body in the direction of evolutionary complexification, in its *own* interests of ever-increasing survivability and well-being.

The operating system of nature is not a drive for survival or domination, but for *evolutionary complexification*. The world is a cosmic Will to Evolve. Evolution is the automatic unfolding of potential in nature. Human beings are a conscious aspect of the process of evolutionary complexification—as such, we are born with a Will to Evolve.

The Will to Evolve is the ability of the human mind to act in accordance with the evolutionary impulse within nature, by steering behavior in the direction of greater complexification, through an ongoing drive for self-development.

Neuromotive Psychology proposes that each of us seek to discover the Will to Evolve, by continually expanding our abilities in pursuit of ever-increasing survivability and well-being.

Maslow's Metamotivation

Thus far, Neuromotive Psychology has much in common with the brand of humanistic psychology espoused by the American psychologist Abraham Maslow, the popularity of which coincided with the emergence of the youthful counterculture of the 1960s, with its dreams of a dawning age of Aquarius.[51]

Humanistic psychology arose as a "third force" alternative to both psychoanalysis on the one hand, and behaviorism on the other. Behaviorism is the psychology which holds that human beings are born as "blank slates," ready to be conditioned by society into anything their conditioners desire. I shall not discuss it further here, suffice to say that my general antipathy toward its sentiments is implicit both in my Neurotypology—stating that human beings are born with unique, neurochemically based temperaments, or personality traits—and my critique of psychological No-Selfism as a convenient justification for mass social conditioning by state, media, or commerce.

To counteract the dehumanizing inferences of both psychoanalysis (man as slave to the unconscious) and behaviorism (man as slave to social conditioning), humanistic psychology stressed the existence of free will, and the need for a positive environment of mutual encouragement, in which individuals may seek and attain fulfillment of their potential as human beings.

Recognizing the existence of higher-level drives beyond those for mere survival and reproduction, Maslow proposed that man possesses a "hierarchy of needs"—a ladder of physical, emotional, mental, social, and spiritual requirements, which he seeks to satisfy sequentially—beginning with the basic (psycho-Darwinist) "need" for survival, and ending with the rather nebulous concept of "self-actualization," which Maslow took from Carl Jung, meaning total fulfillment of potential.

Maslow's model of "metamotivation" continues to be popular today, frequently being taught in psychology courses along with the other dominant models of motivation. The idea of a hierarchy of needs is certainly in accordance with the transhumanist position. It is simply untenable to reduce humanity to the level of an animal which wants nothing other than to "fuck and fight." The transhumanist applauds Maslow's encouragement of human self-development and opposition to Freudian bio-fatalism—the secularization of Original Sin, man as perpetual fallen angel. However, as a Positive Realist, the judicious transhumanist might find Maslovian humanism just a little too naïvely optimistic to adopt as a psychological model of choice.

The Limitations of Self-Actualization

The weakness of humanistic psychology lies in its overidealized notion of what is possible for human beings at this point in their evolution, given their current biological limitations.

As evidence of his "self-actualizers," Maslow openly took the healthiest, happiest, most successful people he could find—with the inference that we could all be like that if we tried. But sadly, this is not the case. There are certain biological obstructions to universal "self-actualization"—tricky obstacles such as an overreactive nervous system predisposing to anxiety neurosis; an overly high basal cortical arousal state leading to excessive introversion; habitually low levels of dopamine, serotonin, or testosterone, causing depression and irritability—not forgetting the small problems of aging and death! The fact is that at this point in evolution innate biological limitations preclude the possibility of total "fulfillment of potential"—the supposed end state of blissful happiness that some adherents to humanistic psychology and its metaphysical sister, New Ageism, appear to believe is within their grasp (perhaps after a few more sessions learning to love themselves).

Better to acknowledge the brutal truth than succumb to wishful thinking. And the brutal truth is that not everyone is born with the luxury of a sanguine and phlegmatic temperament—dominance by the Dopamine and Serotonin Neurotypes—a sound mind in a sound body. And no one *at all* is born with the capacity for *continual* happiness—for our bodies have not evolved to service us with a state of perpetual bliss, but to help us survive, and survival at this point in our evolution requires a certain amount of struggle and strife—the continual will to overcome obstacles in our way.

The belief in the "self-actualization" of man in his present biological form is a myth. We can be better than we are, but we cannot transcend suffering and attain the bliss of nirvana in our current biological state. Such a task would require more radical intervention—through Superbiology.

Positive thinking and encouragement can help raise mood to a certain degree, but it cannot work miracles. While techniques of rational, "positive thinking" can help improve one's attitude, and thus life experience, the individual capacity to maintain a positive mood state is greatly influenced by neurotype. You cannot make a silk purse out of a sow's ear—without science. The only way of enhancing neurotype to a level significantly beyond that which nature has provided is through a future Superbiology of eugoics and Supergenics.

To the positive realist, Maslow's glowing descriptions of his self-actualizers as almost another species in their superiority over the rest of us sounds a certain note of absurdity ("Self-actualizing people, those who have come to a high level of maturation, have so much to teach us that sometimes they seem almost like a different breed of human beings"). One rather doubts that had Maslow returned to his self-actualizers thirty years later, he would have found them in such a blissful state of superhuman health, happiness, and vitality. Quite apart from the trials of life, the process of aging reduces the neurochemistry and physiology of well-being. Nature will have her wicked way with us—if we let her.

From Humanist Idealism to Transhumanist Bio-Realism

The main weakness of Maslovian humanistic psychology is its ignorance of neurobiology. As an example of Maslow's blindness on the biology of behavior, in a footnote to his paper "Metamotivation" he writes: "It is a great mystery to me why affluence releases some people for growth while permitting others to stay at a strictly 'materialistic' level."[52] In other words, Maslow has no idea as to *why* some people are self-actualizers and others not. A neurotypologist would say, "Some people are more psychologically healthy than others, because they have been lucky enough to have been born with an optimum neurotypical balance—such as higher-than-average levels of dopamine and serotonin, producing a sound mind in a sound body." This is way beyond the scope of Maslovian psychology.

Would Maslow have endorsed the possibility of neuroenhancement through Superbiology? He was actually fatalistic about our biological nature in his belief that "peak experiences"—his term for sudden surges of ecstatic pleasure—could not be controlled, but came and went of their own accord. In other words, he accepted mood states as a given—an attitude utterly alien to that of the Neuromotive Psychology espoused here.

In short, the transhumanist vision of human control over our own biological makeup through Superbiology is so far removed from the Maslovian remit that a new approach is needed—a new psychology for a new age of biology and technology.

Here, then, I shall speak not of some slightly smug-superior end-state of fulfilled potential called "self-actualization"—but of a drive for ever-increasing survivability and well-being—the Will to Evolve. My aim is to distinguish the ***bio-realism*** of transhumanist Neuromotive Psychology from the experientialist idealism of much humanist psychology, which is often indistinguishable from the world of New Age unreason in its overidealized notion of what is possible without the bioenhancement of mind and body.

Neuromotive Psychology combines the positivity of Maslovian humanism with the biotech futurism of transhumanism in a model suitable for the twenty-first century.

Neuroanalysis

How could Neuromotive Psychology be applied in practice as a rational tool to increase well-being?

Here I shall speak not of psychoanalysis, or psychotherapy, but of **neuroanalysis**. A future science of neuroanalysis might consist of three elements:

1. Cognitive (psychological: thinking/perceiving):

   Recognize and develop the will to evolve. Raise confidence by encouraging the acknowledgment and development of the innate drive to enhance abilities.

2. Experiential (behavioral: doing):

   Devise a rational plan for creative living. Implement lifestyle changes, based on a plan to attain logically assessed life goals.

3. Neuronal (neurobiological: feeling):

   Prescribe eugoics to enhance temperament. Utilize mood-enhancing drugs to optimize the neurochemistry of well-being.

Cognitive: The Will to Evolve

A future neuroanalysis would emphasize our evolved and evolving nature, and the importance of acting upon our higher, cognitive, or *Nurethic* drives (the Prometheus and Orpheus Drives for Individuation and Integration), both by encouraging us to continually expand our abilities, knowledge, and experience throughout life, and by seeking a sense of unity, or at-home-ness-in-the-world, to replace the feeling of psychological alienation common in depressive states of life devaluation.

A neuroanalyst might stress the importance of the much-maligned concept of "will" to the development and maintenance of well-being. Will is the ability of the mind to steer the behavior of the body in its own interests (hence *Homo cyberneticus*—man the steersman of his own destiny). Failure to do so may result from the nature of the meme map downloaded into the brain, to guide behavior. A meme map riddled with **negamemes** (negative beliefs about oneself or the world) is likely to result in self-destructive behavior. A future neuroanalysis would encourage the attempt to *reprogram the mind* with positive and logical (as opposed to negative and irrational) attitudes—that is, to upload a new meme map into the biocomputer that is the human brain—specifically, the *transhumanist* meme map, based on a powerful **supermeme** (or Big Idea)—the Will to Evolve.

Experiential: Rational Plans for Creative Living

Having emphasized the importance of developing one's innate Will to Evolve (psychology), the next step would be to devise a plan of action (ethics).

Here we might speak of **Rational Plans for Creative Living**.

By Creative Living I mean an attitude of openness to possibilities. Creativity is the ability to construct new patterns out of the stuff of the world. Creative Living is the ability to create new patterns out of the stuff of *life experience*. The opposite of Creative Living is being "stuck in a rut"—acting out the same old routines day after day—without growth, development, progress, or creativity.

The various unpleasant symptoms of anxiety and depression are far more likely to occur if we allow ourselves to get stuck in a rut, instead of living creatively. An example from the animal world is the horrific sight of the caged animal in a zoo, continually pacing up and down, obviously in acute distress, caused by sheer boredom and frustration (a condition to which its "keepers" always seem strangely oblivious. Prison zoos should be banned by any civilized society. A suffering animal in a cage doesn't care about "preserving an endangered species").

Having established the intention to live creatively, based on the underlying principle that we all possess an innate Will to Evolve, the next step would be to determine one's *specific* life goals by developing a Rational Plan for Creative Living—in other words, by working out where one's greatest strengths and weaknesses lie, and devising projects by which to improve both, rectifying weaknesses where possible, while focusing on strengths.

In simple terms, neuroanalysis would encourage you to find out what you do best, and then create a lifestyle that allows you to do as much of it as possible, as *well* as possible—on the wholly logical grounds that the greatest feeling of well-being derives from that which we do best (an idiot won't be happy at Harvard, nor a genius in a chewing gum factory).

Such a tactic might seem like stating the obvious. But how many people actually attempt such a feat? How many people instead end up in jobs they actually despise—effectively wasting their lives away—simply through lack of a Rational Plan for Creative Living?

Neurochemical: Eugoics

So far, neuroanalysis sounds a lot like humanistic and cognitive therapy. However, here we part company. For neuroanalysis would also recognize the importance of *neurotype* on behavior—and the consequent benefits of enhancing neurotypical deficiencies which contribute to behavioral problems, through Superbiology.

For instance, the hypersensitive autonomic nervous system of the Opioid Neurotype increases susceptibility to anxiety and depression, but the nervous system can be strengthened by serotonin-raising eugoics such as Prozac. Similarly, drugs which raise levels of dopamine increase the power of deliberation, volition, or *will*—the ability of mind to control behavior. Enhancing our neurochemistry may thus increase the power of the will—the ability to motivate ourselves in order to control our *own* actions in our *own* interests—rather than those of the selfish genes, which seek to limit us to instinctive patterns of behavior.

By raising levels of neurochemicals such as dopamine and serotonin, the Superbiology of the future may serve to expand the Prometheus and Orpheus Drives for Individuation and Integration, thus increasing our Will to Evolve to ever greater levels of survivability and well-being.

Neuroanalysis would not seek to ignore or repress basic Freudian drives, but rather, like Maslow's humanistic psychology, encourage us to satisfy them, and then move on to higher things. Like cognitive therapy, neuroanalysis would seek to encourage rational, positive thinking. But unlike all three, neuroanalysis would acknowledge, identify, and seek to rectify the neurobiological deficiencies which contribute to frustration and life failure.

Neuroanalysis as Syncretic Psychology

Neuroanalysis, then, would combine elements of different motivational models thus:

- *Psychoanalysis:* Acknowledge the need for satisfaction of basic Freudian drives, then get on with higher things.

Essentially, have regular sex and exercise to prevent physical frustration!

(Freud's sexual theory was influenced by the gynecologist who told him that the cure for female hysteria was "*penis normalis dosim repetatur.*")

- *Humanistic:* Stress the need to act upon the Will to Evolve to avoid *psychological* frustration.

(Maslow: "I warn you, if you deliberately set out to be less than you are capable of being, you will be deeply unhappy.")

- *Cognitive:* Encourage the Will to Evolve by reprogramming the biocomputer to accept only positive, rational thoughts.

(I'd like to quote a famous cognitive therapist, but there aren't any).

- *Neuropsychology:* Employ Eugoic drugs to overcome neurobiological obstacles to well-being.

Perhaps Neuroanalysis might be a more useful method of improving well-being than talking about one's parents, one's relationships, or the daily trivia of one's life for fifty minutes every week.

Freud and the Return of Original Sin

The modern world is deeply infected by the negameme of psychoreductionism introduced into the culture by the Great Pessimists. And none can be held more responsible for the cultural decline into postmodern nihilism than Sigmund Freud.

For unlike Schopenhauer and Nietzsche, Freud was a socially ambitious man, who clearly set out with the intention of making himself famous. Not lacking the art of self-promotion, he delighted in equating himself with Copernicus and Darwin as the final "master of suspicion," responsible for dethroning man from his privileged position of superiority in the cosmos. Where Copernicus destroyed the illusion of man as center of the universe, and Darwin the gulf between man and ape, Freud was proud to consider himself as having put the final nail in the coffin of humanism by destroying the notion of man the master of his fate, captain of his soul—for he was not even in charge of his own mind!

By contrast, though even *more* egomaniacal than Freud, Schopenhauer and Nietzsche were solitary men who never actively sought fame and fortune, while Darwin was so shy about promoting his evolutionary theory that he delayed publication for some two decades!

In one way, Freud's philosophy could be regarded as a "scientific" synthesis of Darwin, Schopenhauer, and Nietzsche—a boiling down of their ideas—combined with some preposterous ones of his own (such as the idea that discovering one has or doesn't have a penis is a cause of universal childhood trauma).

We have Freud above all others to blame for the prevalence of cultural pessimism in the (post)modern world, for it was Freud who deliberately set out to make his philosophy a metameme—the dominant ideology in the meme pool of modern culture.

To Freud, the rational mind could only allow us to *recognize* our unconscious instincts ("where id was, let ego be")—the battle between instinct and reason, body and mind, flesh and spirit could never be *won*—life will *always* be a tragedy because instinct always defeats reason, flesh overrides spirit, body defeats mind. We are biological beings, and the mind is not strong enough to transcend our animal urges. Civilization will continue to be a battleground of irrational impulses, war, aggression, and hostility, barely kept in check by our attempts to impose reason on our destructive urges. Even if we succeed in inhibiting our animal instincts, the result is not satisfaction but neurosis, through the repression of our natural impulse to destroy and be destroyed. The spirit is willing, but the flesh is weak.

In short, Freudianism is a crypto-religious belief system—a form of psychosocial cultural regression to a neomedieval attitude of religious fatalism. Today it's known as "biological determinism." In another age it was known as Original Sin. Its essence is the belief in the inability of man to transcend his animal condition.

Freud and the Postmodern Condition

Freudianism is antihumanism. Antihumanism is antimodernity. Antimodernity is a more accurate way of describing postmodernism. It was not the French "deconstructionists" who were the true founders of postmodern nihilism, but Freud—and his secret mentor, Nietzsche.

The negameme of postmodern nihilism has infected the modern world ever since. It is the reason why young adults in the cinema applaud with glee when a man is tied up and tortured or blown to pieces. It is the reason why a psychologist expresses concern that the constant language of bitchiness and backbiting on TV is turning young people into rude, sneering brats. It is the reason why a young boy has just been arrested for locking his friend up in a hut and burning it down—for fun. For what is benevolent morality if we are slaves to the unconscious id? Nothing but a failed system of ethics.

In Great Britain, in 2004, an "alternative comedian" presented an educational program on Freud for the BBC's *Open University* adult degree course. In his final sentence, he summed up the importance of Freud thus: "Freud teaches us that the only chance we have of staying sane is to recognize that we're all completely mental." This is pure postmodernism—the view of man as a debased creature, at the mercy of irrational impulses, in an absurd universe—but not to worry, we'll just "have a laugh"—at ourselves.

Postmodernism is fiddling while civilization burns. At the dawn of the twenty-first century, much of Western culture seems like a trash-filled graveyard of antihumanistic self-mockery. Freudian pessimism is the psychological underpinning of the postmodern Anticulture of Nihilism.

Beyond Freud

Freud's gloomy pessimism about the human condition was unwarranted. We do not need to remain at the mercy of our animal instincts. We evolve as we increase the use of our brain. The more we do so, the less we lie at the mercy of our genetic programming. Humankind is not a doomed species, trapped between the forces of id and superego, but a halfway house between an animal and a God. For what is a God but one who possesses the power to avoid his own death?

Psychoanalysis was the secularization of Original Sin—the last vestiges of an old religion. In neuroanalysis, the Fall of Man is replaced by the belief in his inevitable transcendence, through Superbiology. We do not need to remain slaves to the unconscious id or selfish genes. We do not need to be chained by Original Sin. We have it in our capacity to transcend the human condition—by increasing *conscious control* over our own destiny.

Where Schopenhauer said man is controlled by the blind Will to Live, Nietzsche by the Will to Power, Freud by the Unconscious Id, and Dawkins by the Selfish Genes, Neuromotive Psychology says we may control our *own* destiny through the power of the conscious, self-recognizing brain—the human mind.

Human drives are not limited to those of simple survival and reproduction. As conscious aspects of an evolving process of complexification, human beings are driven by an innate Will to Evolve which takes two complementary forms—the impulse to expand abilities, and to feel *at home* in the world.

From Eros and Thanatos, to Prometheus and Orpheus.
From Genethics to Nurethics.
From the unconscious will to survive, to the conscious Will to Evolve.
Such is the message of Neuromotive Psychology.

Neuromotive Psychology: Beyond Freud and Maslow

In the modern world, humanistic psychology and psychoanalysis stand side by side as uneasy bedfellows—an idealized vision of human potential fulfilled through personal growth next to a gloomy view of man as the slave of unconscious forces beyond his control.

Neuromotive Psychology rejects both Freudian psychoreductionism *and* the naïve idealism of those who wish to blind themselves to the reality of our biological limitations. It opposes both the cynicism of the inference that man is little more than "naked ape" or "selfish genes"—*and* the overidealized belief that human beings can achieve complete human fulfillment merely by thinking positively and "talking things through."

Neuromotivational Psychology constitutes a balance between psycho-Darwinian reductionism and humanistic idealism that reflects human nature *as it is*—rather than how it appears through the eyes of the temperamental pessimist or optimist.

Thus, in place of both pessimistic Freudian psychoreductionism and idealistic Maslovian humanism, Neuromotive Psychology offers a theory of motivation suitable for the twenty-first century. Neither Freudian cynicism nor New Age naïveté is the answer. Man in his current state is neither doomed nor perfectible. We are limited by our biology—but striving for transcendence.

MAP OF GENETHIC AND NURETHIC DRIVES AND ASSOCIATED NEUROTYPES

	NURETHIC DRIVES (NEUROMOTIVE PSYCHOLOGY)	
PROMETHEUS		**ORPHEUS**
INDIVIDUATION		INTEGRATION
WILL TO GROW	WILL TO EVOLVE	WILL TO LOVE
DOPAMINE	(neurotype)	SEROTONIN

	GENETHIC DRIVES (PSYCHOREDUCTIONISM)	
EROS		**THANATOS**
LIFE INSTINCT	(FREUD)	DEATH INSTINCT
SURVIVAL/REPRODUCTION	(DARWIN)	SELF-DESTRUCTION
WILL TO POWER	(NIETZSCHE) (SCHOPENHAUER)	WILL TO DIE
ADRENALINE	(neurotype)	OPIOID

NOTES

1. Plato, *Phaedo*, ed. David Gallop (Oxford: Oxford World's Classics, 1999).
2. René Descartes, *Meditations* (Cambridge University Press, 1996).
3. Daniel Dennett, *Consciousness Explained* (Boston: Little, Brown, 1991).
4. Susan Blackmore, *The Meme Machine* (Oxford: Oxford University Press, 1999).
5. Eugene W. Holland, *Deleuze and Guattari's Anti-Oedipus: Introduction to Schizoanalysis* (London: Routledge, 1999).
6. Camille Paglia, *Sex, Art, and American Culture: Essays* (New York: Vintage Books, 1992).
7. Kenan Malik, *Man, Beast and Zombie* (London: Weidenfeld & Nicolson, 2000).
8. Hannah Arendt, *The Human Condition* (Chicago: University of Chicago Press, 1958).
9. Ian Hacking, *Rewriting the Soul* (Princeton, NJ: Princeton University Press, 1995).
10. A. S. Byatt, "Soul Searching," *Guardian*, February 14, 2004.
11. Hans Moravec, *Mind Children* (Cambridge, MA: Harvard University Press, 1989).
12. P. M. Thompson et al., "Genetic Influences on Brain Structure," *Nature Neuroscience* 4, no. 12 (November 2001): 1253–58.
13. S. F. Witelson et al., "The Exceptional Brain of Albert Einstein," *Lancet* 353, no. 9170 (June 16, 1999): 2149–53.
14. X. Piao et al., "G Protein–Coupled Receptor-Dependent Development of Human Frontal Cortex," *Science* 303, no. 5666 (March 26, 2004): 1033–36.

15. Peter D. Kramer, *Listening to Prozac* (London: Fourth Estate, 1994).

16. André Nieoullon, "Dopamine and the Regulation of Cognition and Attention," *Progress in Neurobiology* 67, no. 1 (May 2002): 53–83.

17. Ibid.

18. Robert Winston, *The Human Mind* (London: Bantam Press, 2003).

19. C. Corcoran et al., "Bupropion in the Management of Apathy," *Journal of Psychopharmacology* 18, no. 1 (March 1, 2004): 133–35.

20. Stephen Pinker, *The Blank Slate* (London: Penguin, 2002).

21. Kramer, *Listening to Prozac*.

22. "Pramipexole and Escitalopram to Treat Major Depressive Disorder," http://clinicaltrials.gov/ct/show/NCT00086307.

23. L. Pezawas, news release, National Institute of Mental Health, *Nature Neuroscience*, May 8, 2005, reported by Randy Dotinga, "Gene Tied to Depression, Anxiety," Forbes online, May 9, 2005, http://www.forbes.com/lifestyle/health/feeds/hscout/2005/05/09/hscout525611.html.

24. Carl Jung, *Psychological Types* (London: Routledge, 1971).

25. William Sheldon, *The Varieties of Human Physique* (London: Harper & Brothers, 1940).

26. Hans and Michael Eysenck, *Mindwatching* (London: Prion, 1989), chap. 13.

27. C. R. Cloninger, "A Unified Biosocial Theory of Personality and Its Role in the Development of Anxiety States," *Psychiatric Development* 3 (1986): 167–226.

28. John Maynard Smith, quoted in Lewis Wolpert and Alison Richards, *A Passion for Science* (Oxford and New York: Oxford University Press, 1988).

29. James Watson, "The Stuff of Life," *Observer*, April 6, 2003.

30. Jerome Kagan, *Galen's Prophecy* (London: Free Association, 1994).

31. Eysenck and Eysenck, *Mindwatching*, chap. 13.

32. Horace, *The Complete Odes and Epodes* (Oxford: Oxford World's Classics, 2000).

33. Sigmund Freud, *Project for a Scientific Psychology* (1895), in *The Standard Edition of the Complete Psychological Works of Sigmund Freud*, 24 vols., ed. J. Stachey (London: Hogarth Press, 1966), 1:295–387.

34. Sigmund Freud, *Civilization and Its Discontents* (New York: Norton, 1930).

35. Christopher Badcock, *Psycho-Darwinism* (London: HarperCollins, 1994).

36. Lapidus, *In Pursuit of Gold: Alchemy in Theory and Practice* (London: Neville Spearman, 1976).

37. Edmund Burke, *A Philosophical Enquiry into the Origin of Our Ideas of the Sublime and the Beautiful* (Oxford: Oxford World's Classics, 1998).

38. Immanuel Kant, *Observations on the Feeling of the Sublime and the Beautiful* (Oxford: Oxford University Press, 1998).

39. Friedrich Nietzsche, *Human, All Too Human: A Book for Free Spirits*, trans. R. J. Hollingdale (New York: Cambridge University Press, 1986).

40. Friedrich Nietzsche, *The Will to Power* (New York: Vintage, 1968).

41. Richard Dawkins, *The Selfish Gene* (Oxford: Oxford University Press, 1998).

42. Ibid.

43. Malik, *Man, Beast and Zombie.*

44. Freud, *Civilization and Its Discontents.*

45. Freud, *Project for a Scientific Psychology.*

46. Ralph Waldo Emerson, *The Portable Emerson*, ed. Carl Bode (London: Penguin, 1981).

47. Arthur Schopenhauer, *The World as Will and Idea* (New York: Dover, 1966).

48. Nietzsche, *Human, All Too Human.*

49. Friedrich Nietzsche, *Beyond Good and Evil* (London: Penguin, 1988).

50. Freud, *Studies on Hysteria*, in Stachey, *The Standard Edition*, vol. 2.

51. Abraham Maslow, *Motivation and Personality* (New York: Harper & Row, 1954).

52. Abraham Maslow, *Toward a Psychology of Being* (New York: Wiley & Sons, 1993).

Chapter 10
NURETHICS

> The sensible, intelligent animal will promote the contentment of others.
> —J. F. Young, *Cybernetics* (1969)

Ethics are the attempt to answer the question, What is the best way to live?

Clearly, this is the most vital aspect of any philosophical system, for what could be more important than the question of what to do with one's life?

By **Nurethics**, short for neural ethics, I mean programs for behavior constructed by the conscious (that is, self-recognizing), embodied brain we call the mind, to override **Genethics**, the three-part genetic program for survival, reproduction, and self-destruction written by the "selfish genes" in their own survival interests.

Here I identify the evolved, neurobiological basis of moral behavior and prescribe a rational system of benevolent ethics suitable for the modern world.

I address four topics, corresponding to the four basic human functions:

• DOING:	Evolutionary Ethics	Transnaturalism
• THINKING:	Nurethics	The Logic of Love
• FEELING:	The Ethics of Eugoics	The Right to Feel Good
• PERCEIVING:	Positive Realism	From Candidism to Can-Do-Ism

SUMMARY OF TRANSHUMANIST ETHICS

Evolutionary Ethics

The Naturalistic Fallacy is itself a fallacy. Ethics can be based only on nature because man is a *part* of nature. Yet, we do not need to base our ethics on *everything* nature does, but only on those aspects beneficial to human beings (**Transnaturalism**). Evolution is the unfolding process of complexification in nature. Humankind is a conscious aspect of evolution; as such, it is instinctive of human beings to seek to enhance their condition in pursuit of ever-increasing survivability and well-being. The postmodern scientists claim that evolution is directionless (**indeterminacy**); life an evolutionary fluke (**contingency**); man a naked ape (**insignificance**). Rather, evolution proceeds toward ever more complex forms, structures, and operations (**complexification**); the human brain is a predictable outcome of evolutionary complexification (**cerebrelization**); and *Homo sapiens* is the peak of evolutionary complexification on earth (**ascendancy**). As such, we should abandon postmodern self-abasement, begin to feel good about ourselves once more, and seek to improve our condition through the miraculous technology opening up in front of our eyes (**Evolutionary Ethics**).

Nurethics

There are two basic sources of benevolent behavior: the body and the mind (**Felt and Thought Morality**). Bliss chemicals induce empathy to encourage behavior beneficial to survival and reproduction. The evolved mind may further extend the capacity for benevolent behavior by recognizing its logical basis. Consideration for others requires the ability to consider. Thinking of others requires the ability to think. Goodwill-to-all-men is a rational survival tactic (**Sensible Self-Interest**). *Ill* will is illogical, for it merely provokes ill will in return (**Stupid Selfishness**). The **Dawkins Doctrine** of the Selfish Gene inadvertently encourages social demoralization by ignoring the neurochemistry of love. We are not "born selfish" but with a neurochemical instinct for empathy. The meme of the Selfish Gene requires memetic enhancement with the neomeme of the **Benevolent Gene**. Nurethics represents a midpoint between the extremes of Nietszchean egotism (selfishness as a virtue) and Christian/Comtian altruism (the unattainable absolute selflessness of *vive pour autrui* [living for others]), in accordance with principle of **bispectism**; the complementarity of parts and whole; individual and society; self-interest and sociability; strength and kindness. In Nurethics, Prometheus and Orpheus join forces: the Will to Grow and the Will to Love, united in the Will to Evolve.

The Ethics of Eugoics

The emerging science of **neuroenhancement** should not be feared, but welcomed as simply the next, logical step in self-improvement, after exercise, education, and aesthetic surgery. There can be nothing bad about wanting to feel good, and nothing good about feeling bad. A future science of neurochemical enhancement may allow us to increase the neurochemistry of love and well-being allowing us to feel *better than well* (**eugoics**).

Neuromanticism

The best attitude to life is neither irrational pessimism (**Impossibilism**) nor naïve optimism (**Candidism**), but **Positive Realism**—the acknowledgment of our biological limitations, combined with the rational belief that they may be overcome through the power of an emerging technology (**techno-can-do-ism**). We should refute fashionable postmodern cynicism; find **Reasons to be Cheerful** in the miraculous technowonderland of the modern world; and move ahead in a spirit of benevolence, adventure, and a passionate lust for life (**neuromanticism/technoromanticism**).

DOING

EVOLUTIONARY ETHICS

Here I address the question, "What should we *do* with our lives?"

My answer—Live to Evolve! Evolve to Live!

Ethics = Cosmology + Psychology

For one's ethics to be consistent with other aspects of one's philosophical system, they must derive from a combination of one's metaphysics and one's psychology. In other words, deciding how to live (ethics) requires knowledge of what is physically possible, through an understanding both of the fundamental nature of the world (metaphysics, or its scientific equivalent, cosmology) and of human nature (psychology, or its modern, scientific equivalent, biopsychology, or neuropsychology).

Thus, for instance, the essence of religious metaphysics is the belief in the existence of the undetected, immaterial phenomena God, heaven, and hell. The essence of theistic psychology is the belief in the existence of an immaterial entity called the "soul," which inhabits the body and survives its death. Theistic *ethics* follow logically from these two beliefs and can be expressed in six words;

"Life is a preparation for heaven." This is the ethical position of all theists, including young men taught to believe that flying a domestic airplane into a skyscraper will win them the sexual rewards of nubile handmaidens in heaven.

By contrast, transhumanist metaphysics says: The world is a process of evolutionary complexification toward ever more complex structures, forms, and operations.

Transhumanist psychology says: As conscious aspects of evolutionary complexification in nature, human beings are imbued with the innate Will to Evolve—an instinctive drive to expand abilities in pursuit of ever-increasing survivability and well-being.

Transhumanist ethics says: We should seek to *foster* the innate Will to Evolve by continually striving to expand our abilities throughout life. By acting in harmony with the essential nature of the evolutionary process—complexification—we may discover a new sense of purpose, direction, and meaning to life, and come to feel ourselves *at home in the world* once more.

Such is the essence of my Evolutionary Ethics.

The Naturalistic Fallacy

Unfortunately, to propose an ethics based on evolution today is to confront an intellectual taboo which has endured for a hundred years. Its origin was a book published in 1903 by the English philosopher G. E. Moore. In *Principia Ethica*,[1] Moore set out to invalidate attempts to derive ethical principles from the operations of nature. "Borrowing" an idea from the Scottish philosopher David Hume[2] (who denied that reason could be the basis of ethics), Moore refuted the validity of inferring how things *ought* to be, from knowledge of how things *are* in nature. Merely stating how things *are* does not infer that this is how they *should* be as a moral imperative. The practice of drawing ethical conclusions from natural facts about the world Moore called "the Naturalistic Fallacy." It was a catchy meme, and it survived for a hundred years.

Until now. For today, one hundred years after the publication of *Principia Ethica*, the Naturalistic Fallacy must be declared invalid, for the Naturalistic Fallacy is itself a fallacy.

Beyond the Naturalistic Fallacy

The Naturalistic Fallacy denies the validity of drawing ethical conclusions from natural facts. But ethics cannot ignore facts about the world!

To draw an "ought" from an "is" is to recognize biological reality. Fire *is* dangerous to human life; therefore, one *ought* not to set oneself alight. Food *is* necessary for human survival; therefore, one *ought* to eat one's dinner. To bury one's head in a bucket of boiling tar *is* threatening to human life; therefore, one *ought* not do it. Oxygen *is* necessary for survival; therefore, one *ought* to avoid putting a plastic bag over one's head.

Naturalistic ethics can only be refuted if one accepts the ludicrous assertion that self-preservation is not a universal goal. It is impossible to imagine a valid ethical system which ignores the fact that human beings are biological entities programmed to survive, reproduce, and self-destruct; wishing to live, but limited in the freedom to do so by nature. Ethics are limited by what we are, or can possibly conceive ourselves to be, or do. We cannot conceive of an ethics that says, "We should all bounce to the moon on a trampoline," because, sadly, the laws of nature forbid it absolutely. Naturalistic ethics are the only conceivably valid kind, because humanity is a part of nature and thus limited by her laws.

Moore was certainly right to discredit the essentially religious view that *whatever* nature does is good. Clearly, there are aspects of nature that are not beneficial to human beings. Its habit of killing us is one. The view that *everything* nature does is good, despite her being "red in tooth and claw," is as silly as the theist's contention that everything *God* does is good, in the face of disease and disability; tsunamis, hurricanes, and earthquakes; drought and famine; blood, sweat, and tears.

However, one need not positively evaluate *everything* that nature does in order to base an ethics on her *fundamental* mode of operation. If the fundamental directionality of nature *necessarily* benefits humanity, then it is perfectly valid to base one's ethics on *this* aspect of nature and not others.

Moore's attack led to the dismissal of ethics based on facts about nature for a hundred years. I think this is long enough.

I thus declare the return of Evolutionary Ethics.

Transnaturalism

Unfortunately, the term *naturalism* is now so strongly associated with the Naturalistic Fallacy that any attempt to propose an ethics using the word attracts a negative knee-jerk response from the orthodox majority. I therefore propose to use a new term to clarify my position.

The view opposed by Moore, that "good" is *whatever* nature does, I shall call **Metaphysical Naturalism**. By this, I mean all pantheistic philosophies that regard nature as sacred and therefore inviolable; her operations not to be altered in any way.

The type of naturalism that has no ethical inference, but merely sets out to describe nature, I shall refer to as **Scientific Naturalism**. This is the attitude espoused by contemporary science.

Finally, my own position: that it is perfectly valid to base an ethical system on a fundamental aspect of nature that is clearly beneficial to humanity, while opposing other aspects detrimental to human well-being, I shall call **Transnaturalism**.

By Transnaturalism, I mean Scientific Naturalism with an added ethical dimension. Transnaturalism is the belief that by transcending the biological limitations imposed upon us by nature, we are, in fact, acting *in accordance* with her *fundamental* mode of operation; that of *evolutionary complexification*—the continual expansion of structures, forms, and operations in nature, from a state of simplicity to ever greater complexity.

Scientific Naturalism does not tell us how or why we should live, merely "what exists." By contrast, Transnaturalism is the belief that it is *good* to act in accordance with the aspect of evolution beneficial to human survivability and well-being (that is, complexification), while seeking to overcome or transcend those aspects of nature that are harmful to us (that is, by reprogramming our genes to eradicate disease, enhance our capabilities, and defeat death).

Transnaturalism thus implies an ethical belief in the value of extending human capacities beyond those laid down by nature through our genetic makeup.

Transnaturalism does not claim that *everything* nature does is *necessarily* good; rather, that it is perfectly valid to base an ethics on the most *central* aspect of evolution, complexification (the continual increase in complexity of information content in the universe), because such a process is intrinsically beneficial both to survival and to well-being. For expanding capacities increases survivability, which in turn enhances well-being; the weak and dying are rarely ecstatic. At the same time, we may attempt to transcend the limitations nature has imposed upon us in other ways—by seeking to eradicate disease, enhance our abilities, and defeat death.

Evolution is complexification. "Transcendence" is a process of elevation from the simple to the complex. Transnaturalism is the belief in a form of naturalistic transcendence through evolutionary complexification—the Will to Evolve—not "beyond nature," but beyond the limitations of *human* nature; the biological restrictions on human freedom laid down by our genetic heritage.

Evolution may serve as a basis for a system of ethics because its fundamental mode of operation—complexification—is intrinsically beneficial to increasing survivability and well-being—the basic goal of life.

It is perfectly logical to assert that the process of complexification in living matter, if left to continue for a sufficient length of time, will *inevitably* lead to the emergence of the brain and its accompanying faculties (consciousness, thinking, logic, forethought, creativity, and design), because such faculties dramatically increase the ability to survive—and all life is imbued with the innate *will* to survive.

We should act in accordance with evolutionary complexification because it benefits our survivability and well-being to do so. Other aspects of nature detrimental to survival and well-being, such as death, disease, and the limitations imposed on us by our genes, we should seek to transcend through Superbiology.

Transnaturalism is therefore the idea that we may *transcend* the limitations imposed upon us by nature, by acting in *accordance* with her most fundamental mode of operation—the evolutionary complexification of her structures, forms, and operations.

Transnaturalism is the underlying principle behind my Evolutionary Ethics, which I shall now attempt to summarize concisely, in terms of two complementary ideas: the purpose of life for the species and for the individual.

The Purpose of Life 1: The Species
Evolutionary Progress through Conscious Evolution

The word *evolution* means to unfold or unroll, like a carpet or a scroll. Evolution is the unfolding of potential in nature—the continuous complexification of structures, forms, and operations in the metapattern of the world. Evolution is a becoming more, not less.

Human beings are conscious entities, where consciousness means the ability of the embodied information processor we call "the mind" to "re*cognize*" its own existence and that of the world in which it operates. Thus, humanity is a conscious aspect of an evolutionary process of complexification, and the human mind may be regarded as a means by which the evolving process of nature may observe its own existence. *Humanity is the mind of evolution become conscious of itself.*

The human brain is more powerful than that of its ape ancestors. Increasing brainpower increases the ability to survive. Progress means "increasing survivability." Therefore, evolution is progressive. Humanity is a product of evolutionary progress because the greater complexification of our brains over those of our ape ancestors serves to increase our knowledge and abilities, and in turn our ability to survive. The more complex the brain, the better able a species to respond to threats to its survival, such as infection and disease, aging and death. Therefore, the evolutionary complexification of life from ape to humankind constitutes "evolutionary progress," defined as "increased survivability."

Evolution proceeds upward toward ever-higher levels of complexity, where "upward" and "higher" are synonymous with "progress," meaning "increasing survivability." Humanity is the current peak of the evolutionary ladder on earth, because the complexity of the human brain affords the survival advantages of consciousness, language, reason, learning, and creativity to a degree unknown in all other animals.

Through the "Cartesian duality" of mind and body afforded by the emergence of consciousness, we now know what we are—evolved biological survival machines controlled by a genetic program for self-destruction. We do not wish to self-destruct—to suffer the ravages of disease, or the biological limitations of our bodies and minds. We should therefore learn to upgrade our genetic programming in our interests of *ever-increasing survivability and well-being.*

Science is the only method of overriding our biological limitations. Therefore, we should believe in science—and embrace the emerging Superbiology by which to eradicate disease, enhance our capabilities, and ultimately—defeat death.

Rather than viewing man as a debased creature whose higher faculties are mere sublimations of his animal instincts, we should celebrate the powers of the evolved human mind that have enabled us to build the extraordinary technowonderland of the modern world, and now offer the prospect of allowing us to transcend the remaining biological limitations on our freedom.

The goal of humankind is evolution by its own design. The next stage in evolution is the transition from human to transhuman; from *Homo sapiens* to *Homo cyberneticus—from Darwinian to Designer Evolution.*

Homo Cyberneticus

The term *Homo cyberneticus* carries two significant meanings.

The science of cybernetics is the attempt to construct machines based on the same engineering principles as living things. Organisms survive through a process known as *homeostasis*—the ability to self-regulate or alter internal conditions in response to environmental changes. Thus, we sweat when hot and shiver when cold in order to maintain the optimum, stable, internal temperature beneficial to survival; the unhealthy extremes are signified by fever and hypothermia respectively.

The machine equivalent to the automatic temperature control in living things is the thermostat that controls your central heating by automatically switching itself on and off when a certain temperature if reached. In the same way, the human body automatically maintains a temperature of 98.6° Fahrenheit, unless subject to extreme external temperatures, or invasion by a harmful virus, which require that the body heats up in order to fight the intruder.

The word "cybernetics" derives from the Greek *kubernetes*, meaning the steersman of a ship, or the governor of a state. The inventor of the term, Norman Wiener, defined the discipline as "the science of control and communications in the animal and the machine."[3] So the emphasis was placed not, as it could have been, on adjusting oneself to the environment, but firmly on the concept of *control*—that is, self-determination, self-rule, autonomy, independence; in short, the ability of an entity to determine its own behavior.

Today, however, the prefix *cyber* is used in popular culture merely to signify any kind of merger between man and computer—as in cyberpunk or cybercafé. Thus, the etymological link between cybernetics and the concept of control has been lost. However, it returns unabated in the use of the term *Homo cyberneticus* to describe a human species evolved to take control of its own behavior—its own evolution—not through unconscious homeostatic control (the preserve of Genethics, the survival program of the selfish genes), but through *Nurethics*—

behavior governed by the human mind in its own interests of ever-increasing survivability and well-being. It's not homeostasis that Nurethics seeks to override (for it keeps us alive!), but the third command in our genethic programming—the instruction to self-destruct.

The evolution from *Homo sapiens* to *Homo cyberneticus*, man the steersman of his own destiny, will be attained through the use of advanced biotechnology—or Superbiology. This process is likely to involve the use of cybernetics in the *Wienerian* sense through the introduction of artificial, computer-controlled implants to replace worn-out organs and improve general functioning of mind and body.

Hence, the term *Homo cyberneticus* signifies two concepts: the idea of a human species evolved to control its own destiny, and that of a species enhanced by technology.

Homo cyberneticus is *Homo sapiens* evolved to take charge of its own evolution through technology. Where *Homo sapiens* was the slave of its selfish genes, *Homo cyberneticus* will be the steersman of its own destiny.

The Purpose of Life 2: The Individual
Ever-Increasing Survivability and Well-Being

Human beings are the slaves of a three-part genethic program reading "Survive, reproduce, self-destruct." It is the logical response of a conscious entity programmed for self-destruction to seek out methods by which to expand its mental and physical abilities in pursuit of ever-increasing survivability and well-being. Humanity is imbued with an instinctive Will to Evolve.

For the species, this means the replacement of Darwinian Evolution with Designer Evolution—from slavery to the selfish genes, to conscious self-rule by the human mind. For the individual, it means the drive for continual self-enhancement of mind and body throughout life. The purpose of life is the pursuit of excellence. We are here to be the *best* we can be.

Human beings are conscious aspects of an evolutionary process leading to ever-greater levels of complexification in nature. Man is an unfinished symphony. The Book of Life is an unfinished masterpiece. Our goal is to be *better* than we are—to seek More Life, through a Lust for Life. Humankind is the mind of an evolutionary process becoming conscious of itself. Our purpose is to evolve, and through the Will to Evolve we may find our greatest joy.

An Evolutionary Ethics may serve to restore an element vital to the survival and progress of the human species—*self-respect*. We have been taught by the postmodern scientists to believe we are nothing but an insignificant twig on a cosmic bush. But once we recognize ourselves to be conscious aspects of a universal process of evolutionary complexification in nature, we may find ourselves *at home in the world* once more. No more existential alienation in a random universe. No more irrational self-diminution. Human beings are not lowly, worthless creatures, but the peak of an evolutionary process on planet Earth.

For the species, the purpose of life is the evolutionary ascent from *Homo sapiens* to *Homo cyberneticus*; human to transhuman—from Darwinian to Designer Evolution. For the individual, the purpose of life is the pursuit of *ever-increasing survivability and well-being.*

* * *

ANTIEVOLUTIONARY ETHICS: RESPONDING TO THE POSTMODERN SCIENTISTS

> Scientists animated by the purpose of proving that they are purposeless constitute an interesting subject for study.
>
> —Alfred North Whitehead, *The Function of Reason* (1959)

G. E Moore's attack on evolutionary ethics initiated a school of evolutionary skepticism which postmodern scientists continue to endorse today with a strange, nihilistic glee. Compare these passages, the first one by Moore, the second by the most vociferous postmodern scientist of all, the late paleontologist and prolific author of popular science books, Stephen J. Gould:

> What a different thing is being "more evolved" to being "higher" or "better." Mr. Spencer constantly speaks of the process which is exemplified in the development of humanity as if it has all the augustness of a universal Law of nature; whereas we have no reason to believe it other than a temporary accident.[4]

* * *

> We are global accidents of an unpredictable process, with no drive to complexity, not the expected results of evolutionary principles that yearn to produce a creature capable of understanding the mode of its own necessary construction.
>
> We are . . . a small, late blooming and ultimately transient twig on the tree of life, and not the summit of a ladder of progress.[5]

The postmodern scientists gleefully discourage the belief in human progress, significance, purpose, or meaning to life with a threefold nihilistic message:

1. Indeterminacy: Evolution is directionless.
2. Contingency: Life is an evolutionary fluke.
3. Insignificance: Man is a naked ape.

I counter their disturbingly misanthropic claims with my own assertions:

1. Evolution is not directionless but a process of complexification—the continual expansion of structures, forms, and operations in nature.
2. *Homo sapiens* is not a fluke but an evolutionary inevitability. The human brain is the *logically predictable* outcome of a continual process of complexification in matter left to run for a sufficient length of time.
3. Humanity is not an insignificant twig on the bush of life, but the peak of evolutionary complexification on earth, due to the incredible power of the human brain, aided by the manual and vocal dexterity of the human body.

Beyond Indeterminacy: Human Progress Through Evolutionary Complexification

> It is unfashionable in academic circles nowadays to speak of evolutionary progress. All the more reason to do so.
> —E. O. Wilson, *Consilience* (1998)

Postmodern scientists say, "There is no direction to evolution. Man is on a road to nowhere."

They are wrong.[6]

Evolution is a process of complexification—the continual increase in the overall complexity of structures, forms, and operations in nature. The postmodern scientists say complexification is not universal, as some species have evolved into simpler organisms. But only an idiot would deny that evolution *as a whole* is a process of complexification—the continual expansion of information content in the world. Evolution is a becoming more, not less.

Evolution may have simplified some species, but man is not just "some species." Consciousness makes human evolution different from that of any other animal, due to the extraordinary survival value of the properties it affords: reason, learning, forethought, creativity, imagination, and design.

Postmodern scientists say, "Progress is an illusion."

They are wrong.

"Progress" means "increasing survivability." Evolution is complexification. Complexification increases survivability. Therefore, evolution is progressive, where "progress" means an increase in our ability to survive. We are better able to survive than our ape ancestors, with smaller, simpler brains, because our greater intelligence increases our ability to manipulate the environment in our own interests. Increasing brainpower increases survivability. Human brainpower is greater than that of its ancestors. It follows logically that human evolution is progressive.

The human brain is a complexification of the ape brain that preceded it. A bigger, more powerful brain improves the ability to survive. Human evolutionary progress means bigger, better brains increasing knowledge, and so survivability. Man is better able to survive than his ape ancestors subject to the whims of capricious nature by virtue of the greater flexibility of response to the environment facilitated by the complexified human brain. Evolution is progressive because complexification increases survivability.

Every time a postmodern scientist spreads the negameme of the **Progressive Fallacy** (the denial of evolutionary progress) another child abandons the quest to fulfill its potential, safe in the knowledge that society has declared all effort to be futile, for ultimately, human life is without purpose, direction, or meaning. Humanists and transhumanists should no longer stand by and allow the fallacy of human insignificance to pass unchallenged.

Teleology and Teleonomy

Critics argue that Evolutionary Ethics imply teleology, the belief that evolution is being *purposefully* (i.e., consciously) directed, whether by God (if you're a theist) or by nature (if you're a pantheist).

Modern science rejects the idea of conscious directedness in nature toward a preconceived goal as metaphysical wishful thinking. However, observation clearly indicates that there is indeed a definite directionality in evolution—toward complexification—the expansion of forms, structures, and operations in the world, irrespective of the question as to whether anyone is doing the directing.

Keen to avoid accusations of "letting God in by the back door," scientists therefore came up with an alternative term, *teleonomy*, meaning the observed effect of directionality in evolution, without any conscious, intelligent design behind it. Teleonomy means *unpurposefully* directed evolution; directionality without a director; Richard Dawkins's metaphor of nature as a "Blind Watchmaker." Teleonomy is teleology without God.

The concept of a universal Will to Evolve in nature is an example of teleonomy rather than teleology, since no conscious intent or preconceived goal is recognized behind the directionality in evolution (though logically, neither can it be ruled out entirely). Evolutionary complexification is "the nature of nature." However, I also agree with an aside by Richard Dawkins, which he might regret sneaking into the glossary of his book *The Extended Phenotype*: "Essentially, teleonomy is teleology made respectable by Darwin."[7] It is interesting that the world's most prominent advocate of atheism finds it necessary to shake hands with teleology, when scientists have rejected the term as an apology for the existence of God the designer.

In short, there is a definite directionality in evolution toward complexification, irrespective of whether or not it has been "planned" by some unknown controlling intelligence. Complexification for human beings means the enhancement of mind and body, and the products thereof, so increasing our ability to survive. Since survival is instinctively what human beings want most of all (unless they have been meme-washed by religious fanatics) it makes sense to act in accordance with evolution's fundamental mode of operation; a process of which human beings are a conscious (that is, self-recognizing) aspect.

In addition to increasing survivability, evolutionary complexification increases human *well-being*, for two reasons: first, getting what you want makes you feel good, and what we want most of all is to stay alive. Second, increasing mental and physical powers automatically expands the *potential* for greater well-being. The ant is not capable of the ecstasy experienced by human beings through art, philosophy, science, sport, love, or sex because it lacks both a conscious self to recognize its own pleasure and the range of abilities of a human being. The greater the abilities, the greater the capacity for ecstasy—and for agony. The task of transhumanity is to employ its enhanced powers wisely, by cultivating the art of eugoics—the enhancement of the self to ever-greater levels of survivability and well-being.

Beyond Contingency: Human Inevitability through Evolutionary Cerebralization

Postmodern scientists say, "The evolution of man was nothing but a fluke; a chance event in a random universe."

They are wrong.[8]

Humankind is not an evolutionary fluke. The evolution of the human brain is an *inevitable* consequence of the unfolding of potentiality in nature we call evolution.

Evolution is the actualization of potentiality in matter, an automatic process of complexification. The human brain is an evolutionary complexification of the ape brain; the result of genetic mutations preserved over generations for their benefit to survival. It is logically *inevitable* that the complexification of life over millennia should eventually result in the evolution of an organ enabling the processing of complex information, because such a facility dramatically aids survival.

The evolved human brain *necessarily* increases survivability because of the enormous survival value of knowledge and the ability to use it through logical and creative thought. The human mind is the foreseeable outcome of an evolutionary process of life governed by the principles of complexification and survivability. It is logically predictable that the process of complexification in life will eventually result in the emergence of an intelligent species possessing consciousness and its associated abilities of reason, learning, forethought, creativity, imagination, and design.

The existence of humankind is not a chance result in a universe without direction. The human brain is a predictable outcome of an evolutionary process of continual complexification left to run of its own accord. For what else could eventually result from the endless complexification of matter but objects capable of maintaining their own existence—living things—and what better method of increasing their survivability than the development of consciousness and its associated functions: reason, learning, forethought, creativity, imagination, and design.

Human beings are the inevitable product of the evolutionary process. Consciousness is not a fluke but the predictable outcome of evolutionary complexification.

Human life has a definite direction and purpose. The human species is evolving toward ever-greater levels of cerebral control, from Genethics to Nurethics, from *Homo sapiens* to *Homo cyberneticus, from Darwinian to Designer Evolution.*

Beyond Insignificance: Human Dignity through Evolutionary Ascendancy—The 99 Percent Fallacy

Postmodern scientists say, "Man is just a naked ape: 99 percent of his genes are identical to those of a chimpanzee."

This oft-quoted statistic reminds one of the increasing tendency of manufacturers to advertise certain foods as "99 percent fat free." Just as manufacturers want you to think their product has "virtually no fat," so postmodern scientists want you to think that humanity has "virtually no intelligence" above that of chimps. But just as eating a whole package of biscuits that are "virtually fat free" will still make you put on weight (due the effect of the *total* rather than the percentage of fat), so a brain advertised as "virtually human-gene free" will still dramatically increase your intelligence, because of the effect of the total amount of "intelligence genes" supplied by the 1 percent that is uniquely human.

The percentage of genes we *share* with apes is less significant than the nature of those human genes we do *not* share with our ancestors. This relatively small number of genes could quite easily account for the entire development of human civilization, if they were responsible for the evolution of the greater capacities of the neocortex, the seat of our higher functions not shared by our ape ancestors.

Even if we discounted the difference in gene sequences entirely, significant differences in the way human genes *function* are sufficient to explain the vast divergence in abilities between man and ape. As James Watson explains, "The 5 million years of evolution that has separated us from the chimps have led to significant divergencies from chimps as to the exact times at which some human genes function, as well as the rates at which they produce their respective protein products."[9]

The greater capacities of the human brain take us way beyond the capabilities of even our closest relatives. What chimp could have produced Leonardo's *Last Supper*, Michelangelo's *David*, Newton's theories of motion, Bach's fugues, the printing press, radio, telephone, TV, hi-fi, DVD, cell phone, Internet. . . .

> This . . . is . . . the reason humans can do calculus, compose poetry, and build cathedrals, while chimps pick bugs off each other and eat them.[10]

The attempt to debase man by reducing his level of significance in the scheme of things to that of apes is a weapon in the postmodern Meme War against modernity I call the **99 Percent Fallacy**. The repeated, gleeful emphasis on our 99 percent chimp rather than 1 percent human-specific genes serves only to encourage the view that humanity is a lowly, debased, worthless creature, the entire products of its culture—science, art, philosophy—deemed of no greater significance than the ability of an ape to scrape up ants with a stick.

Beyond the Naked Ape

The number of genes we share with our ancestors is a red herring used by bio-Luddites to denigrate the value of our species by implying that human beings are "nothing but mammals." The attempt to degrade humanity by downplaying the vast differences in ability between man and ape might be called **sapienism**. Coterminous with sexism, racism, and the recently discovered medical disorder "homophobia," sapienism is prejudice and discrimination against the human species, an expression of antihumanistic misanthropy. Its latest manifestation is the assertion that humanity must abandon its "last claim to uniqueness"—the ability to create culture—on the grounds that apes have been observed to pass on new skills, such as washing food before eating. Thus, *sapienists* try to persuade us that there is no distinction between the culture of the modern world (source of the Internet, laptop, and video phone) and that of apes who wash bananas.

Sapienism is an irrational instinct for self-abasement; product of Thanatos, the genetic command to self-destruct. The promotion of sapienism contributes to the cultural descent of the modern world into the abyss of postmodern nihilism. We find it echoed everywhere in popular culture; for instance, in the lyrics of the hit pop song "The Bad Touch": "You and me baby ain't nothin' but mammals, so let's do it like they do on the Discovery Channel."

This sentence describes the beliefs and values of the postmodern Anticulture of nihilism in a nutshell.

Ethics cannot be divorced from psychology. What we do is inevitably influenced by what we understand ourselves to be. How we treat other people is strongly influenced by how we think of ourselves. If people are taught by the postmodern scientists to believe they are "nothing but mammals," they will begin to behave that way, by abandoning their specifically human capacity for moral behavior.

We often hear the common myth that other animals are supposedly benevolent in comparison to man, the wicked warmonger. But chimps do not give to charity, help old ladies across the road, or concern themselves with third-world poverty, for they lack the basic capacity for human benevolence—*reason*, the ability to *understand* other people. No chimp is a Good Samaritan because no chimp can maintain an idea in its mind for long enough to understand the concept of "other people's needs," or why it's important to consider them. Consideration for others requires the ability to consider. Thinking of other people requires the ability to think. Conscious, rational thought is the product of our uniquely human genes.

The public have been brainwashed by the postmodern scientists into believing the "nothing-buttery" of their animal nature. But man is more than naked ape or talking chimp. The fallacy of human insignificance is a dangerous lie that would have us all descend into the nihilism of the Postmodern Condition.

Never has there been a more important time for the development of a positive, rational, Evolutionary Ethics.

Huxley's Evolutionary Humanism

One of the few voices to sail against the prevailing antievolutionist tide in the twentieth century was the eminent biologist Sir Julian Huxley, grandson of the nineteenth-century advocate of Darwinism, T. H. Huxley. In a number of popular science books written throughout the twentieth century until his death in 1975, Huxley outlined his own evolutionary ethics as the basis of a philosophy he called "evolutionary humanism"—or "transhumanism," a term he was responsible for coining in the sense used here. The essence of Huxley's evolutionary ethics are expressed in the following passage:

> We cannot say that evolution is purposeful until we are privileged to know what processes occur in the thoughts of God, but we can and must say that it has direction. On average, the upper level of biological attainment has been continually raised. Not only this, but our own standards, moral, aesthetic, and intellectual alike, have been produced by this process, and tend to continue the direction of evolution in the future along the line it has followed in the past—towards "more life." That being so, we must say that those actions which tend to help the advance of the upper level of living matter—today represented by man—along a continuation of the line it has followed in the past, are good; while those that tend to hinder it are bad.[11]

Exactly. Nature is a process of evolutionary complexification. Man is the evolutionary spearhead of life on earth. The purpose of life is More Life. We are here to be better than we are.

Julian Huxley's positive Evolutionary Ethics represented a breath of fresh air in the arid desert of twentieth-century philosophy. Directly following the Second World War, Huxley was appointed the first Director General of

UNESCO, the United Nations Educational, Scientific and Cultural Organization set up in the desire to "build peace in the minds of men."[12]

Today, fifty years on, Professor for the Public Understanding of Science Richard Dawkins chooses to criticize Julian Huxley's attitude toward nature for failing to "admit its unpleasantness."[13] Perhaps he did, but surely not from ignorance, but rather from the conscious intention to focus on the positive; not what nature does wrong, but what she does right, and how we can improve upon her methods in our own interests. What nature does *wrong* is to kill us, to limit our freedom, and to leave us with the vestiges of antisocial instincts that may have been appropriate for the survival of our ancestors, but not for an intelligent species capable of constructing and maintaining a complex civilization. What nature does *right* is to complexify; to allow the unfolding of potential in matter, through increasingly diverse structures, forms, and operations. It is what nature does right that should guide our future behavior, and what she does wrong that we should oppose. Evolutionary complexification is something nature does right. In this, we would be well advised to follow her lead.

To evolve is to be more than we are. Nature is not a being, but a becoming. So let us act in accordance with her fundamental mode of action, and seek to discover the Will to Evolve.

Such is the essence of my Evolutionary Ethics.

* * *

THINKING: NURETHICS
THE LOGIC OF BENEVOLENCE

> If little capacity for intelligent inhibition is built into the machine, its behavior will be describable as stupid selfishness.
>
> —J. F. Young, *Cybernetics* (1969)

Here I address the question of how best to treat others. I identify two sources of benevolent morality, and outline a rational system of ethics representing a midpoint between the Stupid Selfishness of Nietzschean egotism, and the unattainable absolute selflessness of Christian-Comtian altruism.

Summary of Nurethics

Benevolence is the ability to *feel* for or *think* of others. **Felt Morality (Genethics)** is an evolved neurochemical conditioning mechanism, inducing benevolence in the interest of survival and reproduction through feelings of sympathy for others. **Thought Morality (Nurethics)** is the ability of the rational mind to inhibit Stupidly Selfish antisocial instincts in its own best interests. The ability of the mind to *consider* the well-being of others enables a greater capacity for benevolence than is possible through emotional empathy alone. Nurethics allows us to recognize that consideration for others is also in our own best interests. "Goodwill to all men" is a rational tactic for mutual survival and well-being. Benevolence is **Sensible Self-Interest**; malevolence is **Stupid Selfishness**.

Thought Morality avoids the basic limitation of Felt Morality: its proportionality to the overreactivity of the nervous system, such that a neurotic wreck incapable of functioning in the world is likely to possess the most moral feelings of all.

Nurethics represents the advocacy of egoism over the extremes of egotism and altruism. Egotism is a reversal and projection of self-loathing born of social or sexual inadequacy, exemplified by the "master morality" of the hermit, Nietzsche, who titled the chapters of his autobiography "Why I Am So Clever," and "Why I Am So Wise."

Christian ethics are limited by their emphasis on unrealistic, excessive altruism based on empathy (Felt Morality). Turning the other cheek won't stop someone from robbing you. Considering the lilies won't get you off welfare.

But one should not criticize religious ethics without having something better to put in its place. By ignoring the inborn neurochemistry of love, the meme of the Selfish Gene ("we are born selfish") has encouraged the erroneous belief in the absence of innate human goodness and led to social **demoralization**. We are programmed as much for benevolence as for selfishness. Goodness and badness are both part of human nature. Our genes encourage benevolence through neurochemically induced sympathy as much as hostility in self-defense.

We should therefore speak instead of the **Benevolent Gene**, and seek to enhance our capacity for benevolence beyond the level of emotional empathy, by learning to think rationally and inhibiting our Stupidly Selfish impulses to attack others beyond the need for self-defense.

The Evolution of Morality

Morality is the capacity of a conscious agent to determine the nature of its habitual responses toward others. *Benevolent* morality is the willingness to help or avoid harming others.

Human beings are programmed to survive. There thus exist certain moral codes shared by every culture, based on the automatic impulse to stay alive. In this most fundamental sense, "good and bad" means "behavior beneficial and detrimental to survival."

"Morality" in the popular sense of benevolent behavior is instinctive to man, because goodwill benefits self-preservation. Those animals capable of benevolence were more able to survive and pass on their genes than those who acted entirely for themselves. A "bad ape" incapable of sociability was more likely to suffer rejection from the tribe, and thus fail to reproduce—or survive. The phenomenon of Natural Selection through the "survival of the fittest," whereby genetically based characteristics which benefit survival and reproduction are preserved in a gene pool over generations, has thus resulted in a species capable of a high level of cooperation, enabling it to build vast semicivilizations based on mutual assistance. Darwin expressed it thus:

> There can be no doubt that a tribe including many members who, from processing in a high degree the spirit of patriotism, fidelity, obedience, courage, and sympathy, were always ready to give aid to each other and to sacrifice themselves for the common good, would be victorious over most other tribes, and this would be natural selection.[14]

What are the mechanisms behind the capacity for cooperation?

A human being is essentially an embodied mind that evaluates the environment by means of two corresponding functions: feeling and thinking. Consequently, there are two basic sources of human morality: I call them **Felt and Thought Morality**.

Felt Morality: The Neurochemistry of Love

Felt Morality is the oldest, most fundamental source of the human instinct for benevolence.

Feelings are the means by which the body automatically evaluates environmental stimuli positively or negatively, in order to determine a behavior of attraction or aversion to greatest benefit to survival and reproductive success. Feelings serve as a neurochemical, behavioral reinforcement mechanism, conditioning for behavior beneficial to survival and reproduction through the reward and punishment of pleasure and pain. The most basic feelings of happiness and sadness are simply the experience of continual pleasure or pain in response to behavior that helps or hinders our ability to survive and reproduce.

Felt Morality—the "moral sense" in man—is an automatic, neurochemical reward of pleasure or punishment of pain, in response to good or ill will, preserved over generations for the value of benevolence to survival. Nature—that is, our genes—rewards acts of goodwill with a neurochemically induced feeling of pleasure, to encourage behavior that helps us to survive. Helping others makes us feel good because it increases our *own* ability to survive. If you help others, chances are other people will help you in return. Reciprocity, the basis of benevolent morality, is induced through the pleasure of fellow feeling. The ape that scratches his fellow's back does so for the pleasure reward it induces, not because he has reasoned that doing so will increase the chances of reciprocal back scratching. Removing bugs from the skin decreases the chances of infection. It feels good to do good because helpfulness helps us survive.

The pleasure of love and friendship is a neurochemical conditioning mechanism attracting us to those of most benefit to our survival and reproductive success. The human body has evolved to respond to friendly, cooperative, benevolent behavior by releasing a pleasure reward of neurochemicals such as oxytocin (inducing maternal love) and serotonin (inducing brotherly love—or friendliness). Serotonin conditions positively for fellow feeling by inducing relaxed self-confidence—the optimum state conducive to goodwill. Although nervy neurotics may *feel* great emotional benevolence, laid-back, "chilled-out" people are more likely to *act* in a gregarious manner—for they lack a supersensitive fear of others. Oxytocin, secreted during pregnancy and breast feeding, conditions positively for the benevolence of maternal instinct in order to encourage child nurturing. Mothers lacking oxytocin lack maternal feeling. To be kind meant originally to protect one's kin, or *kinder*, the German word for children. Felt Morality is sympathy—an instinctive response to another's needs.

The Limitations of Felt Morality

The limitations of Felt Morality are twofold. First, not all emotions are benevolent, and none are consistent. We are just as likely to fly into an irrational rage as to feel a surge of love for our fellow man. Our ape ancestors, lacking a brain of sufficient complexity to enable the employment of logic, relied on feelings to reinforce benevolent morality. In ape tribes, Felt Morality is apparent in mutually affectionate grooming and distress at the suffering of others. But even in our closest relative, the chimpanzee, there is also violent aggression, sometimes to the point of savage murder of its own kind, and, of course, there is an absence of the complex webs of social rules which allow human societies to function in all their complexity—an extraordinary ability human beings take for granted.

Second, feeling empathy for others requires one to experience pain oneself in response to the suffering of another. It follows, then, that the most moral person is he who experiences the most pain! Strength of feeling is proportionate to the reactivity of the nervous system. Consequently, the weaker the nervous system, the more empathetic one's responses. Emotionally sensitive people feel more sympathy for others than the insensitive. The cause of emotional oversensitivity is an overreactive sympathetic nervous system, which automatically pumps out high levels of adrenaline in response to stimuli. Its neurotypical correlate is the Opioid Neurotype—a personality trait defined by the desire to reduce anxiety caused by excess adrenaline. The body dominated by an Opioid Neurotype releases endogenous tranquilizers—opioids—in the attempt to counteract adrenaline and dampen down a fraught nervous system. So if emotional sympathy was the only source of morality, weakness would have to be considered morally superior to strength, since the weaker one's nervous system, the more hypersensitive one's reactions, and the greater one's sympathy. Ergo, the extreme Opioid Neurotype—a neurotic emotional wreck who overreacts to every stimulus—must be considered the most moral type of all! Felt Morality is the essence of Christian morality—the biopsychological basis of the belief that "the meek shall inherit the earth." One wonders, then, as to the neurotypical profile of its founder.

Should morality really depend on a hypersensitive nervous system? This is the biopsychological basis of the conservative objection to so-called bleeding-heart liberals; the suspicion that emotional displays of compassion are sometimes little more than hypersensitivity; the instinctive response of an over-reactive nervous system to another's suffering. By contrast, actually *doing* something to alleviate suffering requires not public displays of sympathy or grief, but a strong *will*—the ability of the *mind* to rule the body; not "conspicuous compassion,"[15] but action.

Feelings are a vital component of benevolence, but *on their own* they are not sufficient to override antisocial instincts or induce moral behavior. If they were, logically there would be no war, hatred, or crime—in Christian countries at least—since we would all have learned to love one another according to our social conditioning. That centuries of indoctrination with the religious meta-meme of Universal Love have failed so dismally to create heaven on earth suggest that feelings are not enough on their own to reinforce benevolent morality.

So Felt Morality is limited in its effectiveness. The development of human civilization required a greater degree of benevolent cooperation than was possible for our ape ancestors. The faculty that made it possible was reason.

Thought Morality: Sensible Self-Interest versus Stupid Selfishness

The evolved capacity of the human brain for complex reasoning allowed for an extension of benevolence beyond that made possible by Felt Morality alone. It was the ability of the mind to reason exactly *why* benevolence is in our own best survival interests that enabled man to establish our current semi-civilizations based on mutual cooperation.

The conscious mind is able to *think ahead* to the likely *consequences* of an action and prevent itself from reacting with instinctively hostile responses detrimental to self-preservation. The mechanism by which the rational mind is able to prevent the body from acting on Stupidly Selfish instincts is called inhibition. Inhibition is the ability of the mind to prevent a physical impulse from occurring. It was the faculty of inhibition, product of the conscious mind, which allowed humanity to create its present state of semi-civilization by enabling it to avoid the "war of all against all" that would ensue if all were to act on their emotional impulses alone.

Thought Morality is the capacity of the evolved rational mind to inhibit asocial instincts through intelligent self-control. To avoid unreasonable behavior requires the ability to reason. Thinking of others requires the ability to think. Consideration for others requires the ability to consider. No ape can consider the lilies, let alone the needs of its neighbor. Recognizing the survival value of benevolence, the reasoning mind may extend its capacity for goodwill beyond the level induced by neurochemical conditioning alone, in its own self-interest—the maximization of survivability and well-being.

Self-interest is a universal survival instinct. The genetically programmed prime directive of all animals is "Thou shalt survive." But the intelligent animal is able to *consciously* recognize and consider the well-being of others, in the knowledge that benevolence also increases its own survivability and well-being through the principle of reciprocity—you scratch my back and I'll scratch yours. In this, the intelligent animal could be described as acting out of **Sensible Self-Interest.** Other animals display high levels of cooperation—biologists call it eusociality—but none possess the ability of human beings to extend cooperation *willfully*, through *deliberate, conscious* control of behavior. A high level of cooperation requires a high level of intelligence—the general measure of the processing power of the brain, in order to *purposefully direct* behavior to *preconceived* goals, and inhibit the body from responding impulsively to stimuli with Stupidly Selfish instincts.

The unintelligent animal is incapable of inhibiting antisocial instincts that may appear to be of benefit in the short term, but are, by harming others, self-detrimental in the long term. In this, we might say it is being **Stupidly Selfish**.[16] The unthinking brute mugs or murders according to emotional whim, lacking the ability to recognize that malevolent acts serve only to advertise oneself as a threat to others, and thus increase the likelihood of reciprocal malevolence, in self-defense against a threat to survival. Murderers rarely get away with it, because their actions constitute such a threat to the survival of others that they attract reciprocal malevolence in self-defense. Following a murder, the whole society is filled with emotional hatred or intellectual contempt (depending on neurotype), and unites in the attempt to discover whodunnit and withdraw them from society forthwith, whether through incarceration (in civilized societies) or murder by decree (in uncivilized ones).

Neurethics

The recognition of both Felt and Thought Morality, and the need to extend the former through conscious, willful self-control of behavior through the inhibition of antisocial instincts, I call **Nurethics**.

Nurethics means both neurochemical and neural ethics. The essence of Nurethics is the recognition that benevolence is both an innate product of neurochemical conditioning (Felt Morality) and a rational survival tactic, product of intelligent inhibition, made possible by the evolution of the cerebral cortex—the seat of consciousness and the capacity for reason (Thought Morality).

Nurethics is the understanding that neurochemically induced benevolence may be extended by the conscious (that is, self-recognizing) brain in its own rational self-interest. Nurethics tells us that egoism (the ethics of self-interest) and benevolence (the ethics of cooperation) are not incompatible, but rather, a necessary combination for the attainment of maximum individual and social well-being.

Malevolence is **Stupid Selfishness** because advertising oneself as a threat to others decreases one's own chances of survival and well-being. Most murders are committed out of emotional impulsivity in response to a perceived threat. The intelligent animal refrains from responding instinctively at the first sign of feeling threatened, having learned that its own survival interests are not best served by reacting to every act of hostility with reciprocal aggression. The most intelligent animal seeks to surround itself with contentment, through a policy of universal goodwill.

Nurethics tells you why, logically, it is in your own interests to act benevolently:

- If you harm A, then A is likely to harm you back. You don't want to be harmed. Therefore, don't harm A.
- If you are considerate and pleasant, others are likely to respond in kind. Life is far more enjoyable when one is surrounded by contentment. Therefore, it pays to show kindness and consideration to others.
- A baby incapable of smiling in response to its parents is less likely to receive the same level of parental love as the smiling baby who delights in their presence.
- The adult who refuses to be good-natured is less likely to prosper socially than his more gregarious fellows.

Reciprocity. Do as you would be done by. The morality of the Golden Rule. It is the basis of every benevolent moral system. It is sublimely simple and perfectly logical. Expressed in the language of Nurethics; "Good" is the ability of the mind to inhibit Stupidly Selfish instincts; "Bad" is the *inability* of the mind to inhibit Stupidly Selfish instincts. Good is Sensible Self-Interest. Bad is Stupid Selfishness. Nurethics—the morality of enlightened, rational self-interest—is an ethical system suitable for a rational species in a modern world. I think it should be taught in schools.

Teaching Nurethics

An important underlying cause of antisocial behavior is the failure to make the logical causal connection between one's behavior and its likely consequences.

Simply put, many people simply don't *think* about the consequences of their actions. Forethought—the ability to think ahead—is the essence of reason (the ability to determine causal connections between events), logic (the system of rules governing reason), and intelligence (the measure of the capacity to employ reason through logic). Acting unreasonably means acting without reason.

The intelligent individual refrains from dropping litter in the street, not from any great universal love of the species, but by reasoning that the future consequence of such a tactic is likely to be filthy streets for all; as dropping litter merely encourages others to do the same, by increasing the legitimacy of the action. The inconvenience of holding on to one's litter until finding a waste bin is far less than that of being forced to wade through litter-filled streets at a later date. Such a deduction requires *forethought*—the ability to think ahead—to make causal connections between events; in this case, "If I throw litter in the street, others will to do the same; the streets will eventually be filthy; therefore I shall put my litter in my pocket." No emotional empathy whatsoever is required to arrive at this conclusion; it is a wholly logical deduction based on Sensible Self-Interest. One could make the same case for such depressingly common everyday acts of Stupid Selfishness as the depositing of used chewing gum on public transport, or the habit of putting one's shoe-clad feet up on the seat opposite, so the next person to sit down is unknowingly graced with a dusty backside.

Similarly, intelligent individuals refrain from criminal activity or drug abuse not necessarily through any great feelings of empathy for humankind, but because they are capable of *reasoning* that a life of crime or drug addiction is unlikely to be a happy one. Criminals and drug addicts are individuals who failed to *think* of the consequences of their actions—until it was too late. Once the recognition (re-*cognition*) of their mistake dawns upon them, they have already been drawn into a downward cycle of despair from which they are unable to escape. The difficulty of dealing with criminality and drug addiction is the difficulty of changing a fixed habit. The answer lies in prevention, not cure.

Schools would be well advised to start teaching the logic of Sensible Self-Interest. We should tell young people exactly *why* being good will benefit them personally, while being bad—or Stupidly Selfish—will get them nowhere in life. Instead of teaching children (and suicide bombers) to fear hell and seek reward in heaven, we should teach them the *reason* why it is in their *own self-interest* to be good; essentially because a policy of "goodwill to all men" is guaranteed to get them a lot further in life than one of "*ill* will." Successful business leaders should be encouraged to give motivational talks in Life Skills classes, stressing the importance of knowing how to get on with people and maintaining a good reputation through self-control, to their *own* financial and social benefit.

Nurethics is the advocacy of egoism over two extremes: egotism and altruism. Next I shall comment on the ethical systems which most clearly represent them; those of Nietzsche and Jesus.

Nietzschean Ethics: Egotism as Master Morality

"Egotism" means conceited self-obsession and a disregard for others. Essentially, egotism is synonymous with narcissism—the inability to recognize the separate existence of, or value in, other people. To the egotistical narcissist, other people are merely tools for the attainment of one's own goals, not individuals in their own right, with their own needs. The inability to recognize the autonomous existence of others is a recipe for the desperation of absolute loneliness.

The epitome of egotism is the ethics of Nietzsche, who based his system on the dichotomy between two very different concepts of the "good." According to Nietzsche, to the strong, "good" meant "brave and noble," while to the weak, it meant "gentle, friendly, or harmless." He called the two "master and slave morality"—and sided with the former.[17] A world in which kindness and consideration are considered bad would be a quite hideous world in which to live. In practice, however, Nietzsche himself was a gentle, mild-mannered man, who pretended to be wicked in his bedroom by way of compensation. In his syphilis and solitude-induced insanity, Nietzsche reversed and projected his own timidity into a one-sided morality in which "I suffer because I am timid and kind," became "I believe in evil, cruelty, and domination." Nietzsche's egotism was born of thinking too much, and not getting out enough.

In short, egotism is the product of a collapsed ego—a desperate compensation for a lack of self-worth. Egotism is Stupid Selfishness because excessively arrogant, inconsiderate behavior automatically induces hostility in others. Indeed, egotism is but a few steps away from "evil"; the difference being that egotism is out for itself, while evil is *deliberately* out to harm others.

Egotism is not the same as egoism. Egoism—the ethics of self-interest—is necessarily the basis of all moral codes, for we are programmed for self-preservation; the will to live is our very essence. Self-interest is the basis of morality because human biology dictates that it be so. Only the suicidal masochist lives by a code based on something other than self-interest—and then we call it "psychopathy."

The concept of egoism suffers from its association with the Freudian use of the term *ego* (Latin for "I" or "self"). Freud's avowed attitude toward the ego was expressed by his motto "where id was, let ego be;" in other words, identical to the morality of Nurethics: "from control by the selfish genes, to self-rule by the conscious mind." However, in practice, Freud came to believe in the futility of attempting to override "genethic" control of behavior, concluding that the result could only be that of neurosis, through the repression of the basic instincts for sex and aggression. Hence, Freudianism today is associated with the belief in the dominance of id over ego, unconscious animal instincts over the conscious, rational mind. As a result, in a godless age, the new explanation of evil is not supernatural ("the devil made me do it"), but psycho-Darwinian ("the unconscious id or selfish genes made me do it"). In the postmodern age, the Ego has replaced the Devil as the main source of human ills.

The contemporary demonization of the ego derives largely from the common misunderstanding about the distinction between the concepts of "egoism" and "egotism." It is not the ego that is bad, but egotism; not rational egoism, but irrational inconsideration for others; not Sensible Self-Interest, but Stupid Selfishness.

Comtian Ethics: Altruism as State-Enforced Goodness

While egotism is incompatible with the morality of Sensible Self-Interest, so too is the opposite moral code—that of altruism.

Altruism is selflessness—the tactic of living not for one's own goals, but for those of other people. The term was coined by the French philosopher Auguste Comte, founder of sociology; motto, *vive pour autrui* (Live for Others). But the practice of living entirely for the benefit of others exists only in a condition of slavery, and in the imagination of those who subscribe to extreme versions of certain religious and political beliefs. For altruism means putting others first *so that you come off worst*. The absolute selflessness of pure altruism is the equivalent of the dog that lies down submissively with its legs in the air; a practical impossibility for a free human being. No coincidence, then, that Comte's idea of the ideal society was a dictatorship by a secularized catholic church.

By contrast, self-interest is simply the innate will to live. *Not* acting out of self-interest is the definition of masochism—the desire for self-harm. Self-interest must come before selflessness if the will to live is deemed preferable to the will to suffer. Benevolent cooperation is both a necessity and a pleasure. Indeed, its pleasure lies precisely in the fact that it *is* a necessity. The pleasure of kindness is nature's neurochemical reward for living wisely. But absolute altruism—putting others first *so that you come off worst*—cannot logically be placed above egoism—the ethic of self-interest—unless self-harm is deemed preferable to well-being.

Egoism is self-interest—the pursuit of one's own goals. It is not the same as selfishness, which is self-interest *at another's expense*. There is no necessary casual connection between acting out of self-interest and acting in such a way as to harm others. The wise man recognizes that benevolent cooperation is Sensible Self-Interest. This is the meaning of the universal principle I call Bispectism, applied to ethics and politics. There is no moral contradiction between living for your own goals and being kind to others on the way.

Christian Ethics: A Critique of the Mountain Lecture

The archetypal exponent of altruism as a system of ethics is the most influential of all moral philosophers, Jesus. Christian morality is encapsulated in a number of pithy aphorisms expressed in his most popular lecture on ethics, known as the Sermon on the Mount. Here I shall briefly comment on a selection of dominant memes from the mountain, beginning with moral imperatives I consider open to question.

Being Meek

"Blessed are the meek, for they shall inherit the earth" (Matt. 5:5).

The problem with this powerful meme is that it tends to communicate the message that submissiveness is better than self-assertion; that one should not be confidant and outward-going, but rather, imitate the action of a sheep. Good advice for a brute, perhaps, but hardly for young people seeking to make their way in a competitive world.

Of course, the word "meek" can be interpreted as meaning humble as opposed to haughty or arrogant. Hence the first entry in the Oxford English Dictionary reads, "Gentle, courteous, kind, merciful, compassionate." But the subsequent definition reverses the inference: "Submissive, humble. In unfavourable sense: Inclined to submit tamely to oppression or injury." And the definition in the free computer dictionary WordWeb doesn't beat about the bush: "Cowed submissiveness . . . very docile, 'meek as a mouse.' Evidencing little spirit or courage, overly submissive or compliant . . . spiritless . . . tame." Are these really the characteristics of the "good man"?

But let us examine more of the Jesus memes. Much of the Mountain Lecture is simply the propagation of secondhand memes from the Old Testament. Thus, for instance, Psalm 37 informs us that "the meek shall inherit the earth." But Jesus goes on to add his own, unique ideas on how to behave.

Turning the Other Cheek

"Resist not evil, but whosoever shall smite thee on thy right cheek, turn to him the other also" (Matt. 5:39).

Here, in a brilliant, lightning meme attack, Jesus completely overturns the morality of the Old Testament. Memetically manipulated (or "enhanced" depending on your metameme), "an eye for an eye" becomes "turn the other cheek." If someone hits you, rather than defend yourself, you should encourage him (or her) to do it again.

But it's not hard to see how this might be regarded by a young person as something of an uninviting instruction, for it seems to imply that, rather than fighting for respect from one's fellow man, one should allow oneself to be abused. Christian apologists have offered a psychological justification for turning the other cheek on the lines of Confucian psychology, which regards the best form of attack as to absorb a blow, as opposed to responding in kind. This is the basis of the discipline known in English as *Kung Fu*, named after Confucius himself. There is much to be said for it. For instance, one might suggest that the best way to respond to an angry spouse is not to enter into a shouting match, but to calmly and rationally seek to establish and empathize with the underlying cause of anger (though, unfortunately, both the capacity to inhibit Stupidly Selfish impulses and to analyze behavior varies according to neurotype).

Others have proposed that "turning the other cheek" was actually an act of defiance toward Roman guards, before whom one was expected to kneel in submission. But all of this is somewhat hard to believe, given that the next piece of advice offered by Jesus is the proposition that one offer a thief one's cloak, if he steals your coat. This is not the advice recommended by the police.

Considering the Lilies

Other dominant memes propagated in the Mountain Lecture are concerned with the question of material prosperity:

"Take no thought for the morrow . . . consider the lilies of the fields . . ." (Matt. 6:34/28).

"A rich man shall hardly enter into the kingdom of heaven" (Matt. 19:23).

In other words, "It's better to be poor than rich if you want to live after you die, so forget about making life plans, don't bother striving for money or success, and just live for the moment—like a plant." Comforting, no doubt, to an oppressed people two thousand years ago, and to those of us neurotypically predisposed to idleness; but hardly good advice for the high school graduate.

It's all very well to consider the lilies, but to "take no thought for the morrow" is to abdicate responsibility for one's own future. The archetypal exponent of living for the moment is the addict who enjoys the perpetual present of a drug-induced haze. The crack or smack addict certainly pays no thought for the morrow—except when wondering whom to mug to pay for his next fix. Is this the ideal man in Christian ethics? Only the tramp and the drunk and the addict may *truly* pay no thought for the morrow.

The First Shall Be Last

"Many that are first shall be last; and the last shall be first" (Matt. 19:30).

But why, exactly? Is it really justice for the loser in a race to be declared the winner, and the winner the loser? Is this not, rather, absolute *in*justice?

The message appears to be that moral superiority lies in being timid, submissive, and poor, as opposed to strong, confident, and wealthy. If I am one day fortunate enough to have children, I do not think this is the message I shall be trying to instill in their impressionable young minds. I believe I shall teach them that financial independence is vital to well-being; that happiness is much easier to obtain for the rich man than the poor man; and that therefore they should do their utmost to ensure that at the very least they are able to support themselves, by planning for their future career as early as possible—that is, by *thinking about the morrow*.

I shall try to teach them to be confident and self-assertive, *as well as* kind and considerate to others. I shall both encourage them to have pride in themselves (or "self-esteem") *and* admonish them when they're cruel, unkind, or thoughtless. In short, I shall encourage Sensible Self-Interest and seek to deter Stupid Selfishness. I do not believe in the existence of a loving God, and I do not believe I require the assistance of such a being to help me instill benevolence in my children. Should I be condemned as immoral for my beliefs?

Other moral imperatives expressed on the mount have much to offer a young person embarking on the journey of life. I shall now turn to those I regard as the *positive* mountain memes.

Loving Thy Neighbor

The most universal ethical imperative espoused in the **Mountain Lecture** is that of reciprocity.

"All things whatsoever ye would that men should do to you, do ye even so to them" (Matt. 2:12).

"Love thy neighbor as thyself" (Matt. 22:39).

"Love thy enemies" (Matt. 5:44).

In short, "Think how one would like to be treated oneself, and treat others in the same way—even if you don't like them." This is incontrovertibly good advice for those of us who would prefer to see peace on earth—rather than war after war in the name of religion. But it is not exclusive to Christianity. The phrase "Do unto others as you would have others do unto you" not only occurs several times in the Old Testament, but is repeated almost verbatim in the *Analects* of Confucius—the great, secular, Chinese philosopher-prophet—who died in 479 BCE![18]

The tactic of doing as you would be done by is not exclusive to religion. The basis of reciprocity is not "spirituality" or emotional empathy, but *logical* thinking. Jesus is telling us to *consider* how we would like other people to treat us, and to *consciously* treat others in the same manner. Such a tactic requires the capacity to *think rationally* in order to *inhibit* Stupidly Selfish instincts. Apes couldn't do it. Nor could the schizophrenic who has just hacked a man to death in a London street, then boiled and ate his brain for dinner—saying, perhaps, that God told him to do it.

In short, the prime directive of Judeo-Christian and Confucian ethics is simply Sensible Self-Interest.

Morality requires no deity. Love is logical.

Shining Your Light

"Neither do men light a candle and put it under a bushel, but on a candlestick; and it giveth light to all who are in the house. Let your light so shine before man" (Matt. 5:13).

This advice is perfectly in keeping with the Neuromotive Psychology and Nurethics I propose. It is, in fact, a concise exposition of egoism: the belief in the value of *self-confidence* or its fashionable contemporary equivalent "self-esteem." Undoubtedly, there can be nothing bad in feeling good about oneself—providing one is acting according to Sensible Self-Interest, not Stupid Selfishness. On the contrary, self-confidence *increases* the capacity for altruism. The most hostile people are the *least* confident, for hyper-hostility is projected self-hatred, and lack of confidence leads to a profound disgust of the self.

However, not hiding one's torch under a bush is hardly the maxim that most readily springs to mind as the dominant message of the Mountain Lecture, which is mostly the wall-to-wall advocacy of extreme altruism. Nor is it quick to be quoted by defenders of the Christian metameme, who tend to see the Big Idea as loving thy neighbor, not shining one's own light—or blowing one's own trumpet.

Zelig Syndrome

But then, one of the impressive aspects of the Jesus phenomenon is his ability to be all things to all people; a personality trait I call Zelig Syndrome, after the classic film by Woody Allen in which the physical features of an unconfident man literally turn into those of whomever he meets. Thus, Jesus informs us on the one hand that he has "Come not to bring peace, but a sword" (Matt. 10:34), while on the other, "He who lives by the sword, dies by the sword" (Matt. 26:52). And we are told to "love thy neighbor as thyself" (Matt. 19:19) (thus appealing to egalitarians), and yet not to "cast pearls before swine" (Matt. 7:6) (appealing to elitists).

Zelig Syndrome—the tendency to ingratiate oneself with other people by telling each what they want to hear—is the characteristic of the brilliant, wily politician.

New Bottles for New Wine

"Put new wine into new bottles, and both are preserved" (Matt. 9:17).

This maxim from the Mountain Lecture is certainly one that will appeal to the transhumanistically inclined. Indeed, *New Bottles for New Wine*[19] is the title of the book by Julian Huxley that introduced the concept of transhumanism for the first time—in 1957. Fifty years on, in contemporary transhumanism, the maxim can be even more literally interpreted. The new wine is the new belief system or meme map of transhumanism. A new meme map requires an upgraded biocomputer in which to download it. The new bottles are the new bodies and minds of transhumanity—biogenetically enhanced, through Superbiology.

Socialist Darwinism

Christian ethics were designed to alleviate the suffering of an oppressed people two thousand years ago, not through political revolution, but through psychology—by changing the way they felt about their oppression. The ethics of Jesus are essentially a solace for suffering deemed impossible to overcome. The ethics of *transhumanism* are based on the belief in transcending problems in *reality*, not just in the mind, through the effort of *will*—the ability of the mind to guide behavior. There is nothing immoral or malevolent in this belief. Nurethics is the recognition that love is—as a noun, a neurochemical instinct (we are not "born selfish" but with an impulse for empathy); and as a verb, a rational tactic of Sensible Self-Interest (for we help ourselves by helping others).

The recognition of benevolence as a rational survival tactic I call **Socialist Darwinism**, to contrast it with the Stupid Selfishness of the ideology known as *Social* Darwinism, the fallacious belief that one can improve a society by ignoring the suffering of its weakest members. A policy of ignoring the plight of the poor can only conclude in social breakdown, through a collapse in the benevolence necessary for collective survival. Even the richest man cannot be happy wading or driving through streets of beggars and thieves, in perpetual fear of crime. The needs of the individual and the needs of the society as a whole are interdependent. One helps oneself by helping others. By Socialist Darwinism I mean cooperation regarded as an evolutionary adaptation benefiting mutual survival, facilitated by the capacity of the evolved human brain for reason. Socialist Darwinism is philanthropy as Sensible Self-Interest.

Sensible Self-Interest

People are different. The message of the Mountain Lecture may have helped many millions of people live benevolent lives. But it simply doesn't do the trick for everyone. Not everyone can be seduced by an ethic which places selflessness over personal ambition; especially when one sees so many successful people clearly paying only lip service to its principles. "How is the willingness to wage war in the name of Christian values compatible with the principle of turning the other cheek?" a socially disadvantaged young person might say, quite reasonably.

To commit crime takes a certain bravery or daring. Criminals are not generally timid, shy, and retiring. Many young people who turn to crime do so because they are strong or dominant individuals who see no legitimate way of profiting from their dominance, whether due to social disadvantage, lack of intelligence, or self-control. Telling a confident, unruly scalawag that it's good to be poor, meek, and turn the other cheek is unlikely to set him on the path of righteousness. I suggest that the teaching of benevolence through Sensible Self-Interest is more likely to appeal to the type of young person *most susceptible* to crime than a message of unrealistic altruism.

But for those who are upset by such a challenge to their own system of ethics, let me quote some other ethical precepts from the Mountain Lecture, expressed, as always, with brilliant concision:

"Judge not, that ye be not judged" (Matt. 7:1).

"He that is without sin among you, let him cast the first stone" (John 8:7).

"They that take the sword, perish with the sword" (Matt. 26:52).

"Don't admonish the speck in your brother's eye, when there's a the plank in your own" (Matt. 7:3).

I stress that I am in no way attempting to discredit Christian ethics; on the contrary, I consider Jesus to be a quite brilliant moral philosopher, and prose stylist, who expressed his system with unmatchable clarity, concision, and style. But the right to challenge the dominant metameme is a moral imperative in a free society, and Christianity is still the dominant ethical system in the modern world, despite strong challenges from "premodern" religions on the one hand, and postmodern nihilism on the other. My purpose is not to destroy benevolent morality, but to *strengthen* it, by adding to its empathetic admonitions (Felt Morality) a scientific rationality suitable for the modern world (Thought Morality). For those thinking of taking up my critique of Christian ethics, I have a strong message of my own:

Don't criticize religious ethics *without offering a better alternative to put in its place!*

For to publicly attack the main source of moral teaching in society without offering any practical alternative, is to encourage the *demoralization* of man . . .

Beyond the Selfish Gene Meme

The meme of the Selfish Gene is one of the most successful in the modern world. The metaphor is undeniably a powerful one; strings of chemicals in our cells imagined to be living entities, acting in their own interests, by inducing us to "give them a lift" to a new body through sex, that they may survive—while we decay and die. I have incorporated the selfish gene meme into my own philosophy, to help distinguish between my concepts of Genethics and Nurethics: instinctive, unconscious behavior—governed by the selfish genes, and rational, conscious behavior—governed by the thinking mind. However, there is a certain deficiency in the **Dawkins Doctrine** that will simply not go away.

For the propagation of the selfish gene meme has led to an unfortunate consequence: thirty years on from its conception, one need not travel far on the proverbial Clapham omnibus to overhear the blame for some moral transgression or other placed firmly on strange demons living within the body. From adultery to murder the refrain is increasingly familiar: "*It wasn't me, your honor, it was me selfish genes wot dunnit.*"

Defenders of the meme appeal to the Naturalistic Fallacy. One should not confuse biology with ethics, facts with values, they say. How we have evolved to behave does not infer how we *ought* to behave.

But the damage has been done! The **meme-link** between our biological makeup and a condition of innate selfishness has been made. The meme has taken on a life of its own. Whatever the meme of the Selfish Gene was intended to mean, what it has actually been taken to mean by a significant proportion of people in the modern world is something quite different, and, I suggest, something quite dangerous.

For I would argue that the successful propagation of the selfish gene meme over the past thirty years has resulted in a phenomenon quite unintended by the Defender of the Meme: a significant rise in a dangerous belief that human beings are innately lacking in benevolence (incapable of overriding their selfish instincts), and a consequent reduction—not only in the *willingness* to act benevolently—but even in the belief that human beings are *capable* of doing so. The potential negative consequences of such a scenario are incalculable.

"My Selfish Genes"

Unconvinced? To test my hypothesis, I typed "My selfish genes" and "Our selfish genes" into the Google search engine, and examined the results; not from academic papers or book reviews, but from that increasingly popular phenomenon the Weblog, Internet diaries written by ordinary members of the public. The results were much as expected. Here are some of the entries:

Everybody is a creep. We're programmed by ***our selfish genes*** *to be creeps. It's the only thing that makes biological sense . . .*

Due to evolution, ***selfish genes*** *and all that jazz, I do not trust the judgments of people . . .*

Don't blame me, blame ***my selfish genes*** *. . .*

My selfish genes *are telling me not to share . . .*

Maybe I really am just a robot controlled by ***my selfish genes*** *. . .*

We really are the slaves of ***our selfish genes****, and there's no basis for morality other than various forms of tribalism . . .*

Race and racism are in ***our selfish genes*** *. . .*

We are biologically driven by ***our selfish genes*** *. . .*

My ***selfish gene*** *fills up my spleen with bile . . .*

I have a three-year-old daughter. When I look at her asleep in her bed, I love her and want to protect her. Are ***my selfish genes*** *just producing meaningless feelings?*

Are we all inherently evil as someone in this forum has said before because of ***our selfish genes****?*

There is no God, no meaning to life, no purpose other than to pass on ***our selfish genes*** *. . .*

Why don't we heed the call of ***our "selfish" genes*** *and say the heck with it and fornicate like bunny rabbits in the streets?*

Now imagine an entire society dominated by the attitudes expressed above. You won't have to struggle too hard. It's called the "Postmodern Condition"—and it *already* dominates the modern world.

The Postmodern Condition

The Postmodern Condition is existentialism without the pain; the ironic acceptance of life without meaning or purpose in an age with no more ideals. The children of the postmodern age have accepted what the scientists-philosophers have told them. There is no "evolutionary progress." There is no universal purpose or meaning to life. We should pretend to be nothing other than what we are: mindless survival machines made from selfish genes.

The postmodern attitude is one of ironic detachment from all ideals, enthusiasm, and moral codes. Postmodernists wallow in the cheap luxury of cynicism, even as civilization collapses all around them. Irony is the cultural orthodoxy. Belief in the values that built the modern world—reason, science, beauty, truth, and progress—are swept away in a tide of ironic disdain. Only the irrational, the amoral, or the absurd are valued. Ideals and enthusiasm are ridiculed. Absolute relativism rules. No behavior or lifestyle is considered better or worse than any other. How could it be, since we are nothing but the product of the same selfish genes, and there is no ultimate purpose to life?

Descent into amorality is the inevitable result of postmodern relativism. For once truth and meaning are regarded as relative, moral consensus necessarily falls by the wayside.

Born Selfish?

Ullica Segerstrale's book *Defenders of the Truth* provides a definitive account of the Meme War for control of the dominant ideas in contemporary biology. In it, she presents an excellently detailed, balanced view of an ongoing academic argument, based on the sociological implications of biological theories. But she cannot hide her conclusion as to the effect of the selfish gene meme: "Dawkins wants readers to believe in Darwinism as a substitute for myth, but his Darwinism carries the encrypted message that the world is empty of meaning."[20]

I could go further. In *The Selfish Gene*, Richard Dawkins writes the following: "If you wish, as I do, to build a society in which individuals cooperate generously and unselfishly toward a common good, you can expect little help from biological nature. Let us try to teach generosity and altruism, because we are born selfish."[21]

"We are born selfish" is something of an unequivocal statement of belief.

Professor Dawkins is not saying that we *should* act selfishly. Like me, he thinks we should use our *brains* to override our innate "biological" selfishness. This is the meaning of "Sensible Self-Interest"—the overriding of Genethics with Nurethics. But the fact remains that here we have a respected, Oxford University Professor for the Public Understanding of Science, whose claim to fame is a best-selling book for the general public which informs us plainly and clearly that "we are born *selfish*"; a word the professor himself defines as self-interest *at the expense of others.*

Another word for selfish is "wicked."

Telling people they are made from "selfish genes" and "born selfish" is tantamount to telling them that malevolence is natural, inevitable, and thus acceptable. Simply saying that they've misunderstood—that the brain can override our innate selfishness—simply won't do. It's like repeatedly telling your child she's born evil, but should try to be good anyway. How do you *think* she'll behave? Answer: like a lapsed Catholic, by saying, "I'm born bad, so I'll *act* bad—and then feel guilty about it later." What a wonderful moral code! *It is precisely the moral code unwittingly encouraged by the Doctrine of the Selfish Gene.*

Whatever its creator intended it to mean is irrelevant; the Doctrine of the Selfish Gene has served to revive the belief in Original Sin, the myth of innate human wickedness. The result has been a gradual descent into a postmodern Anticulture of Nihilism.

Social Demoralization: The Postmodern Anticulture of Nihilism

In the postmodern Anticulture of Nihilism, trash TV encourages shameless voyeurism, casual cruelty, infantile vulgarity, greed, and humiliation. Semi-celebrities famous for being famous compete for the headlines, while groundbreaking developments in science and technology are routinely ignored or rubbished in hysterical scare stories. Dumbed-down novels are lauded for their fashionable inverse-snobbery. In art, naïveté is preferred to skill, ugliness to beauty. In music, harmony is rejected for dissonance. In cinema, the individual auteur is replaced by mindless movies created by committee. And everywhere the ubiquitous drone of manufactured pop music stuns brainwashed youth into mindless submission.

Such is the "culture" of postmodernity.

The Anticulture sneers at its own civilization. The (anti)ideal attitude is a sort of knowing, smug-cynical ennui known as "postmodern irony." Man is dismissed as a laughable, pathetic creature, devoid of purpose or direction, the only acceptable attitude being one of mockery or self-ridicule.

The values of the Postmodern Anticulture are spread by memetic mutation. The word "concern"—indicating empathy with others—is replaced by "fazed"—indicating solely an imbalance within the self. In youth culture, "bad," "evil," and "wicked" all become terms of approval. Compassion for the weak and unhappy is abandoned. "Sad" becomes an expression not of sympathy, but of mockery for the lonely and unfulfilled. Today, "sad losers" must "get a life."

As a result of memetic invasion by the postmodern metameme, a world-weary mood of quiet desperation has descended on the West. The modern world has seen a collapse in self-respect, vitality, enthusiasm, desire for progress, improvement, or belief in the future; and with it, all possibility of a shared moral consensus. For why be good when the world is meaningless, life an absurd joke?

Irony means saying the opposite of what you believe. To speak in perpetual irony is to reveal a complete absence of belief in anything.

The *ironization* of postmodern culture is leading to the *demoralization* of man: "Life becomes a movie that we watch with the bemused detachment of an absurdist. This is the specter of a thoroughly postmodern morality."[22]

Science versus Philosophy

The underlying problem here is the gulf between science and philosophy.

In the absence of ethical philosophers willing to get their hands dirty in the deep dark waters of science, it has been left to scientists to act as philosophers. But by the very nature of their work (increasingly specialized; analytical not synthetic) scientists are not best suited to the job of philosophers.

Science is—or is supposed to be—the "disinterested" pursuit of objective truth. What society does with this knowledge is the concern of moral philosophers and politicians. Scientists ask "how?" philosophers ask "why?" So when scientists are forced to act as philosophers, the ethical component of philosophy may be downgraded, perhaps unconsciously, in favor of metaphysics and psychology; or its modern equivalents, cosmology and biology. The resultant philosophy may be more descriptive than ethical, brute facts predominating over value judgments.

That is why I consider transhumanism so vital a metameme for the modern world; a philosophy grounded in scientific rationality, which recognizes the need for a strong, clear, positive, benevolent ethic that is absent in much of the scientific philosophizing we see today.

Words matter. What people *think* words mean ultimately matters more than what the author intended them to mean, because of the influence of language on attitudes and behavior. The Selfish Gene is a meme in need of serious, *directed mutation*. Here is how it may be done.

The Benevolent Gene

> When I mentioned *The Selfish Gene*, Popper immediately took issue with the title. Why emphasize selfishness—why not call the book "The Cooperative Gene"! After all, genes also have to cooperate!
>
> —Ullica Segerstrale, interviewing Karl Popper, *Defenders of the Truth* (2000).

The word "selfish" means self-interest *at another's expense*. According to Selfish Gene theory, the chemical sequences which copy themselves to produce human beings are acting in their own interests *against* those of human beings. Clearly, to the extent that genes have us programmed for death once they've moved on to a fresh body, they could be described as selfish.

However, it is also in the interests of genes to encourage human survival for the duration of a life span long enough to permit maximum sexual reproduction.

If a human being dies before his genes have been copied to another body through sex, they will die with him. Benevolent behavior assists mutual survival. Cooperation makes it easier to survive. It is therefore in the survival interests of the genes to encourage cooperative, benevolent behavior. And indeed they do. I have called it Felt Morality. Genes encourage benevolence by releasing neurochemicals that induce feelings of goodwill, compassion, affection, or empathy—otherwise known as love. Genetic self-interest demands human benevolence, because of the survival value of cooperation. Getting on with one another helps us survive.

We are not born selfish, but with the innate instinct for love; an instinct which manifests itself from the moment a baby smiles affectionately at its parents.

In this sense, the Selfish Gene might equally be termed the **Benevolent Gene**, since human benevolence is necessary for survival and reproductive success, and genes neurochemically induce fellow feeling in order to encourage behavior such as nurturing, friendship, cooperation, and love.

Admittedly, the Benevolent Gene may not have sold so many books.

Genes promote human goodwill because it benefits their own survival interests to do so. Particular genes produce particular types of behavioral traits (or "predispositions," as biologists like to call them, careful to avoid accusations of "genetic determinism"). Traits that confer maximum survivability on their owner are more likely to be passed on through sex, and so preserved in the species, because you're likely to have more sex the longer you live and the healthier you are. Those traits of no benefit or detrimental to survival are more likely to die out with those who possess them, because you can't have sex if you're dead. Survivability is increased by benevolence. This is the message of Nurethics. We have evolved to possess an instinct for goodwill because goodwill aids survival, which in turn increases the likelihood of our passing on our genes through sex. Genes need human beings to survive for the duration of their reproductive life, if they are to survive and prosper themselves.

Thus, the interests of man and his selfish genes coexist. For the genes to move house, the body in which they currently abide must survive long enough to take care of the removal process—through sex. Genes favor human benevolence because cooperation assists survival. Genes are selfish in programming us for death, but benevolent in ensuring we survive long enough to reproduce. They do so by encouraging us to cooperate, through the neurochemically induced fellow feeling we call friendship, affection, or love. In the words of James Watson:

> I think the morality comes from human nature.
> I think we were born to care for one another . . .
> It gives people pleasure to help each other.[23]

Love is genetically induced benevolence; a neurochemical pleasure reward for fellow feeling, induced by chemicals such as oxytocin (maternal and romantic love) and serotonin (brotherly love). Maternal, romantic, and brotherly love are the products of the Benevolent Gene.

We are not composed of "selfish chemicals" at all. Our genes continually flood our bodies and minds with the neurochemistry of love. I call it Felt Morality. Human beings are not "born selfish," but born to love. So let us speak instead of the Benevolent Gene:

> While kids still help old ladies
> With their cases, at the station,
> I know that the meme of the Selfish Gene
> Is pure imagination.

The injection of the meme of the Benevolent Gene into the Selfish Gene is the memetic engineering required to enhance a powerful but defective meme into a Supermeme. The ethical principle of Sensible Self-Interest tells us to recognize the logic of benevolence. "Goodwill to all men" is not a religious dream, but the optimum tactic for mutual self-preservation among coexisting organic "survival machines" (Thought Morality). The Benevolent Gene meme tells us that our bodies induce benevolent feelings from birth, through the release of Bliss Chemicals—the neurochemistry of love (Felt Morality).

We are not born bad, but rather with the neurochemical instinct to be good, and the potential to be bad if we override it. It is up to our conscious, thinking minds to act as steersman, guiding our behavior toward the good and away from the bad in our own self-interest, since cooperation is a rational tactic of mutual benefit to survival and well-being. So let us seek to enhance both the neurochemistry of love and our understanding of its rational basis in mutual self-interest.

Felt Morality is the essence of Christian ethics. Today we can identify its neurochemical basis. Add to it the Thought Morality of Sensible Self-Interest, and one gets Nurethics—Thought and Felt Morality. There is no need to attack Christian ethics, only to enhance them through scientific rationality.

So let us strive together to develop both the feeling and the logic of love.

Aggression

Some say, "The aggression of our animal ancestors is proof of the inherent maleficence of man."

Aggression is an evolved, instinctual response of the nervous system to a perceived threat to survival.

A certain amount of aggression is a "necessary evil," for a species devoid of the capacity to respond in kind to a threat to survival would not have lasted long in the evolutionary journey. However, the animal incapable of refraining from aggression *beyond* the need for self-defense *decreases* its chances of survival. The ape that breaks the rules by attacking its opponent with excessive force embarks on a perilous path, for aggression sparks reciprocal aggression in self-defense.

The intelligent animal learns to inhibit excessive aggressive impulses in the interests of self-preservation. The capacity of the evolved human brain for the intelligent inhibition of stupidly selfish impulses allows for a far greater level of benevolence than is possible for apes.

So whatever our ape ancestors may have done in the way of aggression is irrelevant. Our evolved intelligence allows us to recognize that malevolence is Stupidly Selfish, and benevolence Sensible Self-Interest.

Evil

Some say, "Evil is caused by the devil within."

Evil is not the product of an invisible demon inhabiting the body, but a malfunction of an evolved, neurochemically induced instinct for aggression in self-defense against a perceived threat to survival.

Hostility is cognitive aggression—a conscious aversion response to a perceived threat. Evil is hostility experienced as pleasurable. Why should hostility be pleasurable? Because the capacity for a certain amount of instinctual aggression is necessary for survival. The lion is better able to survive than the lamb when competing for food or territory. This explains the existence of the phenomenon known in German as *schadenfreude* (literally, "damage to joy"), a word for which, significantly, there is no equivalent in English. Schadenfreude is pleasure in the misfortune of others—the secret delight experienced by the failure of our friends, the downfall of a celebrity, or politician. So contradictory to Christian morality is schadenfreude, that the English never even acknowledged its existence by inventing a word to describe it! But exist it does, and it is the basis of evil.

Schadenfreude is Evil Lite.

Mockery

The difference between pleasure in another's misfortune and pleasure in harming another *oneself* is only a matter of degree. That is why mockery must be considered one of the worst of all vices, because delight in belittling others is but a few steps away from delight in torturing or killing them.

The willingness to commit murder begins with the acceptance of mockery, when open hostility to others is experienced as pleasurable—even sociable. It would not be entirely overstating the case to say, *he who mocks a friend today may murder an enemy tomorrow*. A kinder, better world would ensue if this maxim became common knowledge.

Sadly, the taste for mockery (in colloquial English, euphemized to "taking the piss," so "watering down" the severity of the crime) is ubiquitous. Mockery of others is the basis of most humor, when it is not *self*-mockery. The best kind of humor acknowledges our universal, mutual imperfections; the worst targets those of individuals or minority groups. To mock an individual or group for their distinctive characteristics is the beginning of barbarism. It is precisely the social acceptability of group-ridicule that allowed a nation to turn a blind eye to attempted genocide by its leaders. For mockery or ridicule devalue and degrade the object of their attack. We should laugh lightly at ourselves as a species, but think carefully before laughing at others for being different from ourselves.

However, this principle should be the subject of personal, moral choice—*not* state coercion, else in our attempts to "build peace in the minds of men," we succeed only in building iron curtains by which to imprison them.

The Bio-Defense: On Free Will and Moral Responsibility

Some say, "Knowledge of the biological influence on behavior will be used as an excuse to absolve people of personal responsibility for their actions, by implying that there are certain genetic impulses they are powerless to resist."

The possible effect of neurotype on moral responsibility will become an increasingly important subject as the public understanding of biopsychology grows. Already defense lawyers in criminal trials have begun employing the argument that innate biological predispositions provide a mitigating circumstance for criminal actions—an argument that might be called the **bio-defense**. My position on the effect of biopsychology on moral responsibility for one's actions may be summarized thus:

1. Free will is proportional to the mind's ability to override the genetic predispositions of the body.
2. The capacity for free will differs between individuals, because bodies differ in:
   - *The intensity of the sympathetic nervous system's response to stimuli*
     A weak (that is, overreactive) nervous system leads to greater and so less-controllable levels of emotional impulsivity.
   - *Hormonal predispositions to aggressive impulsivity*
     Greater habitual levels of adrenaline and testosterone, and, correspondingly, lower levels of dopamine and serotonin, increase the propensity for aggressive impulses.

And brains differ in:

- *Susceptibility to social conditioning through education and discipline (Doing)*
  The behavioral psychologist Hans Eysenck demonstrated that be-

havior is more easily conditioned in introverts than extroverts, suggesting that a higher "basal cortical arousal state" increased the reactivity of their brains to stimuli. In other words, introverts shy away from stimulation because their brains are habitually "turned up" higher than those of extroverts, causing greater sensitivity to external stimuli, thus making it easier to indoctrinate their brains with instructions on how to behave. In short, introverts are more susceptible to mind—or meme control. This is why a boisterous boy is so much harder to control than a shy, nervous one—because the brains of extroverts are harder to condition.

- *Ability to inhibit antisocial instincts by employing logic to determine behavior most beneficial to self-preservation (Thinking)*

  Differences in brainpower (that is, intelligence) depend upon the structural and operational complexity of the cerebral cortex. The greater the capacity for rational thought, the greater the ability to deduce rational tactics for successful living—benevolent cooperation being one of them.

- *Neurochemically controlled capacity for emotional empathy (Feeling)*

  Capacity for fellow feeling probably depends on habitual levels of two neurohumors: serotonin—the neurochemical correlate of brotherly love, resulting from a decreased fear of others, produced by a reduction in the brain's responsiveness to stimuli; and oxytocin—the main neurohumor responsible for inducing motherly love—the nurturing facility.

3. A neurotypical susceptibility to asociality no more excuses criminality than poverty excuses theft. The purpose of punishment is deterrent. Punishment is a behavioral conditioning mechanism "dis-*courage*ing" asociality through *fear* of the consequences.
4. The sensible response to asociality is not one of emotional hostility, but the rational attempt first to understand its causes, then to prevent it through a combination of nurture and nature; conditioning for benevolence through the teaching of Sensible Self-Interest, and enhancing the neurochemistry of love through Superbiology.

Biology Is No Excuse to Be Bad

Thus, while the evolved capacity for Thought Morality is universal (psychopaths excepted) it must be recognized that nature has not provided everyone with an absolutely equal predisposition to benevolent behavior. Some people have a greater neurological predisposition for cooperation than others, by virtue of variations in both the neurochemistry of benevolent feelings (Felt Morality) and the capacity to inhibit antisocial impulses (Thought Morality).

"Loving thy neighbor" (reacting to high levels of endogenous "love drugs") and "resisting temptation" (inhibiting stupidly selfish instincts) come more easily to some than to others. "Loving thy neighbor" means responding to neurochemicals such as oxytocin, which induce feelings of benevolence, originally to assist child rearing. Humans differ in habitual levels of oxytocin. "Resisting temptation" means inhibiting impulses induced by the limbic system, the part of our brain we share with our ancestors. Humans differ in the capacity of the cerebral cortex to inhibit impulses induced by the limbic system.

But the conventional morality of reciprocity is not contradicted by biopsychology. The recognition of the neurochemical basis of morality in no way negates the justification for punishment. The main purpose of punishment should not be revenge (a primitive instinct to attack a perceived threat), but social conditioning for benevolence. Society must punish antisocial behavior in order to deter against it. Punishment is a behavioral conditioning mechanism discouraging asociality through fear of the consequences. Justice is the consistent application of punishment in proportion to the severity of the malevolent act.

Recognition of the biological influence on criminality does not absolve moral responsibility; it simply means that some people have to try harder than others to be good; something most of us come to realize through life experience.

Biopsychology no more absolves moral responsibility than drunkenness excuses murder. Biology is no excuse to be bad. But the recognition that neurotype affects the capacity for social behavior may serve to alter our attitude toward asociality from one of impulsive emotional hostility to a rational understanding of the biosocial factors which underlie it.

This is the logical basis of Christian compassion, expressed in the words of Jesus: "Forgive them, for they know not what they do."

* * *

FEELING: EUPHILIA
THE ETHICS OF EUGOICS, THE RIGHT TO FEEL GOOD

> We are on strike against the dogma that the pursuit of one's happiness is evil.
> —Ayn Rand, *Atlas Shrugged* (1957)

> If Nietzsche were alive today, he could be pitching anti-depressants for Pfizer.
> —Jan Morris, *Guardian* (March 27, 2004)

Here I address the question of the best way to *feel*.

By **euphilia** I mean the love of feeling good. The eugoic revolution will soon force society to face the question, Is it good, or right to neurochemically enhance our mood states *beyond* the level of "normal" well-being? Transhumanists say an unqualified "Yes!" There is no good reason for a taboo against the biological enhancement of well-being. The pursuit of happiness is both innate and morally virtuous.

Which of us has the right to tell others they may not use the technology available to improve their condition? Which of us would seek to deny the weak the means to be strong? Why should *biological* self-improvement be considered wrong, when self-improvement through exercise and education are deemed virtuous?

If we don't like the way we look, we are free to spend our disposable income on aesthetic surgery if we so choose. If someone doesn't like the way they feel, why should they not be free to take eugoics—chemical mood-enhancers—to make them feel better?

The bio-Luddites disagree. For them, suffering is a part of human nature we should not seek to alter. But the desire to *preserve* states of misery and distress might be considered pathological; perhaps born of a neurotypical predisposition to melancholia, in which the glass is always seen as half empty: "I'm not gloomy, I'm melancholic. . . . With Prozac, I think there are a lot of people who should be off it and suffering and living with depression," says the novelist and self-proclaimed Grumpy Old Man, Will Self.[24]

Here, then, I respond to the critics of eugoics and applaud the pioneering **New Alchemists** for their private experimentation in neuroenhancement.

The Cause of Suffering

What is the cause of human suffering?

Theists say God included suffering in the world because it increases human compassion; or if that seems a little unlikely— because he moves in mysterious ways.

Sorry, can't go for that one.

Those who believe in reincarnation say disease and disability are a punishment for misdemeanors in a past life.

Charming.

Transhumanism says, "The cause of suffering is biological; the product of a blind, unplanned process of natural evolution. Deficiencies in our biological makeup can increasingly be rectified through science. We should embrace such knowledge unequivocally, allowing individuals the freedom to enhance their emotional makeup if they so choose, just as they may enhance their physique through exercise, or their brain through education. We should be no more afraid of temperamental self-improvement than educational, vocational, or psychological self-improvement. If someone doesn't like the way they look, they are free to improve their appearance through aesthetic surgery. If someone doesn't like the way they feel, they should be equally free to take eugoics to make them feel better."

The bio-Luddites disagree:

- THEISTS say, “Sadness is part of God’s world. We should learn to accept ourselves as we are.”
- MISERABILISTS say, “Neurochemical happiness would make us bland and uninteresting.”
- EVOLUTIONARY PSYCHOLOGISTS say, “Depression is a necessary deterrent against maladjusted behavior.”
- LIBERALS say, “Eugoics will be limited to the wealthy, increasing social inequality.”
- DOOM MONGERS say, “The drugs won’t work.”

I shall respond to each argument in turn.

Unhappiness as Natural

Theists say, “Misery is a natural, God-given part of life. We should learn to accept ourselves as we are.”

Such critics may have been lucky enough to be born with temperaments beneficial to happiness and emotional and mental stability. But not everyone is so fortunate. Certain “God-given” temperaments dramatically increase susceptibility to mental-emotional disorders that can ruin the life-chances of innocent sufferers. A temperamental predisposition to anxiety, depression, or general emotional instability is no joke.

It’s no use the priest or therapist telling an emotionally impulsive, introverted Opioid Neurotype suffering from severe anxiety and depression to “pray” for the strength to cope with adversity, or to “look on the bright side of life”; for the neurochemistry responsible for depression also reduces the enthusiasm and willpower necessary to overcome it.

The concept of altering our neurochemistry to make us feel better is nothing new. We do so every time we take a drink to calm our nerves. Psychopharmacology and a future genetic enhancement (eugoics and Supergenics) are merely more effective extensions of something the vast majority of us do already: use drugs such as caffeine or alcohol to improve our mood. But maybe God wants to ban us from drinking coffee, too, because caffeine alters our God-given temperament?

If someone is born with one leg, we don’t think twice about fitting a false one, so why should we think twice about fitting a new personality if blind, butterfingered nature has let us down? So long as we harm no one in the process, let us be free to do as we please in the effort to improve our well-being.

Consciousness is the last great taboo; the last part of nature we daren’t control through fear of God. Yet mind-altering drugs have been a part of the human experience throughout history. One might ask the theist why an omnipotent, benevolent deity would bother creating mind-altering drugs if he did not wish us to use them.

Oh, I forgot; He gave us free will—together with cancer, heart disease, Parkinson's, Alzheimer's, Huntington's, diabetes. . . . What a gent.

If God gave us the capacity for reason that enables us to develop technology, why would he want us not to use it to cure disease and enhance our abilities? The only possible answer is that God is a malicious trickster who takes great delight in giving us things, then punishing us if we use them to ease our suffering and improve our quality of life!

Is there anything quite as nonsensical as religious belief?

Happiness as Bland

Miserabilists say, "Replacing unhappiness with a state of continual bliss would turn human beings into dull, bland, idle Lotus Eaters."

Strange how often one hears this illogical argument. Rather than "ironing out personality," the eugoics of the future will enhance the capacity for creativity, empathy, and happiness! If you had the choice of feeling any way you wish to feel, would you choose a passive state of drugged-up docility or one of maximum mental and physical dynamism, drive, energy, and enthusiasm for life?

For those who choose the former—let them! It's a free world!

Depression as Deterrent

Evolutionary psychologists say, "Depression is a necessary deterrent against behavior detrimental to survival."

In other words, we need misery to make us abandon hopeless plans. Without the emotional safeguard of depression to deter us, we would refuse to give up on lost causes.

There is probably much truth in this theory. Depression might be regarded as a neurochemical conditioning mechanism evolved to prevent the continuation of hopelessly unproductive behavior, by making us recognize our limitations—and know when to quit. Thus, if we are feeling depressed and the cause is not obvious, such as grief for a loss, we would be wise to think to ourselves, What behavior is my body here encouraging me to abandon as unsuccessful? The answer is most likely to be some failed enterprise in relationships or career, caused by an overestimation of one's own abilities, or underestimation of the problems involved in realizing one's goals. An anxious, neurotic Opioid Neurotype of low IQ who persists in the attempt to become a brain surgeon or airline pilot is likely to experience depression as a result.

However, it is quite wrong to regard the concept of "depression as an evolved deterrent against overambitious behavior" in a *positive* light; for depression puts a stop not only to *unproductive* behavior, but to any behavior at all! Clouded by negative emotion, deficient in the chemicals necessary for feelings of well-being, energy, vitality, confidence; the depressive is a slave to sloth. Contrary to the view that we should listen to the voice of depression, it is in the states of greatest well-being that we make our best decisions; in our most feeble-minded states, the worst.

The pursuit of happiness is not just valuable for one's own sake, but for the effect one's own mood and behavior has on others. So much antisocial and criminal behavior is linked to the state of life-devaluation and unhappiness we call depression. Sufferers of depression typically report feelings of hostility toward other people they feel powerless to overcome. Many people confuse the characteristics of depression with those of an antisocial personality. But the depressive does not *choose* to have a negative attitude toward the world; it follows causally as a consequence of an imbalanced neurochemistry. If one feels bad, the world seems bad, so we tend to treat others badly. By contrast, happiness increases our ability to treat others well. In unhappiness we treat others worst; in happiness, best.

Further, "the happiness habit," like gloom, tends to "rub off" on those to whom we expose it. If it is better to "spread a little happiness" than a little gloom, then it must be morally good to *increase* one's level of happiness—artificially if necessary.

It is one thing to recognize our innate limitations as they exist today, but quite another to accept the principle of human limitations as an eternal given. Today's neurotypical temperamental deficiencies are potentially rectifiable tomorrow through Superbiology. The pursuit of happiness is an ethical enterprise. Raising happiness is a moral act. So let us choose happiness.

Eugoics As Socially Divisive

Egalitarians say, "The benefits of eugoics will be limited to the wealthy, increasing social inequality."

On the contrary, the main effect of ever-increasing public accessibility to eugoics will be increased *equality* through the *leveling up* of society.

Nature is no egalitarian. She has always preferred the inequality provided by a balance between the socially dominant and submissive, probably because, for our ape ancestors, mutual survival was facilitated through social hierarchies. A society composed exclusively of dominant apes would be likely to destroy itself through infighting. We differ in levels of social dominance because that is how our ape ancestors survived best. Hierarchy maintained relative stability. Both dominants and submissives have been preserved by natural selection because inequality benefited social order on the planet of the apes.

But in a *Transhuman* species leveled *up* by Superbiology, with a dramatic *decrease* in the numbers at the mercy of socially debilitating conditions such as timidity, anxiety, depression, aggression, or even low intelligence, social stratification will inevitably reduce. No matter that "those at the top" will also continue to expand their abilities. For the neurological deficiencies that allowed the weak to be downtrodden by the strong will be no more. The dream of the religions and revolutionaries will be attained through technology. Through Superbiology, the meek will—if not inherit the earth—be made strong enough to demand their fair share of the spoils.

The modern world is undeniably keener than ever for greater equality. Material prosperity has afforded ever-greater levels of power to the disadvantaged. The acceptance of class divisions is inversely proportional to the economic power of the poor. It's not unusual today for a plumber to earn more than a college lecturer. As a result, people are less prepared than ever before to accept the old class-based social deference; they naturally expect to be treated as equal citizens. Equality is an instinctive desire for those who have been forced to defer to their supposed "social superiors" on economic grounds.

And lo and behold, what do we see in a society looking for greater equality? —the appearance of the "leveling-up drug" serotonin; increasing both social confidence and empathy in Ecstasy-experimenting youth and Prozac-prescribed adults alike. Across the generations, the goal is the same: the instinctive desire to increase self-esteem and sympathy; confidence and sociability—the basic twin drives of a conscious entity; Prometheus and Orpheus; Individuation and Integration; the Will to Grow, and the Will to Love.

Neuro-Love

If we are all made confident and benevolent through eugoics, the age-old dichotomy between dominant and submissive, master and slave will decrease. One does not fear one's equal. As neurochemical enhancement allows us to become more confident and empathetic, and our happiness levels increase, we may cease to fear the world and begin to love and appreciate life and each other, as never before.

If we are all neurochemically empowered, perhaps we may *really* be able to start "loving thy neighbor as thyself." As we become able to create the chemistry of enlightenment through biotechnology, perhaps we may *all* come to feel the mystic's bliss of universal love. Through Superbiology, perhaps we may all eventually attain the kind of universal empathy dreamed of only by the religions.

Some say, "Rather than increasing our ability to love, eugoics will be used to make people more evil, devoid of benevolent feelings."

But it is a mistake to think that, given the choice, people will choose to make themselves evil! We are programmed to seek pleasure. People prefer drugs like Ecstasy because of the pleasure of benevolence. They do not seek out drugs that make them hateful, for it is simply not part of human nature to seek unhappiness, and hatefulness is always a state of profound unhappiness. One of the first benefits of Superbiology will be the ability to increase the human capacity for empathy by enhancing the neurochemistry of love. Transhumanity will make itself *more* benevolent, not less.

If "spirituality" is available in pill form, and the fear of death is as remote as the fear of life, what then for religion? Men need religion only to fight malevolence, suffering, and death—the products of our human-all-too-human condition. Once empathy, happiness, and immortality are attainable through Superbiology, the need for religion will disappear.

In the future, the tasks of decreasing human suffering and increasing universal love will not be those of a "new church," but of neurochemists. Universal love will be attainable not through religion or politics, but biotechnology. Not New Age enchantments but neuroenhancement—therein lies the future.

The Drugs Don't Work

Doom Mongers say, "The drugs won't work. They always result in unwanted side effects and reduced effectiveness due to the buildup of tolerance."

Tolerance is the body's ability to automatically adjust its own neurochemical levels in the attempt to maintain the state of neurochemical balance prescribed by the genes. Because of the phenomenon of tolerance, eugoic drugs can, indeed, only be a short-term measure. The Real McCoy will be the future development of the Superbiology that allows our bodies to *naturally* produce the ideal neurochemical balance for maximum well-being. Until such a point, however, eugoic drugs will be the only method of effecting temperamental enhancement. As such, the ongoing development of eugoics is vital, not only for short-term benefits, but in order to increase our knowledge of the neurochemistry of mood and behavior, as a precursor to the development of advanced forms of neuroenhancement in the future. Such techniques will probably include the use of stem cell therapy, genetic modification (Supergenics), and, eventually, nanomedicine—the miniaturization of biomedical processes.

Of course, pills are no quick fix for every medical condition. If you have been diagnosed with high blood pressure, you take no exercise and drink a bottle of scotch a night, you would be morally and practically advised to take exercise and cut down on alcohol, rather than bother the overburdened medical profession with endless tests and drug regimes. But one must also ask, What exactly is it that predisposes some people and not others to take no exercise and drink too much? An Opioid Neurotype characterized by a hyperreactive (or "weak") sympathetic nervous system is at far greater risk than others of succumbing to alcohol and drug problems, and equally more likely to display an aversion to physical activity—and hence vulnerability to conditions such as high blood pressure, heart disease, and stroke. In the long term, Superbiology promises to solve the problem by rectifying the genetic deficiency at the source. But in the short term, Eugoic drugs can be of enormous value in enhancing a vulnerable neurotype.

For instance, dopamine-raising drugs such as those used to treat Parkinson's disease, and those currently being prescribed as an aid to give up smoking, are likely to be extended in their use and prescribed as general mood-brighteners, helping to turn anxious-depressive Opioid Neurotypes into their lively, confident dopaminic equivalents.

Similarly, serotonin-raising eugoics such as Prozac, which serve to reduce brain sensitivity to stimuli and dampen the "flight or fight" response of the sympathetic nervous system, may help to reduce anxiety, raise mood, and increase self-confidence. By enhancing self-esteem and interest in life generally, such drugs may serve to increase the drive and willpower to, say, take up sport or exercise and cut down on alcohol.

The Drugs Are Dangerous: The Prozac Scare

Some say, "Prozac the wonder drug turns out to increase suicidal or violent behavior."

The **bio-defense** argument has already been employed in court in the attempt to reduce the conviction of the Prozac user, on the grounds that "the pills made me do it." But to distinguish the truth from media hysteria requires an understanding of the way in which Prozac works, something that has not yet been established by the medical profession.

In my **Reduced Sensitivity Hypothesis**, raising serotonin levels decreases anxiety and general emotionality by reducing the activity of the nervous system. It is not hard to imagine that the resultant effect on a minority of anxious depressives might be an increase in suicidal feelings, either from a certain dulling of the emotions, or an increase in aggression resulting from reduced levels of anxiety, and hence absence of fear of others. But such effects are only experienced by a tiny minority! To dismiss Prozac because of its negative effect on a few is like seeking to ban pecan pie because some people suffer from a nut allergy.

Designer Drugs

Nobody is saying that existing eugoic drugs are a miracle cure for temperamental disorders. We are only at the very start of a revolution in our understanding of neurochemistry and the development of mood-enhancing drugs. The poor reputation of neuropharmacology as a positive tool for improving mental states stems largely from the primitive nature of the first eugoics. But this is no argument for abandoning the effort! It would be stupid to reject the potential benefits of neuropharmacology merely because the early attempts have been crude.

The early pharmaceuticals used to treat psychological personality disorders were primitive "dirty drugs"; that is, those not designed for a specific purpose, but merely discovered by chance to have a beneficial effect for a particular disorder. Such drugs often had side effects worse than the original symptoms. Lacking the subtlety that comes from sophistication, dirty drugs were the equivalent of throwing a pot of paint at a canvas on a wall and hoping that at least some of it hits the spot. Inevitably, much of it covered the wall as well.

The difference between dirty drugs and the designer drugs of the future is the difference between an artist who throws paint at the canvas and hopes it looks interesting, and the products of a Leonardo or Michelangelo, artists who deliberately and painstakingly work toward a specific goal. Side effects are proportional to lack of design sophistication. The first flying machines didn't get too far off the ground because their design techniques were primitive, but the first aeronautical engineers did not simply give up and say, "We should abandon flying machines because they crash." They learned from their mistakes and went on trying, improving their methods and knowledge, until now, we have cheap global air travel and the prospect of space tourism in the twenty-first century.

The future lies not in dirty drugs but Designer Drugs; those deliberately created to do a specific job by affecting specific neurochemicals in specific ways. The first great success story has been the class of mood-brightening eugoic drugs—SSRIs (selective serotonin re-uptake inhibitors), most famously, Prozac—trade name of the chemical fluoxetine. Where previous types of antidepressant acted on several types of neurotransmitter at once, and pharmacologists frankly weren't sure which—the equivalent of throwing a pot of paint at a wall—SSRIs were specifically designed to raise levels of serotonin, essentially by "putting a plug in" to the relevant brain receptors, to stop the chemical draining away after use.

As scientists learn more about the behavioral effects of specific neurotransmitters, new classes of mood-brightening eugoics will appear. Already we have seen the emergence of SNRIs, targeting the excitatory chemical, noradrenaline, of which dopamine is the chemical precursor. Now that the importance of the "dopamine-noradrenaline pathway" has been recognized as a source of mood-enhancement, we will begin to see new eugoics targeting both dopamine/noradrenaline *and* serotonin in an attempt to produce the optimum state of well-being—*mens sano in corpore sana*—a sound mind in a sound body.

But in the long term, drugs will be obsolete, as the emerging Superbiology of biotechnology and nanotechnology combine to permanently optimize our neurological level of well-being.

Drug Addiction as Self-Medication

Until now, experimentation with eugoic drugs has been left to the street—and thus to the weakest members of society. Much drug addiction is simply self-medication; the attempt to improve a deficient neurotype. Drugs such as heroin, crack cocaine, and alcohol achieve their effect by raising the levels of neurohumors such as opioids, dopamine, or GABA. Drug addicts are often self-medicating against neurochemical personality disorders that have made successful social functioning impossible.

And who can blame them? Sometimes drug addiction may be a helpless response to intolerable life experience, such as poverty or physical abuse, and that's a sociopolitical problem, which can be solved if we have the compassion and good sense to care about people at the bottom end of the social scale. But sometimes people are just born with a neurotypical predisposition to be melancholic, depressed, apathetic, uptight, nervy, or painfully shy.

The development of eugoics will provide an alternative option to the present policy on drug abuse: repeatedly arresting addicts for their daily crimes committed to pay for their next fix, or handing out free drugs—such as daily doses of the heroin substitute methadone—thus perpetuating the misery of addiction.

The New Alchemists

The widespread belief in eugoic as opposed to street drugs is illustrated by the countless private experiments in neurological self-enhancement taking place around the world.

Increasingly, individuals armed with the ever-expanding oracle of information that is the Internet are scouring scientific research papers and exploring the potential of nutrients and neurotransmitters as yet ignored by the mainstream medical profession, but available for private purchase, often from offshore net-pharmacies, in the quest to enhance mental and physical functioning beyond the levels of "normal" well-being.

The situation is reminiscent of the medieval and Renaissance alchemists—the first neurochemists—who devoted their lives to the quest for the elixir of life; the chemicals that would turn base metal into gold, and make man immortal.

Today's **New Alchemists** are psychonauts and cybernauts surfing the Web in search of new possibilities, armed with a vision of a higher state of well-being.

Governments may align themselves with drug companies in seeking to prevent such private experimentation, by outlawing "nonmedical" eugoics they consider dangerous (or effective?). But as one is removed from circulation, another appears—in the unregulated global marketplace of the Internet.

Neuromanticism

The difference between the use of psychedelic drugs by the hippies of the sixties, and the exploration of neurotransmitters and amino acids by the New Alchemists of the twenty-first century, is the difference between an *anti*modern mentality which sought to abandon (or "blow") the rational mind, and the **New Modernity** of transhumanism, which seeks to enhance rather than reject the capacity for reason.

In this sense, transhumanists might be described as **neuromantics**: passionate believers in human transcendence—not through mysticism—but technology; the product of the *rational mind.*

Where the counterculturalists wished to jettison the scientific rationality they viewed as a product of the "patriarchal Western ego" in favor of getting "far out," the transhumanist neuromantics seek to employ the cutting-edge technologies of Superbiology to enhance the capacity for both emotion *and* reason, feeling *and* thinking. The ideal state might be called one of **Fully Firing on All Four Cylinders**, the maximum operation of all four faculties: sensing, feeling, thinking, and perceiving.

The essence of Romanticism is imagination and passion. Imagination is the creative power of the mind. Passion is the active power of the body. **Neuromanticism** is the belief in enhancing our capacity for positive feeling, by means of the rational mind—through science and technology. Neuromanticism is a synthesis of romantic passion and "classical" scientism; the union of Dionysus and Apollo; feeling and thinking.

Freedom to Experiment

As with the development of all new technologies, there is a risk involved in experimenting with eugoic drugs; the risk of self-harm resulting from unexpected side effects. But equally, there is a tendency of the state to treat the public like children who must be protected from dangers to themselves they are considered insufficiently wise to recognize.

Similarly, there is a tendency for the state to act in the interests of pharmaceutical companies seeking to prevent consumer access to nonmedically sanctioned eugoics, in order to avoid reduced profits from their own drugs.

Though it is sensible to support both public health education and the invaluable work of pharmaceutical companies in developing new drugs to alleviate suffering, we should oppose attempts by the state or drug companies to prevent access to nutrients and supplements such as amino acids, proteins, and neurotransmitters that are currently the source of untold private eugoic experimentation.

Let us believe in freedom of choice. Let the state treat the public like adults, not children; with respect, not patronizing condescension. The role of the democratic state is not that of parents, there to stop us from hurting ourselves, but servants, there to put our will into practice. In the words of the pro-eugoics group CERI, "We support the right of adults to decide for themselves what is in their best interest."[25]

A revolution is about to occur in society; we are beginning to recognize that we need not accept the mood states offered us by nature. Our stoical acceptance of unhappiness will one day be regarded as the behavior of a primitive species. In the future, we will look back at a society which allowed the mentally or emotionally vulnerable to crawl around dingy rooms sticking needles into their arms in order to avoid conditions of emotional distress with the same horror that we look back at slavery today.

In the future, we will no longer need to fatalistically endure negative mood states or feeble brainpower. We will increasingly be capable of enhancing our capacity to feel and think in pursuit of ever-greater levels of well-being. The only thing holding us back is fear of the unknown. Fear can never be the basis of ethics.

The eugoic revolution has only just begun.

* * *

PERCEIVING: NEUROMANTICISM THE TRANSHUMANIST ATTITUDE TO LIFE

> And now we are on the brink of the can-do era, when the leopard will be able to change its very spots.
> —Robert Ettinger, *Man into Superman* (1972)

> It became no longer the thing to mope.
> —Peter Medawar on the dawn of Renaissance humanism,
> *The Strange Case of the Spotted Mice* (1996)

Here I address the question of how to perceive the world—that is, the attempt to answer the question, What is the best *attitude* to life?

I describe the transhumanist attitude to life as neither irrational pessimism nor naïve optimism but **Positive Realism**, summed up by the phrase **from Candidism to Techno-Can-do-ism**. I refute fashionable Antimodern cynicism, and offer **Reasons to Be Cheerful** in the miraculous technowonderland of the modern world.

The Meaning of Attitude

An "attitude" is a condensed summary of one's world picture, belief system, philosophy—or meme map. Attitudes consist of two elements—beliefs and values—corresponding to the two main areas of a philosophical system. Beliefs correspond to metaphysics or cosmology, describing the nature of the world, as we perceive it. Values correspond to ethics, and describe how best to *act* in the world. Beliefs correspond to Being, values to Doing. Beliefs are "is" statements (the world is round—or flat). Values are "should" or "shall" statements (thou shalt not kill—or suffer a witch to live).

Values depend on beliefs. How we think we ought to behave, or what we should do with our lives, necessarily depends upon our understanding of the nature of the world in which we are to act. If we believe that the world is ruled by God or the Devil, or is made of green cheese, we are likely to have rather different values than if we believe that the world is a natural process of unfolding evolutionary complexion. This is the basis of my refutation of the "Naturalistic Fallacy," the assertion that facts do not determine values. On the contrary, facts *necessarily* determine values, except for the insane, who base their moral codes on delusions about the nature of the world.

The English serial killer known as the "Yorkshire Ripper" murdered prostitutes because he believed that God wanted him to do so. His moral code was based on his beliefs about the nature of the world; namely, that its ruler dislikes prostitutes and wished him to kill them. Facts determine values. The Yorkshire Ripper's facts about the world were wrong. Had they been correct, he would not have gone out at night with a sledgehammer in his van.

One's general attitude to life depends upon the meaning or significance one *perceives* in the world, where perception is used in the popular sense of the word, synonymous with "interpretation." How we perceive a thing is our interpretation both of its nature (an "is" question, about Being) and its value (as "ought" question, about Doing). The original Latin meaning of "perception" is the ability to grasp, seize, or take hold. Thus, perception is the ability of the mind to grasp or apprehend the world, by piecing together the objects identified by the senses into a coherent pattern—or a meme map. The question of how we perceive the world is thus synonymous with our understanding both of its physical (or metaphysical) nature, and our ethical *evaluation* of its nature. How we "see" the world depends upon two elements: what we understand the world to be like, and how we evaluate what we think it's like. In short, ethics follow metaphysics, and one's attitude to life is a condensed summary or reflection of both—a personal world picture, or meme map.

Transhumanist ethics are based on transhumanist metaphysics, or cosmology, which describe the world as a harmoniously unfolding process of complexification in nature, leading to ever-greater human knowledge, abilities, and well-being. Knowledge increases well-being by allowing us to improve our physical and material condition, for example, by building warm, comfortable homes, instead of living in caves; or by preventing and curing disease, through antibiotics, surgery, or Superbiology.

Candidism

> "What is optimism?" said Cacambo. "Alas!" said Candide, "it is the mania of maintaining that everything is well when we are wretched."
> —Voltaire, *Candide* (1759)

> But against the palpably sophistical proofs of Leibniz that this is the best of all possible worlds, we may even oppose seriously and honestly the proof that it is the worst of all possible worlds.
> —Arthur Schopenhauer, *The World as Will and Idea* (1819)

The transhumanist thinker Max More has called the transhumanist attitude to life one of "Dynamic Optimism," an expression that conveys well the belief both in positivity over negativity and activity over passivity: "A dynamic optimist is not someone who sits back and believes that all things will work out fine regardless of their effort. It is someone who says that the universe is open to intelligence, to possibility, to the application of our own capabilities. We just have to make things better, applying our will and intelligence."[26] Here, however, I shall avoid the term *optimism*, conscious of the need to deflect unjustified but common criticism of transhumanism as ingenuousness—rather than ingenuity.

The term *optimism* originated in the eighteenth century to describe the philosophy of Leibniz, who famously reasoned that this world must be "the best of all possible worlds," since any other was logically incompatible with God's omnipotence and benevolence. In other words, the real meaning of optimism is the belief that this world is the "optimum" that one could conceive of creating. The French rationalist philosopher Voltaire duly devoted an entire book to lampooning this piece of illogic. In his classic novella *Candide, or Optimism*, a naïve young man befalls a series of increasingly disastrous misfortunes, each of which are justified by his mentor, Dr. Pangloss, in the unshakable conviction that, despite the endless suffering of life, "everything is necessary in this best of all possible worlds."[27]

In its original sense, then, the word "optimism" implied a *naïvely* positive attitude. And such an inference is still carried today; optimism means hope *in spite* of a poor prognosis. A cinematic representation of the true meaning of optimism is the ironic image of a crucified Christ being encouraged by his fellow martyr to "Always look on the bright side of life." The power of this secular icon lies in the universality of the attitude it represents: the naïveté of optimism in the face of brute reality. The transhumanist attitude to life I espouse is not the naïve optimism I call, after Voltaire, **Candidism**, (pronounced "con-deed-ism").

But neither is it the equally irrational negativity that is "pessimism." The word is often used to describe the attitude of the nineteenth-century German philosopher Arthur Schopenhauer, who famously responded to Leibniz by asserting that this world must rather be considered the *worst* of all possible worlds, since any *more* suffering would make the very existence of life an impossibility![28] Pessimism, the tendency to believe things will turn out badly, is based on an underlying attitude one might call **negativism**—a predisposition to negative evaluations. Negativism is negativity as a habitual, temperamental predisposition. To hold a negative attitude toward a modern world overflowing with plenty is simply churlish; the brattish response of a spoiled kid.

The Positive Realism of Techno-Can-Do-Ism

The sensible attitude to life is neither the naïveté of optimism (Candidism) nor the irrationality of pessimism (negativism), but rather an intermediary position I call **positive realism**: a blunt recognition of the biological limitations of life, combined with the positive conviction that such problems will be gradually overcome through the application of science and technology; genetic, rather than social engineering; evolution not revolution.

Positive realism is the recognition of fundamental biological limits to human progress, combined with the conviction of their inevitable eventual transcendence through Superbiology. We *can* improve our condition, but only if we are prepared to make use of the greatest power tools at our disposal: science, technology, and the power of the rational mind.

The transhumanist's rejection of the naïve optimism of utopian idealism for positive realism could be summed up in the phrase **from Candidism to can-do-ism**. Positive realism is not a naïve optimism based on wishful thinking, but a logical deduction that the most likely future outcome for humankind is the gradual reduction of biological limitations through the continual increase in abilities made possible by technology. In short, positive realism is a twenty-first-century **techno-can-do-ism**.

The belief in gradual improvement through effort as opposed to the extreme idealism of optimism or cynicism of pessimism is sometimes called **meliorism**, defined in the Oxford English Dictionary as, "The doctrine, intermediate between optimism and pessimism, which affirms that the world may be made better by rightly directed human effort. As used by some writers, the term further implies the belief that society has on the whole a prevailing tendency toward improvement."

The positive realist rejects both the pessimism of postmodern nihilism and the naïve idealism of Candidism for a twenty-first-century techno-can-do-ism based on meliorism: the belief in the inevitability of gradual progress through technology, leading to a positive attitude toward life, and the hurdles it places in front of us.

To the positive realist, a problem is not an unbreachable barrier to success, but a hurdle to be leaped.

Reasons to Be Cheerful

The Positive Realism of Techno-Can-Do-ism provides an alternative to the directionless vacuity of ironic nihilism that is the hallmark of the Postmodern Condition. Of course there are failures, mistakes, things we would love to see improved in the modern world. But why always focus on the negative side of the balance sheet? Is there no evidence to support a belief in the modern world? Yes, of course, providing you're not an ostrich; one who would bury his head in the sand rather than experience any kind of joy or pleasure in living.

Next I shall briefly set out a few of the positive features of the modern world; features that are conveniently overlooked by those with a vested or temperamental interest in downplaying human achievement in favor of the dubious pleasures of postmodern cynicism.

I shall call it Reasons to be Cheerful.

In Praise of Progress

> The list of diseases we have conquered in this country is a long list: malaria, yellow fever, bubonic plague, polio, cholera. Measles, 20 years ago 500,000 annual cases, in 1993 there were 301 cases. That's progress.
>
> —Ed Regis, author, *Mambo Chicken and the Transhuman Condition,* HotWired (November 15, 1996)

In the post (or anti)modern age, ironic condemnation of one's own civilization has become a mark of social acceptance.

I say, baloney! The modern world really *is* best, and it's not difficult to state why:

- We have dramatically increased the average standard of living.
- We are developing the most astonishing technology, which will soon allow instant access to the sum total of human knowledge at a click of a button.
- We are fast learning how to defeat disease.
- We no longer throw Christians to the lions.
- We do not amputate the limbs of criminals, or stone them to death.
- We do not hang, draw, and quarter our enemies.
- We extend basic rights to animals.
- We believe in equality between sexes and races.

What I'm describing here is a common phenomenon. It's called human progress. It's deeply unfashionable at the present, and it should be made fashionable again. Why? Because those who think things won't get better, stand and watch as things get worse. Cynicism is merely sloth, an excuse not to try.

In Praise of Technology

Ever-advancing technology has produced a modern world positively overflowing with extraordinary inventions: from advanced surgical techniques, to the humble washer, dryer, dishwasher, microwave, telephone, Internet, personal computer, CDs, laptop, cell phones, DVDs, PDAs, MP3s . . . the list goes on and on. We now live in the sort of technowonderland our ancestors could only dream of, but which most of us take entirely for granted. And incredibly, most of it has come into being only over the past few decades—indeed, much of it only in the last few years!

The spread of technology to the home now enables all of us to be artists, musicians, writers, doctors, and psychologists, at the click of a mouse. We can record our own CD, write and print our own books, magazines, pamphlets or photos. Design our own notepaper or business cards. Become video artists. Learn more about medicines than our general practitioners, and share that knowledge with others from around the globe. Sell products to the other side of the world without leaving one's home. Learn about any subject or communicate with anyone across the planet . . . the possibilities are endless.

Superbiology—the enhancement of mind and body through biotechnology —will represent one more Wonder of the Modern World.

We can choose to embrace the Modern Age, or succumb to **cyber rage**; a state of technophobic anger and hostility toward new technology not yet understood. The choice is ours.

In Praise of Prosperity

In the postmodern age, it has become popular to refute the notion that material goods and services are an accurate indicator of standard of living. The benefits of the modern world are supposedly counterbalanced by undesirable side effects such as "soulless" materialism, stress, or pollution. Yet in the modern world, more people have a higher standard of living today than at any point in human history. Shops are overflowing with choice and increasingly fabulous design, a witness to the wonder of human creativity.

Yes, of course there's still poverty and disadvantage in the world, but only a fool expects utopia overnight. Progress toward a world of plenty takes time. *Homo sapiens* is a young species; a mere child of evolution. Think what we have accomplished in fifty years through technological production, and imagine what miracles of abundance await us in the future.

Are technophobic negs really going to dismiss the extraordinary opportunities made possible in an evolutionary wink of an eye by science and technology on the grounds that everyone on earth cannot immediately share them? What illogical nonsense. Despite the whining of the doom mongers, most people in the modern world have access to the basic material necessities of life, and there is a higher provision of universal healthcare, wider educational opportunities, and generally greater freedom from want and suffering for more people in the world than ever before. We should evaluate the affluence of the consumer society positively, in terms of profusion, abundance, plenty; not negatively, in terms of overconsumption, materialism, and waste. Plenty is better than scarcity, yet the neg-ridden Luddites act as if it were somehow morally superior to identify oneself with poverty, scarcity, want, and states of distress.

Simply identifying oneself with disadvantage is not morally virtuous. One thinks of the inverted snobbery that is **Cheap Chic Syndrome**, a condition in which identifying oneself with poverty (perhaps by dressing like a tramp) is confused with the grim reality of poverty itself. Yes, let us tackle poverty in the world, but let us not bite the hand that feeds! Let us be proud of the magnificent civilization we have created; a civilization now in serious danger of being overthrown by one of greater self-confidence that laughs in the face of our chronic lack of self-belief.

In Praise of "Consumerism"

The negative connotations of the term *consumerism* derive from the original Latin meaning of the word "consume": "to use up completely," with the implication of wastefulness. By their propagation of the "consumerist" meme, anti-modernists have managed to convince the public that a continual increase in the available goods and services in society is somehow a dreadful thing, since it means "using up resources"; a brilliant meme attack indeed!

According to the antimodern metameme, decent people should feel guilty about their prosperity and live in "voluntary simplicity"—euphemism for abject poverty. I would like to wage a Meme War against the idea that the continual increase of goods and services in the modern world is somehow morally reprehensible. We are living in a wonderland of prosperity. We should rejoice; it's a human miracle.

The fact that at this point in history everyone on earth does not yet share the prosperity of the modern world should no more negate our ability to appreciate its value than the fact that we cannot all be Beethoven devaluates our ability to appreciate the Ninth Symphony. You don't have to hate yourself and your society in order to show sympathy with the disadvantaged. Wearing sackcloth and ashes won't feed the poor. The way to end world poverty is by *extending* the consumer society, not destroying it through a perverse guilt trip.

To regard the use of products and services as "using up resources" is the nadir of the neg. Instead of the "Consumer Society," we should call it the "Enhancement Society" for we make goods and services to enhance our lives!

In Praise of Growth

Eco-fundamentalists say, "The greedy West should abandon economic growth."

But growth is the very essence of life, and economic growth is no different than any other kind. It's good to want more. It's instinctive to want more. To want more is what humans do. It's the very fiber of our being. When Oliver Twist asked for more, we applauded him, because we knew intuitively it was the right thing to do; to be given only just enough to survive is wrong. Why? —because to grow is in our very nature. This is what I mean by the Will to Evolve: The conscious instinct to want more; to be more; to improve; to develop; to be the *best* one can be, is not only good—it is the very essence of what we are as human beings!

Wanting more is not the same as greed. The desire to grow to the fullest possible extent is not the same as the belief that "greed is good." Greed implies wanting more at the expense of another. To harm others in the pursuit of more is Stupid Selfishness, for malevolence breeds malevolence in return. But nobody really wants *less*. Even those who drop out of the rat race for a quieter life do so to increase their survival chances by reducing levels of stress. What "downgraders" are really looking for is *more life*, not fewer material goods.

That which does not evolve, stagnates and dies. Most of us know of someone who gave up on a dream or plan, lost enthusiasm, and "went downhill" emotionally and physically thereafter. Or one who retired early and promptly dropped dead. The reason? Abandoning the Will to Evolve can kill, for life *itself* is evolution; a process of becoming, not being.

When we cease to grow, we begin to die. Americans know this instinctively, which is why they are a youth-oriented country like no other. The powerful belief in constantly striving for self-improvement through cheerful effort unto the grave is one reason why we live in the age of the American Empire, and the United States is by far the strongest nation on earth. The motivation behind "the pursuit of happiness" is the Will to Evolve, for happiness is nature's reward for behavior beneficial to survival, and the Will to Evolve is the drive to increase survivability by extending capacities.

Let us abandon the view that it is somehow wrong to want more. To want more is the basis of progress, and progress is the very essence of modernity—the philosophy that built the technowonderland of the modern world.

A civilization which no longer wants more but confines itself to critical self-examination is a civilization in decline, ripe for replacement by some other, more assertive culture, with altogether different ideals . . .

In Praise of Modernity

Postmodernists say, "The ideology of modernity is 'Eurocentric,' concerned only with European and North American interests."

But the belief in social-material progress through techno-industrialization is no longer uniquely "Western," but shared around the world. Japan is a shining ambassador for the values of the modern world, while China is busy modernizing as fast as she can, having woken up to the fact that prosperity through modernity is better than poverty through ideology.

Modernity is not tyranny, but techno-industrial progress for anyone who wants it. Most people *do* want it, because they recognize that it brings freedom and prosperity. At a time when our freedom is severely under threat from pre-modern fundamentalism on the one hand and postmodern authoritarianism on the other, belief in the values of modernity is even more important than ever.

What is needed now is a renewal of self-belief in the modern world before it's too late.

Transmodernity

At this point, the judicious agnoskeptic might be forgiven for asking, "How is it possible to simultaneously declare one's love for the modern world, and condemn its domination by an Anticulture of Nihilism?"

The answer is simple: one may love the cultural artifacts of a society without loving its metameme.

The word "culture" signifies both products *and behavior* in a society. Cultural artifacts refer to its *products*—such as personal computers, the Internet, cell phones, digital radio, satellite TV. Cultural *behavior* within a society is based on its dominant metameme—the beliefs and values held (and propagated) by the dominant members of the society.

The dominant metameme of the modern world is currently that of postmodernism. But the cultural artifacts of a society bear no necessary affinity with its currently dominant metameme. The cultural artifacts of the modern world are dominated by Information Technology, but there is no necessary connection between an IT-dominated culture and the postmodern metameme.

In fact, the logical ethical correlation to an IT culture is not postmodernism at all, but the beliefs and values of modernity—or rather, the **New Modernity** depicted in the Transhumanist metameme—**Transmodernity**, an extension of the benefits of techno-industrialized society to all who desire its pleasures.

The logical ethical response to the technowonderland of the modern world is not postmodern nihilism, but *Transhumanist e-phoria*!

The Happiness Habit

> Pessimism is partly a matter of bad experiences or/and hormone shortages. These can be remedied, if you can hang on a while.
> —Robert Ettinger, *Man into Superman* (1972)

It's not always easy to look on the bright side. Life is tough. Accumulated negative experiences can gradually reduce the capacity to maintain a positive attitude and drag you down. Yet who could argue that it is better to *try* to feel good, rather than to wallow in misery; something all good parents teach their children?

Much depression is simply adult sulking; aggression an adult tantrum. A susceptibility to depression and aggression could be significantly reduced by good parental conditioning in childhood. The failure of parents to stamp out the self-pity of sulking and the selfishness of tantrums is one of the secret causes of adult depression and aggression.

And here the experientialists are right. The brain *is* malleable—if caught young enough. Habitual levels of the neurochemicals necessary for well-being can be reduced through continual negative or aggressive attitudes and experiences.

The message for parents is clear: if you want your child to grow up to be a happy, successful adult, let neither sulking nor tantrums become habitual. Teach the happiness habit, and if you can't, because you're depressed or short-fused yourself, go to your doctor and ask for eugoic mood-brighteners until you can.

There can be no doubt that a general predisposition for pessimism is neurochemically based. Just as the drinker's gradual slump into melancholic, maudlin self-pity is the result of collapsing neurochemical levels artificially raised by alcohol, so some people are born with a neurotypical susceptibility to negative mood states resulting from a hereditary deficiency in bliss chemicals. The sensible response, however, is not merely to accept one's melancholic pessimism as a given, but to fight it, both with positive thinking and with modern, mood-enhancing medical drugs—eugoics.

Happiness may be a habit, but the happiness habit is easier to learn with a top up of neurochemical bliss. And as the science of eugoics develops in the twenty-first century, it will become easier still. So let us reject the ironic cynicism of the Postmodern Condition for the belief in **New Modernity**; a renewal of belief in the inevitability of human progress through reason, science, technology—and the Will to Evolve.

NOTES

1. G. E. Moore, *Principia Ethica* (Cambridge: Cambridge University Press, 1962).
2. David Hume, *A Treatise of Human Nature*, ed. Ernest C. Mossner (1739–40; repr., London: Penguin, 1985).
3. Norbert Wiener, *Cybernetics* (New York: Wiley, 1948).
4. Moore, *Principia Ethica*, p. 49.
5. Stephen Jay Gould, "Life's Grandeur," in *Full House: The Spread of Excellence from Plato to Darwin* (New York: Harmony Books, 1996), p. 216; and "Three Facets of Evolution," in *How Things Are: A Science Tool-Kit for the Mind*, ed. John Brockman and Katinka Matson (New York: William Morrow, 1995).
6. "Evolution Has Direction," *New Scientist,* January 18, 2003.
7. Richard Dawkins, *The Extended Phenotype* (Oxford: Oxford University Press, 1982).
8. "Why Evolution Will Always Create Intelligence," *New Scientist,* November 16, 2002.
9. James Watson, interviewed at the Future of Life summit, hosted by *Time* magazine, Monterey, California, February 2003.
10. Dean Hamer and Peter Copeland, *Living with Our Genes: Why They Matter More than You Think* (New York: Doubleday, 1998).
11. Julian Huxley, *Essays in Popular Science* (London: Chatto & Windus, 1926).
12. *About UNESCO*, http://www.unesco.org.
13. Richard Dawkins, *The Devil's Disciple* (London: Weidenfeld & Nicholson, 2003).

14. Charles Darwin, *The Descent of Man* (London: J. Murray, 1871).

15. Patrick West, *Conspicuous Compassion* (London: Civitas, 2004).

16. J. F. Young, *Cybernetics* (London: Iliffe Books, 1969).

17. Friedrich Nietzsche, *Beyond Good and Evil* (1886; repr., London: Penguin, 1988).

18. Confucius, *The Analects*, trans. D. C. Lau (London: Penguin, 1979).

19. Julian Huxley, *New Bottles for New Wine* (London: Chatto & Windus, 1957).

20. Ullica Segerstrale, *Defenders of the Truth* (Oxford: Oxford University Press, 2000).

21. Richard Dawkins, *The Selfish Gene* (Cambridge: Cambridge University Press, 1976).

22. Robert Wright, *The Moral Animal: The New Science of Evolutionary Psychology* (New York: Pantheon Books, 1994).

23. James Watson, Skeggs Lecture, Youngstown State University, Ohio, 2003.

24. Will Self, *Sunday Times*, April 24, 1998.

25. "Evolution Has Direction," *New Scientist.*

26. "Dynamic Optimism: An Extropian Cognitive-Emotional Virtue," http://www.maxmore.com/optimism.htm (May 10, 2005).

27. Voltaire, *Candide, or Optimism*, ed. John Butt (1759; repr., London: Penguin Classics, 1990).

28. Arthur Schopenhauer, *The World as Will and Idea*, trans. E. F. J. Payne (1819; repr., New York: Dover, 1966).

Chapter 11
NEUROPOLITICS

> Who then is free? The wise man who can govern himself.
>
> Your own safety is at stake when your neighbor's house is in flames.
>
> One wanders to the left, another to the right. Both are equally in error, but are seduced by different delusions.
>
> —Horace, *Odes* (23 BCE)

Political philosophy is the attempt to answer the question, What is the best way to organize society?

I answer, "According to the principle of 'Maximum individual freedom compatible with social cohesion.'"

Here, I address two subjects: the nature of past societies, and the nature of the ideal society.

- *The Bodypolitic*—the authoritarian state as corporate body
- *The Neuropolitic*—the **Neurosociety** or **Self-Enhancement Society** as a blueprint for the ideal society

SOCIETY

By **Neuropolitics** I mean the vision of an ideal society in which behavior is determined not by an unthinking public body controlled by head of state or other corporate body (**the Bodypolitic**, **Marionette Politics**)—but by a voluntary collaboration between conscious, autonomous, logically thinking, individual minds, each acting out of Sensible Self-Interest, through the freely tendered mutual encouragement of **Reciprocal eugoics**—that is, self-development, self-enhancement, or personal growth. In Neuropolitics, just as the autonomic nervous system works automatically to maintain a state of homeostatic balance in the body by allowing the free activity of individual parts within limits required for its harmonious functioning as a whole, so, too, conscious human brains (that is, minds) work together in their own mutual self-interest to maintain a state of *social* homeostatic balance between extreme states in the **Neuropolitic**. Neuropolitics is thus the democratic politics of individual freedom and social concern combined. The Neuropolitic is a **meritocracy** in which individuals are allowed the freedom to pursue the dictates—not of the Bodypolitic, via **groupthink**—but of their innate, individual Will to Evolve—the drive to be the *best* they can be, in the recognition that the best way to be all that they can be is to cooperate benevolently with *other* individual "Wills to Evolve" in the neuropolitic, each seeking the same goal.

SUMMARY OF TRANSHUMANIST POLITICS

The ideal society is not an unconscious public body controlled en masse by an authoritarian head of state or corporation (**Bodypolitic**), but a harmoniously functioning, voluntary collaboration between autonomous, self-governing, individual "Wills to Evolve," each acting out of Sensible Self-Interest in pursuit of ever-increasing survivability and well-being, for the good of self and society alike, through a deliberate policy of continual, mutual encouragement (**Neuropolitic, Self-Enhancement Society, Reciprocal eugoics**).

The Meme War between the politics of the Left and Right is a product of **Dichotomal Thinking**. The belief in individualism is vital to an evolving society that wishes to avoid the descent into authoritarianism, stagnation, and decay (against **Groupthink**). No one has the right to deny others the freedom to improve their bodies or minds as they see fit **(Bio-Libertarianism)**. At the same time, complete absence of social concern in favor of absolute individual freedom is self-defeating (**Stupid Selfishness**). No man wishes to wade through streets full of beggars and thieves, in constant fear for his life and property (**Soft versus Hard Libertarianism**). A society is a homeostatic organism requiring continual regulation between extreme states for maximum survivability. The best way to evaluate political and ethical questions is to apply the principle of "maximum individual freedom compatible with social cohesion" or harmony **(Bispect Politics)**. Soft Libertarianism is neither the laissez-faire "Social Darwinism" of the extreme Right, or the laissez-faire *social* Liberalism of the Left, but the politics of the sensible center (**Progressive Centrism**).

THE RESURGENCE OF ETHICO-POLITICS

Today, the freedom of the individual is under threat from a two-pronged attack by both sides of the democratic political equation. The bio-Luddism of both the Religious Right and the Liberal Left is one example of a dangerously disturbing trend toward the acceptance of authoritarian attitudes in modern politics. The phenomenon is occurring, I suggest, for two interrelated reasons: the increasing prosperity of the modern world and, as a result, the emergence of a greater economic and political consensus between Left and Right.

The political consensus in the modern world is that of a dynamic state of balance between market forces and state intervention, expressed most recently by Prime Minister Tony Blair and President Bill Clinton as "The Third Way." (Not a new idea, this; the concept of "convergence politics" was much discussed during the cold war, as a method of breaching the Iron Curtain that separated the Capitalist West from the Communist East).

If there is less argument on socioeconomic issues because Left and Right broadly agree, then what's left to argue about in politics, except ethical issues that were previously held to be apolitical? If both parties move to the center of politics, then those who hold strong opinions on moral issues will begin to spot an opening in political life for the promotion of their views. Defenders of Recessive Memes will suddenly wake up to the realization that a recessive meme can become a dominant meme, in the right meme pool, at the right time.

Thus, as a result of economic prosperity and political convergence, moral issues have begun to enter into politics to a greater degree than ever before. Moral dilemmas such as the life-and-death issues of abortion, euthanasia, and animal rights are now at the forefront of politics, where for previous generations, all that really mattered was jobs and taxes. As I write, fox hunting has just been outlawed by the British Labor government—an act that would have been unthinkable for any previous administration.

Unfortunately, the greater one's moral convictions, the less likely one is to be tolerant of those who disagree with them. Moral righteousness tends to breed political authoritarianism. The seepage of authoritarian policies through the back door by way of ethical issues is one of the dangers facing the modern world in the twenty-first century. The task of the Neuropolitic will be to attempt to avoid descent by slippery slope into either the "egalitarian" authoritarianism of the Left, or the evangelical authoritarianism of the Right, spurred on by the sight of a populace newly eager to embrace **ethico-politics**—the heady Bodypolitics of moral fervor.

The way to avoid a return of the perils of the authoritarian Bodypolitic is to embrace the vision of the **Neuropolitic**—society regarded as a multiplicity of autonomous, individual wills-to-evolve, freely tendering Reciprocal eugoics in their own Sensible Self-Interest—instead of a gluttonous Bodypolitic, ruled by its head—the politics of the marionette, whose strings are pulled every which way—but loose.

BODYPOLITICS

> Look sharp, or they'll pull wool over your eyes.
> —*Spirit of the Times* (September 29, 1842)

Politics is social ethics: the application of a system of moral codes to the organization of society. The essential question in political philosophy is, What is the proper role of the state? I suggest that the evolution of the Bodypolitic is a natural process of development from massification to individuation—from control by the head of the Bodypolitic, to ever-increasing self-control by the individual.

Past generations feared that the new technology would lead to mass unemployment by replacing jobs. But as usual, the fears of the Luddites proved unfounded. In fact, new technology has *increased* employment by creating a myriad of new jobs, from the humble call center to the highest of high-tech IT. By increasing employment, productivity, and prosperity, new technology results in ever-greater freedom from state control, and ever more "power to the people"—one by one.

We should encourage the process. For if wealth is power, poverty is impotence; if happiness is freedom to be oneself, misery is slavery to the diktats of the state. I sum up the idea of political evolution from centrally imposed groupthink to independent-minded self-rule as a transition **from Bodypolitic to Neuropolitic**.

Where the Bodypolitic has always been implicitly a symbol of authoritarian rule—the *body* of men and women controlled by the *head* of the state or corporation—the *Neuropolitic* is explicitly a symbol of independence; society seen as a **metamind** emerging automatically from the voluntary cooperation between autonomous, self-governing minds.

The human mind is a neural net of individual neurons, collaborating to produce the phenomenon of consciousness, or self-recognition. Where the Bodypolitic acts *un*consciously, behavior having been conditioned by the head of state or corporation—the Neuropolitic is self-aware, self-conscious—"sussed"; each member of the Neuropolitic has *nous*—Greek for "mind," "intelligence," "common sense," or "gumption."

In the Bodypolitic, the people operate en masse, as unconsciously as an arm or a leg to the mind which controls them, while the Defenders of the Meme are safely ensconced at the *head* of the Body, more often than not, invisible to the body they control. In the Neuropolitic, the people "know the score"; they know what's going on—they have "savoir faire," or "savvy"; you just can't "pull the wool over their eyes."

Thus, where the Bodypolitic is synonymous with coercion, repression, subjugation, dependency, totalitarianism, absolutism, tyranny, autocracy, dictatorship, and war—the Neuropolitic is synonymous with independence, self-government, self-rule, self-determination, liberty, freedom, and peace.

To employ the body as a symbol of society necessarily implies centralized control by the head of state or corporate body. In the language of memetics (which I predict will come to be the dominant social science of the future), "Bodypolitics is meme control through meme washing by defenders of the meme." In other words, authoritarian politics is the control of attitudes and behavior through brainwashing by those seeking to impose their worldview on society.

Bodypolitics is **memetic determinism**—the control of society by Defenders of the Meme—their dictum: "Whatever the dominant meme, thus shall be the actions of the people." *For he who holds the power, is he who controls the meme.*

The Philosophical Bodypolitic

The Bodypolitic is a metaphor, or figure of speech, in which a society (or group within society) acting with a single will is compared to the organs of a human body acting for the good of the body as a whole. The Oxford English Dictionary defines the Bodypolitic thus: "An artificial person created by legal authority for certain ends; a corporation. . . . Always with (the) defining adjective *body corporate, body politic*."[1] The word "corporate" means "embodied." Thus, the concept of the "body corporate," or corporation, is synonymous with the body politic. The comparison of the public body—or state— to an individual human body has a long history.

Plato

Plato includes the analogy of the Bodypolitic several times in *The Republic*[2]—his attempt to describe the ideal state. For Plato, the Bodypolitic symbolizes the need for (as we would say today) homeostatic balance between the various parts of society, as between the parts of a human body:

> And, as in a body which is diseased the addition of a touch from without may bring on illness, and sometimes even when there is no external provocation a commotion may arise within—in the same way wherever there is weakness in the State there is also likely to be illness, of which the occasion may be very slight, the one party introducing from without their oligarchical, the other their democratical allies, and then the State falls sick, and is at war with herself; and may be at times distracted, even when there is no external cause. (Book 8)

In the language of memetics, or **memespeak**, just as a human body falls ill through infection by a virus, so the insemination of **rogue memes** into the dominant metameme of the state, whether "oligarchal" (by the few) or "democratical" (by the many), can lead to sickness in society. Plato continues with the analogy of the sculptor:

> Why do you not put the most beautiful colors on the most beautiful parts of the body—the eyes ought to be purple, but you have made them black?—to him we might fairly answer, Sir, you would not surely have us beautify the eyes to such a degree that they are no longer eyes; consider rather whether, by giving this and the other features their due proportion, we make the whole beautiful. . . . And therefore we must consider whether in appointing our guardians we would look to their greatest happiness individually, or whether this principle of happiness does not rather reside in the State as a whole. But if the latter be the truth, then the guardians and auxiliaries, and all others equally with them, must be compelled or induced to do their own work in the best way. And thus the whole State will grow up in a noble order, and the several classes will receive the proportion of happiness which nature assigns to them. (Book 4)

In other words, just as a sculptor does not make the eyes of a statue more beautiful than any other part of the body, but rather, strives to attain a harmonious balance of the whole, so, too, happiness resides not in the life of the individual body, but in the body of the state, or Bodypolitic. Therefore, no individual should seek happiness above his station.

Thus, Plato's ideal was a classically authoritarian Bodypolitic, in which the happiness of "the State," imagined to be a living entity, was of greater importance than the happiness of the individual. How many, reading that sentence, do not receive a certain shiver of recognition at the obvious association with the belief of a certain dictator in the subordination of the individual to the state, empire, or—"Reich."

Aristotle

Where the idealist Plato delights in the poetry of metaphor, his successor, the pragmatic Aristotle, is typically more blunt. In his treatise on *Politics*, the authoritarian implications of the Bodypolitic analogy are made explicit; "Thus the State is by nature clearly prior to the family and the individual . . . for example, if the whole body be destroyed, there will be no foot or hand . . ."[3]

Civitas, Republica, Commonwealth, State

Immediately, then, the authority of the Bodypolitic over the individual is clearly established in Western thought. The authoritarian Bodypolitic is implicit in concepts such as the Roman *civitas* (civic state) and *res republica* (literally "public thing"); in the English term *commonwealth* (from "common weal," meaning "good," as in "weal and woe"); and in the concept of the state defined in the Oxford English Dictionary as "The body politic as organized for supreme civil rule and government."[4]

Hobbes

The seventeenth-century English philosopher Thomas Hobbes famously compared the Bodypolitic to the biblical sea monster Leviathan, to whom individuals willingly surrendered their freedom in order to avoid a life that is "Solitary, poor, nasty, brutish, and short": "For by art is created that great Leviathan, called a commonwealth, or State, in Latin civitas, which is but an artificial man. . . . The multitude so united in one person, is called a Commonwealth. . . . This is the generation of that great Leviathan, or rather . . . that mortal god, to which we owe under the immortal Gods, our peace and defence."[5]

Once again, the bodypolitical analogy is used to justify an essentially authoritarian politics. For "The multitude so united in one person" is the definition of a dictatorship, whether benevolent—or not ("Ein Reich, ein Volk, ein Fuehrer").

Bodypolitics as Political Holism

All these concepts are based on the view that a collective body of individuals acting together constitutes an entity in itself—something greater than the sum of its parts. In short, *Bodypolitics is political holism*—the belief that collective activity is intrinsically superior to that of individual activity. So Beethoven's Ninth Symphony would have been much improved had he asked for some help with the melody.

The weakness of Bodypolitics is the weakness of holism. For if Bodypolitics is rule by the general will, this begs the question, Who exactly is it that *determines* the general will? The idea that every single member of society may be persuaded to act in perfect agreement on all issues at all times occurs only in the power fantasy of the "mad dictator." In short, holism ignores the parts for the whole—the individual for the state.

By Bodypolitics, then, I mean a society, or group within a society, organized according to the principle that its leader ("head of the Bodypolitic") is expressing the general will of all, when in *fact* determining rules of behavior for him- or herself.

"Will" is the ability of the mind to determine behavior—the control of the body by the mind. Bodypolitics assumes that everyone in society shares a single will—a single mind, or meme map—such that whoever assumes the role of dictating the general will of the Bodypolitic is justified in doing so, because he (or she) is acting on behalf of the wishes—or will—of each and every citizen. But the idea of the collective subordination of all to the rules of a head of state deemed to be automatically acting on behalf of the general will of the people as a whole is complete anathema to the modern world, in which democracy is universally regarded as preferable to the authoritarian control over individual behavior, whether it be imposed by Roman Imperialism or British Imperialism; Italian Fascism or Italian Catholicism (the word "catholic" being derived from the Greek "oleos" meaning "whole"); German National Socialism, or Russian *inter*national socialism—that is, Communism.

Bodypolitics as Memetic Determinism

A conscious human being is not ruled by the body, but by the mind. When the body rules the mind, we call it "lacking self-control," or worse, "out of control," and implore the person concerned to "pull yourself together"—that is, to willfully coordinate the activity of the mind; to stop being, literally, "scatterbrained."

Of course, the Dionysians among us will reply, "But I value my physical and emotional impulses; they make life worth living." Very well, but only the fool, the adolescent, the hippie, or the foolish adolescent hippie believe you can organize a *society* on the basis of emotional impulses instead of reason. They tried it in the sixties. As Camille Paglia has written, it ended not in heaven on earth, but in murder by Hell's Angels at the Rolling Stones concert in Altamont, December 1969,[6] which suitably signaled the end of the Aquarian dream—the dream of feeling unregulated by thinking; love untainted by logic; passion untamed by reason; compassion unchallenged by common sense.

So where organized behavior is concerned, the body must be ruled by the mind. Therefore, to follow the metaphor through, a Bodypolitic is not ruled by the people on their collective behalf—by the organs and limbs of the Bodypolitic—but by its mind—that is, by the *head* of state—he (or she) who is afforded the highest *status*. Ergo, Bodypolitics does not represent equality and democracy—the rule of all, or the many—but rather, oligarchy or monarchy—rule from above by the few, or the One. In short, Bodypolitics is authoritarian absolutism—pressure from the top.

Societies based on the Bodypolitic are inherently authoritarian because the body is ultimately ruled by the mind. An arm or a leg do not act of their own accord, but according to the dictates of the mind. When they don't—when an arm or a leg acts on its own impulses—it is because the brain lacks dopamine, the neurochemical correlate of will (the power of the mind to determine the behavior of the body). It's called Parkinson's disease. My mother suffered from it for over thirty years. Her limbs would either jerk around violently of their own accord, or stay immobile when she (her mind) told them to move.

In a disease-free body, the mind controls the actions of the limbs, not the other way around. Hence the message of Jesus, "If thy right hand offend thee, cut it off" (Matt. 5:30). Jesus is not advocating amputation as a punishment for carnal desires. He's telling us we should be responsible for our *own* actions, because we're able to control the behavior of our bodies, with our *conscious minds*. In other words, it's the advocacy of free will. (Interestingly, the movement of the right hand is dominated by the left side of the brain, which controls language, self-consciousness, and reason.)

By contrast, the barbaric amputation of limbs we see in certain theo-Fascist Bodypolitics is punishment for the failure to act according to the *general* will of the *Bodypolitic*—laid down by the Defenders of the Meme. Jesus notoriously rejected the idea of political revolution, a policy summed up in the maxim, "Give unto Caesar that which is Caesar's" (Matt. 22:20). Hence, where for Jesus, benevolent morality was a question of individual conscience, for a theo-Fascist dictator, it's a matter of law. Jesus speaks of amputation as a moral metaphor—theo-Fascist dictators, as a political policy. Here, then, Jesus represents the values of Neuropolitics—the belief in personal responsibility for one's own actions—while a theo-Fascist dictator represents Bodypolitics—the belief in forcing others to do, and think, what they're told, by law.

The Religious Bodypolitic

Jesus sometimes endorses the values of an independent-minded Neuropolitic, as opposed to authoritarian Bodypolitic—and sometimes not—at least, not here in the Gospel according to John, where he talks about himself a lot: "I am the bread of life . . . I am the light of the world . . . I am the door . . . I am the good shepherd." Which of course begs the question, if Jesus is a shepherd, what does that make the people? (John 6:35, 8:12, 10:9, 10:11).

His followers were not so ambivalent. The absolutism of the *religious* Bodypolitic is illustrated in the Christian concept of *corporis Christi*, or "the mystical body of Christ," a concept derived from the account of the Last Supper, in which Jesus offered bread and wine to his disciples on the eve of the crucifixion, as symbols of his body and blood. Saint Paul's first epistle to the Corinthians makes the Bodypolitical analogy explicit: "For as the body is one, and hath many members . . . so also is Christ. . . . And the eye cannot say unto the hand, I have no need of thee: nor again the head to the feet, I have no need of you. . . . Now ye are the body of Christ" (1 Cor. 12:12).

Thus, in the central rite of Christianity (the Eucharist, Holy Communion, or Lord's Supper), the consuming of bread and wine symbolizes not only the sacrificial death and resurrection of Jesus, and the rather dubious concept of the value of suffering, but crucially, the unity of the church, through the infusion of its members with the "holy spirit." By consuming the blood and body of Christ, they become one with him. Thomas More (the English chancellor beheaded by Henry VIII for refusing to accept him as head of the church—instead of the pope) explains succinctly, "He doth incorporate Christian folk and his own body together in one corporation mystical."[7]

Thus, in Christian Bodypolitics, the ultimate act of obedience to the head of the Bodypolitic is one of symbolic cannibalism. As I eat you, I become one with you. Anthropologists of the occult call it "magical thinking." In the language of memetics as social science, to be "filled with the holy spirit" is to be "inseminated by the Defenders of the Meme with the theistic metameme, for the purpose of attaining meme control through meme washing, thus maintaining control of the dominant metameme, in a continual Meme War against other, now recessive, but potentially dominant metamemes."

A mystical union of all who inhabit the Bodypolitic is a rather convenient means of ensuring that the Defenders of the Meme remain in complete authority over the meme map of the Bodypolitic. Hence, in the influential sixteenth-century law dictionary, *Termes de la Lay*, John Rastell tells us that "Bodies Politique are Bishops, Abbots, Priors, Deanes, Parsons of churches, and such like, which have succession in one person only."[8] The Church was indeed the "*corpus mysticum et politicum*"—religion and state united under an authoritarian *Body*politic.

The analogy of the Bodypolitic, like the meme of the Holy Spirit, is a Weapon of Memetic Destruction employed to maintain control of the metameme. The propagation of the meme is vital for any budding memetician who seeks to impose his own meme map on the meme pool, for *he who holds the power is he who controls the meme*: "He that receives the seed into the good ground is he that hears the word and understands it; which also bears fruit, and brings forth some hundred-fold . . ." (Matt. 13:23).

So let us be sure to sow the seeds of love, not war—freedom, not tyranny.

The Royal Bodypolitic

An early mention of the Bodypolitical analogy occurs in Shakespeare's description of the English monarchy in *Henry VIII*: "This realm of England is an Empire . . . governed by one supreme Head and King . . . unto whom a Body politick . . . is bound to bear a natural and humble obedience."[9] Thus, an absolute monarchy unregulated by parliamentary democracy was known as a Bodypolitic. The monarch symbolized or *embodied* the state as a whole, so justifying his absolute power.

Of course, the Bodypolitical absolutism was not confined to the English monarchy. It's most famous and pithy description is to be found in those three, chilling little words of Louis XIV (ruler of France during the late seventeenth and early eighteenth centuries): "L'etat, c'est moi." The State? That's me!

The attempt to justify absolute monarchism was based on the medieval concept of "the divine right of kings," by which the authority of command over the populace was deemed to have been granted by God himself. The eighteenth-century English poet Alexander Pope wittily summed up this rather convenient justification for the totalitarianism of the absolutist Bodypolitic as "The Right Divine of kings to govern wrong."[10]

The Imperialist Bodypolitic

The term *imperialism* derives from the Latin *imperium*, meaning "empire"; *imperare*, to "command"; and *imperator*, "emperor." Imperialism is the attempt to expand the Bodypolitic beyond its natural borders by force. A suitable analogy, then, is the unfeasibly obese man who persists in gluttony, despite all warnings as to the effect on his health. The collapse of an empire, or Bodypolitic, results from the inability of the centralized state—the mind of the Bodypolitic—to control its colonial territories—or "spare ribs." In other words, empires fall when they take on more than they can chew.

Roman Imperialism

The absolute Bodypolitics of the Roman Empire is suitably signified by its official insignia, the *fasces*: an axe head imbedded in a bundle of twigs—so binding them together by force (one recalls the etymology of the word "religion": *ligare*, "to bind"). The Roman system was not known for its great liberality, despite the odd placating nod to democracy such as "bread and circuses" for all: "(Caesar) Augustus never went so far as to restore genuinely free elections or the organs of popular government. He kept the crowd happy with chariot races, gladiatorial contests, and the dole of bread."[11]

British Imperialism

British Imperialism suffered from the same problem: it got too big for its boots, until the feet rebelled from the mind and took a different route—independence. Thus, the British Empire became the British Commonwealth—voluntary cooperation replacing coercion. The Commonwealth is one example of a *non*authoritarian Bodypolitic, in which the controlling head or brain is content to let the various parts of the body act of their own accord, through the automatism of instinct, not the authoritarianism of absolute control. However, a Bodypolitic lacking a brain is unlikely to be a great deal of practical use.

The Fascist Bodypolitic

Italian Fascism

Named after the symbol of Roman authority, Fascism was Mussolini's attempt to repeat the absolutism of the Roman Empire in twentieth-century Italy. The organizational foundation of Fascism was the concept of corporatism, or the corporate state. The word "corporate" means "embodied." "Corporation" originally described what we today would call "physical constitution." Hence, this rather charming quote from the eighteenth-century English satirical novelist Tobias Smollett, from his novel of 1753, the delightfully entitled *The Adventures of Ferdinand, Count Fathom*: "Sirrah! My corporation is made up of good wholesome English fat."[12]

Thus, the term *corporation* means the physical body—and Italian, Fascist corporatism meant the submission of the individual body to the body of the state, as Mussolini made crystal clear: "The keystone of the Fascist doctrine is the conception of the State, of its essence, its functions, and its aims. For Fascism the State is absolute, individuals and groups relative."[13] Hence, in 1929 the *Times Literary Supplement* (of July 11) spoke of "The Fascist corporate State";[14] and by 1935 the *Economist* magazine was informing its British readers: "The Italian newspapers announce that 'Corporatism' has become a reality."[15] In short, the political corporatism of Italian Fascism is another example of absolutist Bodypolitics—the body of the people, controlled by the head of the state.

German Nazism

Inspired by Mussolini's neo-Roman totalitarianism, Hitler's National Socialism was the attempt to establish a new German Empire on the lines of previous models (the Holy Roman Empire and the Prussian State of Frederick II) through an expansionistic, militaristic dictatorship—"the Third Reich." Hitler's emphasis on the need for the complete submission of the individual to the Reich, and the defense of German "blood and soil"—"a code word implying the protection of a real personality" according to the author of a study on twentieth-century Ecoism[16]—reveals Nazism to be another example of authoritarianism founded on Bodypolitics—the reduction of the individual to the level of a limb being controlled by the mind of another, as opposed to an independent, autonomous, freethinking human being. Bodypolitics is **Marionette Politics**.

The Communist Bodypolitic

Communism

The Union of Soviet Socialist Republics (USSR) was based on a system of locally elected councils (or "soviets") acting under the authority of a permanent, one-party central state. In Communism, the Roman-Fascist symbol of axe through twigs, representing the authority of the emperor, was replaced by the symbol of the hammer and sickle, representing the authority of the workers through control of industry and agriculture.

One suspects that the word "sickle" derives from the ancient Hebrew word "shekel"—first a unit of weight, then of economic currency—due to the obvious causal link between the weight of crops (harvested with a sickle) and economic prosperity. Thus, the symbol of hammer and sickle also represents the idea of power attained through *economic* control, by controlling "the means of production and distribution."

The moral justification for the Communist state lay in the idea that local councils (or soviets) were elected democratically, while a powerful central state was required to regulate the Bodypolitic as a whole. In reality, of course, it didn't quite work out like that, as this quote from the *Times Literary Supplement* of 1930 reveals: "The chairman of the village soviet may in theory be master in his own limited sphere; in practice he is the servant of a Communist 'cell.'"[17]

The word "cell," here, is notable. The members of the local soviet council resembled not so much *brain* cells in a conscious mind (neurons in the Neuropolitic, determining behavior collaboratively through rational argument based on Sensible Self-Interest) but *infected blood* cells, in a body that has ceased to grow, ruled by the puritanical mind of a central state that cares not for the body, only for its own survival.

Interesting that the authoritarian Plato, who's ideal Republic was a benevolent dictatorship ruled by "philosopher kings" (a pleasant fantasy indeed), was a despiser of the body in preference for the "soul" (that is, the psyche, or mind). How much of the instinct for the coercive propagation of one's memes in the meme pool is in reality a sublimation of the failure to succeed in the *genetic* equivalent is open to question. The Freudian psychologist Wilhelm Reich certainly thought he'd discovered the cause of authoritarian instincts in sexual repression, and advocated something called *sexpol* as the cure—orgies, in other words. Mind you, he was also something of a crackpot, who thought he'd discovered the secret life force of the universe, and captured it in a big black box.

"Pure Communism," as its remaining followers are quick to inform us, has "never been tried." Rather, the Russian, Chinese, and other models are examples of absolute State Socialism, the realizable "precursor" to the "stateless State" predicted by Marx, after the inevitable "withering away" of the desire by Defenders of the Meme to hold on to power over the Bodypolitic. Unfortunately, this dream neglects one small truism, expressed rather well in a famous maxim on the human condition: "Power corrupts, absolute power corrupts absolutely."

Plato describes his ideal republic as a "whole organism which binds body and soul together into the unitary system managed by the ruling part of it." A few paragraphs later in his train of thought, we get the logical conclusion to the Bodypolitical analogy: "the greatest good for our city (is) the common ownership of wives and children . . . they must have no private houses or land or any private possessions."[18]

In the Bodypolitic, no man is an island, because "everyone belongs to everyone else"—as Aldous Huxley described life in the state dictatorship of the *Brave New World*.[19]

Corporate Socialism

Modern corporate, or municipal, socialism reflects the Russian model of soviets—locally elected district councils supposedly acting on behalf of the citizens. However, how many of us have never felt a certain twinge of exasperation in the suspicion that our communities are being controlled by petty council bureaucrats who serve their own ideological interests, instead of responding to social concerns raised by local people? Charles Dickens, writing here in a letter of 1866, expresses a feeling most of us know too well today: "I sat pining under the imbecility of constitutional and corporational idiocy."[20]

In corporate socialism, the authoritarian Bodypolitic rears its power-hungry head once again.

The Socialist Bodypolitic

Socialist Bodypolitics begins with *The Social Contract*, the famous treatise by the eighteenth-century philosopher Jean-Jacques Rousseau. In this quote, Rousseau directly employs the term *Bodypolitic*, to describe his belief in the transference of power away from the individual to the collective or "general will" of the people: "In place of the individual person . . . this act of association creates an artificial and collective body . . . that body acquires its unity, its common ego, its life and its will. The public person thus formed by the union of all other persons . . . is now known as the . . . Bodypolitic."[21]

Rousseau spoke in glowing terms of the joys of abnegating one's will to that of the Bodypolitic:

> The passing from the state of nature to the civil society produces a remarkable change in man; it puts justice as a rule of conduct in the place of instinct, and

> gives his actions the moral quality they previously lacked. It is only then, when the voice of duty has taken the place of physical impulse and right that of desire, that man, who has hitherto thought only of himself, finds himself compelled to act on other principles, and to consult his reason rather than study his inclinations. . . . His faculties are so exercised and developed, his mind is so enlarged, his sentiments so en-nobled, and his whole spirit so elevated that . . . he should constantly bless the happy hour that lifted him for ever from the state of nature and from a narrow, stupid animal made a creature of intelligence and a man.[22]

On the surface, Rousseau's sentiments appear to be in accordance with the transition advocated here, "from Genethics to Nurethics"; that is, from control by the selfish genes through instinct, to self-rule by the conscious mind through reason. However, for Rousseau, control by the rational mind did not mean self-rule by the *individual*, but rule by the *collective* mind or "general will" of the people—determined by the state.

Rousseau believed that human beings were born naturally good, but required a benevolent state to *bring out* their goodness. He did not advocate that we abandon society and live in the woods, as the above quote clearly demonstrates. He saw the state as the *embodiment* of human goodness—the means by which human beings "born free" could expresses their true humanity, instead of descending into the barbarism of animal instinct. His mistake was that of all socialists—the failure to recognize, with Aristotle, the potential tyranny of a "benevolent state."

The underlying reason why an initially benevolent, democratic government with a large majority tends to end up as an authoritarian Bodypolitic controlled by a dictator, is the failure of Nurethics to transcend Genethics—that is, the failure of the rational mind to replace rule by the selfish genes. Evolved animal instincts impel the potential leader to seek as far as possible the primacy of the "alpha male," in order to minimize threats to survival from others within the tribe. In the euphoria of power, rationality tends to "wither away," animal instincts overriding Sensible Self-Interest and compelling the head of the Bodypolitic to seek total domination, as head of the tribe, leader of the pack, lord of the jungle—king of the hill.

The Postmodern Bodypolitic

The flaw in the ideology of state socialism lies in the reduction of freedom resulting from centralized control of society. Simply put, the greater the enforcement of the "general will" by the state, the lesser the freedom of the individual. *Absolute* control by the "general will" means the tyranny of Fascism or Communism (that is, national or international socialism).

After the Second World War, the Postmodern Left came to adopt this critique of Rousseauian socialism, summed up by the blunt assertion of the philosopher Theodore Adorno, "Enlightenment is tyranny." In other words, the concept of progress through the collective will of the people, whether imposed by Right or Left, fascists or communists, always leads to the same thing—totalitarianism—the absolutism of the Bodypolitic.

The "dictatorship of the proletariat" on the extreme left is little better than the dictatorship of the pseudo-uberman on the extreme right. But the radical solution proposed by the Postmodern Left—the abandonment of totalized belief systems altogether—was no solution at all. As I have suggested, the effect of this extreme knee-jerk reaction—the advocacy of absolute relativism—is the cultural descent into nihilism and demoralization through the absence of belief in anything whatsoever.

In fact, the Postmodern Left don't believe in nothing whatsoever; they believe in what might be called Cultural Communism—the pursuit of equality not in the economic but in the cultural sphere, through a policy of absolute ethical and aesthetic relativism. This is the cause of that most distressing of contemporary phenomena, "dumbing down"—the ideological attempt to impose cultural equality by lowering intellectual and aesthetic standards to appeal to the least able or educated. Thus, high- and low-brow newspapers merge into one; exams become evermore easy to pass; arts and entertainment become "Dumb and Dumber."

The cultural Communism of the Postmodern Left is the latest in a long line of authoritarian Bodypolitics in which ideology (the metameme) is imposed from the top in the name of "the people." From "bread and circuses" to Big Brother, the message is always the same—"Let them eat cake."

The Democratic Bodypolitic

When the Postmodern Left stress the virtues of "democracy," they use the word as a euphemism for socialism, to mean "rule by the majority, for the benefit of the disadvantaged." This may seem like a jolly good idea. However, majority rule on behalf of the many, or the few, instead of on behalf of *everyone*, is no panacea for all ills. For democracy can mean the *tyranny* of the majority, as Aristotle pointed out: "Democracies . . . may be founded on violence, and then the acts of the democracy will be neither more nor less acts of the State . . . than those of an oligarchy or of a tyranny."[23]

In fact, to Aristotle, democracy meant precisely what it means to the socialist: rule by the poor ("Where the poor rule, that is a democracy").[24] But the tyranny of the poor is no better than the tyranny of the rich. If a man is oppressed by his neighbor, he cares not one jot if his neighbor is rich or poor. Thus, Aristotle described democracy as "one of the perversions": "The perversions are as follows: of royalty, tyranny; of aristocracy, oligarchy; of constitutional government, democracy. For tyranny is a kind of monarchy which has in view the interest of the monarch only; oligarchy has in view the interest of the wealthy; democracy, of the needy: none of them the common good of all."[25]

Aristotle's assertion that democracy is the tyranny of the needy, not the common interests of all, may be shocking to many people today, so thoroughly has the democratic meme been disseminated in the meme pool of the modern world. But the sentiment behind Aristotle's critique was the same as that behind that famous quote by Winston Churchill, spoken in Parliament after the Second World War:

> Many forms of Government have been tried, and will be tried in this world of sin and woe. No one pretends that democracy is perfect or all-wise. Indeed, it has been said that democracy is the worst form of government except all those other forms that have been tried from time to time.[26]

Perhaps a democratic "tyranny of the needy" might have prevented Churchill from persuading his country to embark on the perilous journey to defeat Nazism, on the grounds that war served to detract economic resources from the needs of the poor.

In short, it is a mistake to think that the majority is always right by virtue of greater numbers. The majority supported Hitler and Mussolini. Instead of worshiping the will of the majority, we should begin to think about the autonomous will of the individual, governed by the benevolent morality of Sensible Self-Interest. The Neuropolitics I propose provides an alternative to the tyranny of the powerful or of the poor, which may result in the prosperity of both.

The Conservative Bodypolitic

Of course, authoritarian Bodypolitics is not the exclusive preserve of the Left.

"Where there is discord, let us bring harmony." Thus spoke the British prime minister Margaret Thatcher, quoting Francis of Assisi on the doorstep of 10 Downing Street on the day of her election in 1979. A few years down the line, and the country was deeply divided—miners versus state, people versus "poll tax," yuppies versus hippies, defenders of the bomb versus protesters for peace.

No doubt the prime minister was sincere in her words. But power corrupts . . . the velveteen debutante quickly became the Iron Lady. Whatever the arguments as to the long-term benefits of her policies, her abrasive style caused severe division in the Bodypolitic, which made life in the eighties sometimes seem like a battleground.

A society is indeed like a body—a body controlled by a brain. The body is an automatic homeostatic regulation mechanism—it naturally seeks to restore order, balance, or harmony within; the immune system automatically seeks to heal itself from the *dis*order of disease. The *good* leader is like a brain that follows the natural inclinations of the body, by seeking to maintain order and balance within. The good leader does not act *against* the Bodypolitic, but in accordance with its innate impulse for self-regulation, by seeking to maintain cohesion or stability in society. The *bad* leader is one who seeks to rend the cohesion of the Bodypolitic through force—the mind acting *against* the wishes of the body—in his (or her) *own* interests.

The fault of Bodypolitics lies in the tendency of the head of the Bodypolitic to convert to the puritan religion, by despising the body it was elected to serve. "With my body I thee worship" goes the wedding vow. But the head of the authoritarian Bodypolitic does not *worship* the body—it seeks to control it, and duly severs the harmonious union between the two by its actions. For the head of the Bodypolitic, head and body are in a state of perpetual war, requiring the domination of one over the other. By contrast, in a Neuropolitic, each individual member of society is an autonomous, thinking agent, free to determine behavior for itself. The minds of the Neuropolitic are not so Stupidly Selfish as to act against the interests of the body as a whole, for they recognize the essential interdependence of self and society, mind and body. They seek to control the body only when it threatens to follow the dictates of the Stupidly Selfish leader who leads it toward self-destruction, just as the individual mind, following a program of Nurethics, seeks to override the program of the Selfish Genes which command its body to wither and die.

The political propaganda of the dictator in a Bodypolitic is the memetic equivalent of gang rape—the forceful insemination of selfish memes into the brains of an unthinking Bodypolitic. Just as soldiers who lose self-control under the intolerable stresses of war sometimes resort to gang rape to retain a feeling of power, so the political head of the Bodypolitic, when faced with the intolerable stresses of power, resorts to the forceful insemination of memes into the meme pool, saying, with every political spin of the meme, "I know you want it, really."

* * *

NEUROPOLITICS

FROM BODYPOLITIC TO NEUROPOLITIC, MEMETIC DETERMINISM TO FREE WILL

Socrates described the ideal politician as a "physician of the bodypolitic"—and decried the failure of what we now call "spin doctors" to live up to the task! Politicians in an authoritarian Bodypolitic are like Molierian doctors who ignore the power of the body to heal itself, and insist on administering quack remedies against the patient's will, which only make him worse. The English were recently agog at the habits of "Doctor Death" Harold Shipman, probably the world's worst serial killer, who put other dodgy doctors in the shade with his penchant for murdering elderly patients by morphine injection. To double (but not mix) metaphors, the administration of the political spin doctor's quack medicine into the life blood of the Bodypolitic signifies the dissemination of negamemes, or rogue memes, into the meme pool of the meme culture, as deliberate Weapons of Memetic Destruction in a Meme War for control of the Metameme.

By contrast, in Neuropolitics, society is regarded as a *self-regulating* or homeostatic organism, which maintains its own inner organization or "health" without the need for authoritarian control by an external force. In the Neuropolitic, the lawgiver is not the state but the individual citizen, through autonomous self-rule guided by Nurethics—the benevolent morality of Sensible Self-Interest. A Neuropolitic is a *brain* as opposed to *body* corporate—a neural network of individual wills to evolve. A multiplicity of logical, independent minds each thinking for itself, each acting out of Sensible Self-Interest for the good of both self and society as a whole, results naturally in the emergence of a Self-Enhancement Society based on Reciprocal eugoics—the mutual, freely tendered encouragement of personal growth. Domination by a centralized state is neither required nor desired. The transhumanist politics I espouse rejects the authoritarian Bodypolitics of Roman Imperialism, British Colonialism, Italian Fascism, German Nazism, Russian Communism, Corporate Socialism, American Imperialism, and PC Postmodernism for the twenty-first century politics of individual self-rule—Neuropolitics.

In Neuropolitics the Corporate Bodypolitic ruled by the state in the interests of the "general will" is replaced by a neural network of freethinking, freely collaborating "wills to evolve"; a **Neuropolitic** of individual minds united by the shared pursuit of continual self-development—the Self-Enhancement Society—or **Neurosociety**.

How we picture our society and species matters. The fifteenth-century German poet Sebastian Brant satirized human beings as seamen on a "Ship of Fools" (Das Narrenschiff) bound for "Narragonia"—the "fool's paradise." By replacing the Bodypolitic with the Neuropolitic, Bodypolitics with **cyberpolitics**, we may transform this ship of fools into a cybernetic or self-steering Ship of the Wise—a cybership, bound not for a fool's paradise, but for cyberia—a land in which human beings are masters of their fate, captains of their soul—*steersmen of their own destiny.*

Neuropolitics as the Freedom to Be Yourself

Of course, the Neuropolitic is not a perfect metaphor. Individual neurons, or brain cells, are not imbued with individual self-consciousness, or rationality.

That's not the point. The point is to distinguish between a society in which the attitudes and behavior of the majority are determined from the top, by the head of state, party, media, church, or corporation (that is, a Bodypolitic), and a society composed of individuals who think for themselves, and so determine their *own* behavior, without the need or desire to perpetually be told what to think or do (that is, a Neuropolitic).

The autonomic nervous system automatically maintains a state of homeostatic balance by allowing the free activity of individual parts within limits required for its harmonious functioning as a whole. In the Neuropolitic, self-conscious human brains—minds—work together out of Sensible Self-Interest, to maintain a state of *social homeostasis* balance between extreme states in the Neuropolitic.

The vision of society as a **metamind**—a multiplicity of minds each thinking for themselves—is, I suggest, preferable to the vision of society as an unthinking, unconscious body controlled from on high by an autocratic brain. Neuropolitics is the politics of individual autonomy, freedom, self-determination, independence, self-rule—in short, the freedom to be your*self.*

The belief in sociopolitical evolution from Bodypolitics to Neuropolitics mirrors the psychological and ethical belief in the evolution of consciousness from Genethic to Nurethic control—from the Bodypolitics of Eros and Thanatos, to the Neuropolitics of Prometheus and Orpheus.

Beyond the Fulfillment Society

> Happiness depends upon being able to develop to the limit of one's abilities.
> —Peter Medawar, *The Strange Case of the Spotted Mice* (1996)

Every political philosophy has its vision of an ideal society. The biologist Julian Huxley, grandson of the eminent Victorian Darwinist T. H. Huxley, advocated what he called a "Fullfillment Society" committed to the full development of human potential: "We should set about planning a Fulfillment Society, rather than a Welfare Society, or an Efficient Society, or a Power Society. Greater fulfillment can only come about by the realization of more of our potentialities."[27]

It was Huxley who was responsible for coining the term *Transhumanism* in his 1957 book, *New Bottles for New Wine*, as an alternative expression for his philosophy of "evolutionary humanism."

Huxley's philosophy was much like the psychology of his friend, humanistic psychologist Abraham Maslow, who described the attempt to satisfy a "hierarchy of needs" beginning with survival, food, and clothing and ending, if successful, in *self-actualization*—his term for the ideal state of well-being. Huxley and Maslow represent the populist postwar, humanistic psychology that dominated the latter half of the twentieth century, in which the "fulfillment of potential" was synonymous with happiness.

The idea of a Fulfillment Society is much in accordance with present-day transhumanism. However, strictly speaking, Huxley's term *Fulfillment Society* suffers from the same fault as Maslow's concept of "self-actualization"—for it implies an end state of satiety which is simply not possible, given the current biological limitations of the species. "Self-actualization" and "fulfillment of potential" fail to take into account the limitations of the human mind and body, which preclude the possibility of absolute satisfaction—at least at this point in human evolution.

Beyond Self-Actualization

Frankly, to the Positive Realist, as opposed to the naïve optimist, Maslow's concept of the "self-actualized" individual sounds a certain note of absurdity. He describes them thus: "Self-actualizing people, those who have come to a high level of maturation, have so much to teach us that sometimes they seem almost like a different breed of human beings."[28]

So Maslow's focus of interest lay in superior human beings, deemed superpeople by virtue of their ability to "actualize their potential." But what if one's potential is not up to much, because of an unlucky spin of the genetic dice? What is "actualization of potential" for the Opioid Neurotype incapable of inhibiting life-crippling feelings of anxiety and depression, or the Adrenaline Neurotype incapable of inhibiting aggressive impulses—an acceptance of neurosis and thuggery? You can't ascend your "hierarchy of needs" if your Neurotype holds you back.

Thus, the transhumanist ethic I espouse is simply the belief in continually striving to enhance one's mind and body throughout life, as opposed to a belief in the attainment of the sort of smug-superior state of pseudo-perfection that tends to ooze from the faces of narcissistic New Agers who've learned to love themselves—and can barely disguise their sense of superiority over those poor souls who've failed to follow them on the "Road Less Traveled."

Ultimately, humanistic psychology is an ideology derived from an over-idealized Enlightenment notion of the Perfectibility of Man. Failure to acknowledge present human limitations, weaknesses, and failings is a dangerous mistake which in the past led to the guillotine, gulag, and gas chamber, as the postmodernists have pointed out. However, the extreme knee-jerk response of the postmodernists to the danger of the belief in "perfectibility through politics"—to banish universal belief systems altogether—was not only absurd, but as authoritarian as the metamemes they sought to destroy. The postmodern "Road to Nowhere" is no better than the New Ager's "Road Less Traveled"—both lead up your own backside.

In the Self-Enhancement Society—or neurosociety I espouse—both the naïve optimism of social utopianism and the nihilistic cynicism of postmodern relativism are rejected for the balanced, Positive Realism of meliorism; the belief not in an end state of pseudo-perfection, or a nihilistic road to nowhere, but in ongoing, unending human progress through reason, science, and technology—the evolution of knowledge and abilities, for the benefit of individual and species alike.

Beyond the Self-Diminution Society

While the concept of the Self-Enhancement Society represents a rejection of the overidealization implicit in much humanistic psychology, it also seeks to counter the precise opposite trend in contemporary culture—toward a Self-*Diminution* Society. What is lacking in society today is mutual encouragement. Too often we put each other down when we should be building each other up.

The postmodern metameme of cultural relativism has led to the emergence of a Self-Diminution Society of mutual self-deprecation, in which the greatest virtue is a negative—*not* to do or say anything which could possibly be regarded as evidence that one considers one's own beliefs, values, or attitudes in any way preferable to those of anyone else. The sensible motto "Each to their own," taken to extremes in the postmodern age, has led to the imposition of an absolute cultural relativism by which the villain is no less revered than the hero—because they're both just "doing their thing."

The virulent spread of the antimodern metameme has led to the downgrading of self-worth and achievement through fear of looking superior to anyone else. In the postmodern anticulture, young people are actively being conditioned to hide their intelligence for fear of ridicule. Skeptical? Here is Theodore Dalrymple, writing about his experiences as a psychiatrist in the inner city of London. It is, I feel, worth quoting at length:

> It takes a special kind of perversity for students at the high school situated 400 yards from my hospital to say to one of their colleagues, who took an overdose because of the constant bullying to which she was subject: You're stupid because you're clever. What did they mean by this apparent paradox? They meant that anyone who tried hard at school and performed well was wasting his time, when he could have been engaged in the real business of life, such as truanting in the park or wandering downtown. Furthermore, there was menace in their words: If you don't mend your ways and join us, they were saying, we'll beat you up. This was no idle threat: I often meet people in their twenties and thirties in my hospital practice who gave up at school under such duress and subsequently realize that they have missed an opportunity which, had it been taken, would have changed the whole course of their lives much for the better. And those who attend the few schools in the city which maintain very high academic standards risk a beating if they venture to where the poor white stupids live. In the last year, I have treated two boys in the emergency room after such a beating, and two others who have taken overdoses for fear of receiving one at the hands of their neighbors.[29]

This is not a description of Victorian Britain, but the Britain of today, here and now, in the twenty-first century. One almost daren't contemplate the full, devastating consequences to modern civilization of the inverse snobbery of postmodernism which threatens to dumb down the West into an idiot culture of epsilonic infantilism—though Aldous Huxley tried: "Then came the famous British Museum massacre. Two thousand culture fans gassed with diclorethyl sulphide."[30]

In the Postmodern Anticulture of nihilism, the dumb are getting dumber, and the dumbest is the best of all.

Epsilonia

Postmodernism threatens to create a dumbed-down **Epsilonia**—a culture dominated by the attitudes of its least-educated members. In the Postmodern ethic, the values of modernity—reason, progress, science, intelligence, endeavor, the quest for knowledge—are ridiculed in favor of irony and "playfulness." In the words of the late postmodern philosopher Jean-François Lyotard, "Stupidity is instead something which we wish to claim for ourselves . . . being right is not important, singing and dancing is what matters."[31]

One doubts he'd get very far in computer science.

From Mutual Self-Deprecation to Reciprocal Eugoics

The transhumanist journey is neither the postmodernist's idle dawdle along a Road to Nowhere nor the therapist's narcissistic parade down the Road Less Traveled in the Emperor's New Clothes, but a noble adventure across the tightrope between human and transhuman. The beginning of the journey is a transition from a postmodern Self-Diminution Society based on mutual self-deprecation, to a Self-Enhancement Society based on the mutual encouragement of ever-increasing well-being. The transition has already begun. We already see evidence of belief in the Self-Enhancement Society emerging from the ashes of postmodern nihilism. No longer "satisfied with dissatisfaction," those dreaming of greater physical, emotional, intellectual, or creative fulfillment are now beginning to seek it without shame.

The desire to be all that one can be is not the product of selfish egotism but an inborn Will to Evolve. In the words of my father, the cybernetician J. F. Young, "The sensible, intelligent animal will promote the contentment of others. By doing so it will promote its own contentment." The Self-Enhancement Society is not a Stupidly Selfish society but an intelligent society which recognizes that benevolence is Sensible Self-Interest. In the Self-Enhancement Society, you encourage one another to be the best you can be, because you know it makes sense. The practice of mutually increasing well-being I call **Reciprocal eugoics**.

From mutual self-deprecation to Reciprocal eugoics—a multiplicity of individual "Wills to Evolve" committed to the mutual encouragement of personal growth. That is the vision of the Self-Enhancement Society I espouse.

Not "greed is good," but "growth is good"—that is the motto of the Self-Enhancement Society. And "personal growth" includes the enhancement of body and mind through Superbiology.

Bio-Libertarianism

Central to the Neuropolitics of self-enhancement is a concept I call **Bio-Libertarianism**.

Bio-Libertarianism is the belief in the freedom of individuals to seek biological self-improvement in any way they wish, provided that such action does not threaten other individuals, the society, or the species. Bio-Libertarianism is a political attitude for people who want to be the *best* they can be, and expect freedom from government interference to do as they wish with their own bodies and minds.

As with any other service in the modern world, bioenhancement above the level of treatment for ill health will be the financial responsibility of individuals—a service to purchase if they choose—or can afford. Consequently the wealthy will gain better initial access, thus attracting strong disapproval from the political Left, who believe that the prime function of an elected government is not the establishment and maintenance of a stable socioeconomic framework in which individuals can flourish according to their abilities and efforts, but the redistribution of income and power.

Just as not every human being can afford a facelift, breast enhancement, private yacht or jet, but only the hardworking, intelligent, naturally gifted, or fortunate, so, too, the benefits of Supergenic enhancement will, at least initially, be enjoyed most by the most privileged members of society. This is called living in the free world. The basic alternative to consumer-controlled bioenhancement is to either ban the technology outright out of spite for the better off or to hand complete control of it to the state, to distribute as it wishes. The former is akin to banning automobiles because some people can afford a Rolls Royce, the latter a recipe for tyranny, called eugenics. Ideologues, however, should note that aesthetic surgery, once the preserve only of the superrich, is now increasingly something generally affordable by those who may choose to spend their disposable income on improving their appearance, rather than on a holiday, new car, kitchen, or satellite TV.

As economic growth continues, through the benefits of a free-market economy, an ever-increasing number of people will be able to afford the advantages of bioenhancement, and Superbiology will gradually spread throughout society. There will still be rich and poor, success and failure, but the general direction will be one of leveling *up* instead of down, the result of the much-maligned "trickle-down" effect of free-market enterprise, based on the simple principle that the more affluent a society becomes, the better off its least affluent members.

As one wag put it in an Internet reply on the question of future **immortologies** (technologies to cure aging): "I hope the wealthy get to use it first, then we'll all have it when it becomes cheap—instead of state bureaucrats deciding who they think deserves it!"

Meritocracy against Groupthink

> It is individuals and not the society that "make" history.
> —Arnold Toynbee, *A Study of History* (1934–61)

A world without individualism is a world without innovation—a world of bland, drab conformity. The transhumanist politics I espouse is passionately opposed to the repression of individual freedom of thought and action, whether by state, religion, business, or media. Bio-Luddites like to equate the transhumanist agenda with a state-controlled Brave New World. But the transhumanist's *greatest nightmare* is a society of mindless automatons conditioned by groupthink. Freedom and individuality are the sources of the greatest human achievements. What links the likes of Leonardo, Shakespeare, Newton, Mozart, Beethoven, Darwin, and Einstein is their unique individuality. What sets genius apart is not socially enforced groupthink, but the kind of individuality capable of thriving only in societies which allow maximum individual liberty. We can't all be Einsteins, but we can at least establish societies which allow individuals the freedom to develop their capacities to the full.

In other words, the Neurosociety is a Meritocracy—a society which rewards according to individual merit, where merit equals abilities plus effort.

The endorsement of the Neuropolitics of individual freedom is consistent with my ethical advocacy of Nurethics—the deliberate control of behavior by the mind, rather than the blind instincts induced by the selfish genes. The more the mind controls behavior, the more individual a person becomes. The body is controlled by universal, evolved instincts, but the mind is relatively free to think and direct behavior as it wishes. Man increases his individuality in proportion to his ability to think for himself. It is the mind which enables individualized action.

In a colony of ants, without the freedom afforded by self-consciousness (the recognition of one's own existence) there can be no individuality, only collectivity. Collectivity without individuality is tyranny. In the ant colony, each member plays a narrow, limited, wholly determined role. There is no scope for change, creativity, individuality, *freedom*. Freedom is the autonomy of the thinking mind. The political equivalent of the ant colony is the kind of extreme leftism which takes the view that collective action and thought is always better than individual action or thought.

While social compassion is both morally good and practically necessary for mutual well-being, the goal of humanity should not be seen as some static state of universal material comfort and the absence of conflict—the totalitarian's dream—but one of continual expansion—the open-ended development of individual and collective abilities.

Social harmony occurs naturally in society when human beings of differing goals and talents are free to organize their behavior among themselves. This a fundamentally beneficent view of human beings, as opposed to the view that man is so wicked as to require the state to force him to be good against his own will.

The New Invisibile Hand

The view that social order will generate itself automatically is sometimes called Spontaneous Order, a theory espoused in the twentieth century by the Austrian economist Friedrich Von Hayek,[32] as a defense of free-market economies against the kind of centralized state planning that came to dominate postwar Europe in the twentieth century.

Hayek posited two types of rationality—practical, and evolutionary, relating to two types of order—planned and spontaneous. Central planning was appropriate for individual businesses, but no single business or government could predict or control the entire flow of resources in society, which was governed by a natural, complex process of Spontaneous Order, by which parts naturally came together as organized wholes through the action of mutual self-interest.

The concept of Spontaneous Order is popular among libertarian-minded transhumanists who base upon it their belief in individualism and free-market politics. Antistatism is essentially an instinctive distaste for coercion. Underlying the theory of Spontaneous Order is the intuition (that is, unconsciously reasoned belief) that in general, things and people tend to work best if left to operate by themselves in their own interests, rather than be forced into actions they do not wish to pursue.

As we shall see in my metaphysics, this insight is fully in harmony with the principles of the emerging science of Complexity—the attempt to identify the mathematical laws governing complex natural structures which display the property of self-organization—the automatic formation of ever more complex patterns, which maintain internal order while existing in a state of continual dynamic fluidity—a process of perpetual change and self-regulation—or Spontaneous Order.

Essentially, of course, the theory of Spontaneous Order is a neo-biologized version of the Enlightenment philosopher Adam Smith's notion of the "invisible hand," which guides individual self-interested action to the automatic benefit of society as a whole, without the need for centralized control. **The New Invisible Hand** is the hand of nature operating through the laws of Complexity.

In the politics of the New Invisible Hand, the free exchange of goods and services in society results in the spontaneous appearance of order in the Neuropolitic by the actions of autonomous, individual minds in the metamind—or Neuropolitic—acting out Sensible Self-Interest for the common good—just as the body *automatically* seeks to maintain internal stability through a process of homeostatic self-regulation, without the need of the conscious mind to control it.

The Neuropolitic thus has twin aspects—on one hand, it mirrors the relationship between the human mind and the self-regulating autonomic nervous system of the body. In this respect, the citizens of a Neuropolitic do not require a "head of state" to control completely the flow of resources through society, because the process unfolds automatically, just as the conscious mind allows the body to get on with the job of regulating basic functions. On the other hand, in a Neuropolitic, citizens think for themselves, each determining behavior according to the instinctive drive of a conscious entity for self-development, in the interests of ever-increasing survivability and well-being.

So in a Neuropolitic, the citizens let "nature take its course" with regard to the free flow of goods and resources, without excessive state control and regulation, while thinking for themselves with regard to their own, individual goals, rather than drifting along according to instinct, or following blindly the dictates of state, party, church, corporation, or media. The result is a harmonious multiplicity of conscious, self-governing, individual "wills to evolve"—the Neuropolitic.

Attempts to restrict the natural process of social cohesion occurring through the free action and communication between individual "wills to evolve" results only in social stagnation, frustration, and deterioration. Such an idea recalls the theories of the cultural historicists Arnold Toynbee and Oswald Spengler, both of whom saw cultural decline as a descent from creativity, to stagnation, to stasis—and finally to authoritarianism.[33]

A gradual, continual improvement in the human condition will occur naturally if we let companies provide services that people want, because they serve to enhance survivability and well-being. Excessive centralized state planning is neither necessary nor desirable. To this extent, instead of advocating a Fulfillment Society, with its distant echoes of state-planned idealism ("vote for us—we'll provide for your every need"), I espouse a Self-Enhancement Society, composed of autonomous, independent-minded, *free* human beings committed to the mutual encouragement of ever-increasing survivability and well-being.

Creative Evolution

Is a rigidly geometrical garden more beautiful than a natural landscape? The intuition that says not is the awareness of the underlying geometry of nature, which spontaneously creates order more complex—and hence more beautiful to the mind—than simple repetitive patterns of unchanging regularity.

The geometry of nature is one of flexibility rather than rigidity. The beautiful patterns in nature—a coral shell, a cloud, an oak tree—involve a greater complexity than is attainable through linear connectivity in which one part is added to another like a pile of bricks. Rather, the formation of patterns in nature involves *exponential* growth—a branching out in multiple trajectories. Such is Spontaneous Order—a sort of Creative Evolution.

The political correlate of Spontaneous Order is that of classical eighteenth-century Liberalism (that is, individualism) as opposed to social statism. If the centrally planned socialist state is like a pile of bricks at the Tate Gallery, the society creatively evolving by means of Spontaneous Order is like an organic life-form—the Bodypolitic in a state of continual, self-directed, Creative Evolution. This is the meaning of Designer Evolution—evolution not "planned by the state," but self-directed by the creative designs of individuals operating in their own self-interest, in a free society.

When I speak of the gradual evolution of the species resulting from the widespread application of Superbiology, I do not mean a process of state planning, but of Spontaneous Order. Designer Evolution is not a master plan by mad scientists, but a naturally occurring process resulting from the free actions of free individuals in a free world.

In the transhumanist vision of the future I espouse, an ever-increasing plurality of individual Wills to Evolve, freely making their own choices in bioenhancement, will eventually result in the evolution of the human species to a stronger, happier, higher state of being. No central controller is required—or desired.

In Designer Evolution, the designer is not the state, or the scientist, but the individual.

Designer Evolution is evolution not by social planning, but by self-design.

Paglia, Politics, and Nature

Human beings are fundamentally biological entities. Economics and politics are the products of their behavior. Therefore, economics and politics may be considered extensions of biological phenomena, and so are subject to the laws of nature. There is clearly a tendency in nature to gravitate *automatically* toward states of order rather than disorder. Therefore, a libertarian politics is preferable to one of collectivist planning. Generally, things and people work best if left to their own devices. The state should not seek to enthusiastically intervene wherever it can in people's lives, but rather as little as possible. The proper role of the state is not the engineering of society, but the maintenance of social cohesion through the establishment of a stable, basic framework of laws, taxes, and services—and then the willingness to leave people alone to flourish in their own way.

The "postfeminist" thinker Camille Paglia has expressed the essence of the attitude behind Libertarian Bio-Politics: "Biology is our hidden fate. . . . I believe history is in nature and of it (so) I tend to be far more cheerful and optimistic than my Liberal friends. Despite crime's omnipresence, things work in society because biology compels it. Liberals, following Rousseau, believe man is free, but everywhere he is in biological chains."[34]

The judicious reader will note here my source of inspiration for the opening line of the Transhumanist Manifesto. Paglia shares my bio-realism ("everywhere in biological chains"); my equal respect for the powers of rational intellect and biogenetic instinct ("I honor Apollo and Dionysus equally, as the Sixties did not do"); and the need to at least *try* to free ourselves from the limitations of our biological programming ("my libertarian view . . . is that we have not only the right, but the obligation to defy nature's tyranny. The highest human identity consists precisely in such assertions of freedom against material limitation."). But Professor Paglia does not believe that we can actually *succeed* in overthrowing nature's tyranny. Like most lapsed Catholics, she is ultimately a bio-fatalist, who thinks we should try and try, knowing we shall fail and fail—being subject to the Fall of Man. She is, in short, a Dionysian who, with a nod of admiration to Apollo, Prometheus, and Icarus on one hand, and Buddha on the other, accepts the ultimate dominance of nature, and man's tragic-comic attempts to overthrow her bodypolitical rule.[35]

At first sight, there may appear to be a contradiction between the recognition of Spontaneous Order in nature leading to the *rejection* of the belief in *social* engineering through state-planned (or -designed) economies—and the transhumanist advocacy of *intervening* in nature through *genetic* engineering—or Designer Evolution. How then is the politics of Spontaneous Order compatible with the transhumanist belief in the ultimate interference in nature's operations—taking control of evolution?

As outlined in my ethical concept of transnaturalism, we should follow the laws of nature *where they are clearly beneficial to increasing human survivability and well-being,* and otherwise seek to alter nature's operations where necessary in our own best interests. There is no more contradiction in such an attitude than in saying, for instance, "I'll eat the soft-centered chocolates I like, and leave the hard ones I don't." Just as you don't have to eat a whole box of chocolates to demonstrate a love of chocolates, so you don't have to conform to *all* of nature's demands to love nature. And indeed, we *do* love nature—for her beauty and magnificence—we are merely not fool enough to believe that she is entirely *benevolent* or worthy of *worship*—for we remember the praying mantis! Let us love and honor nature, but not be so foolish as to worship and obey one who ultimately has it in for us.

Socialist Darwinism

Having set out a basic defense of Libertarianism, I must now temper it by rejecting its extreme version.

While social systems have a tendency to self-regulate if left to their own devices, we should not take the theory to its extreme—the complete abdication of state responsibility for the organization and maintenance of society, and the protection of its weakest members. Such a politics, of course, would be little more than a straight rehash of nineteenth-century *laissez-faire* Social Darwinism—and just as cruel to those left behind in the process by which only the "fittest" survive.

Nature may demonstrate the ability to spontaneously create order, but society demonstrates the ability of the weak to fall by the wayside into drug addiction and crime. Abandoning the weak to their fate is not a recipe for Creative Evolution, but for social breakdown. For even the richest man cannot be happy wading through streets full of beggars and thieves, in continual fear for his life and property.

Just as, despite its miraculous powers of automatic self-regulation, the human body cannot survive without control and guidance from the conscious brain, so, too, a certain amount of centralized state regulation is necessary in the Neuropolitic. Just as the brain cannot ignore the health of even the most apparently insignificant part of the body when the automatic homeostatic regulation mechanism breaks down, as it does from time to time, so a certain amount of state regulation is required to ensure the stability or cohesion of society.

Thus, the proponents of Spontaneous Order should be careful to temper their healthy individualist and antistatist instincts with the recognition of the need for social compassion expressed through sensible state intervention, and hence limitations on absolute freedom.

The advocacy of individualism need not be equated with Social Darwinism—a self-defeating policy of Stupid Selfishness. The essence of Social Darwinism—the recognition that competition is intrinsic to life—should be balanced by an attitude I call **Socialist Darwinism**—the recognition that collective assistance for those in need is beneficial to individual and society alike—for benevolence is Sensible Self-Interest. A society which abandons the responsibility for others is like a mind that decides not to bother about the body—it won't be long before it withers away—not into the blissful anarchy of a Communist utopia, but into the total stagnation of stasis and death.

Against Hard Libertarianism

> The essence of civilization is that the strong have a duty to protect the weak.
> —George W. Bush (March 31, 2005)

Thus, I do not endorse the attitude I call **Hard Libertarianism**—the kind of *extreme* individualism that sees *any* state restrictions on the individual as unjust.

Hard Libertarianism—essentially the view that only the individual matters—is the political equivalent not of the proud lion, but of the solitary ape forced to leave the tribe due to its inability to cooperate with others. The extremism that is Hard Libertarianism or any other form of "neo-Social Darwinism" is not only cruel, barbaric, and unworthy of a civilized, intelligent species, but also detrimental to the welfare of the individuals who espouse it.

Ignoring the needs of others less able than oneself is simply **Stupid Selfishness**. Mutual goodwill and cooperation are necessary for the smooth functioning of society. Social responsibility through collective concern is a prerequisite for mutual survival. John Donne's dictum "No man is an island" is a truism precisely because it is true. A sensible degree of collective responsibility is simply common sense. The imbalanced attitude that is self at the expense of society, organism at the expense of environment, species at the expense of the ecosystem which supports it, is detrimental to the survival of all concerned. The intelligent animal would be as unwise to ignore the survival needs of the environment which supports it, as to ignore its own, *individual* needs.

Applying this principle to politics, a certain level of benevolent action by the state on behalf of the society as a whole is simply Sensible Self-Interest. Ignoring the plight of the poorest members of society does not benefit the most wealthy, unless it is deemed desirable to live surrounded by an ever-increasing sea of poverty, under an ever-increasing threat of crime.

Such an attitude is consonant with the universal principle I call **Bispectism**—the recognition of the complementarity of parts and whole in any natural system. Applied at the social, biological, and ecological levels, Bispectism asserts that the most effective survival behavior is that which addresses the needs of both self and society, organism and environment, species and ecosystem.

Soft Libertarianism

In contrast to extreme or Hard Libertarianism, I call my political position Pragmatic or **Soft Libertarianism**.

Soft Libertarianism is a practical center ground between the extreme individualism of Hard Libertarianism on the one hand, and the authoritarian statism of the extreme Left on the other. Soft Libertarianism means individual freedom tempered by pragmatic social concern, but not to the overextent that precious individualism is lost to the herd instinct of authoritarian groupthink.

Soft Libertarianism takes the Nurethics of Sensible Self-Interest as its starting point. The most basic human drive of all is the individual survival instinct. Correspondingly, self-interest is the natural instinct of a human being. The basic principle of political ethics should be one of freedom of action, unless such action threatens social cohesion. Each individual should be free from coercion by state, business, media, religion, or anyone else, to live life as they wish, provided that their behavior neither contravenes the right of others to do the same nor threatens the stability of society as a whole. The English Enlightenment philosopher John Stuart Mill put it rather wittily: "The liberty of the individual must be thus far limited; he must not make himself a nuisance to other people."[36]

But Soft Libertarianism tempers this natural, healthy self-interest with the required reciprocal altruism necessary to maximize mutual survivability and well-being.

For an uncaged animal, total selflessness is a contradiction in terms, for life itself is self-interest. The principle of putting others first knowing one will come off worst is a practical possibility only for the slave. But equally, it is the height of Stupid Selfishness to live by a creed of Hard Libertarian individualism, which ignores the need to maintain social stability through a measure of collective, benevolent action, for the good of all.

Soft Libertarianism recognizes that self-interest is an innate product of our genetic programming for survival, and that social responsibility is a necessary aspect of self-interest. In a nutshell, self-interest and social responsibility are complementary not contradictory.

The Fundamental Principle of Soft Libertarianism

The fundamental principle of Soft Libertarianism could be stated as a formula: *maximum individual liberty compatible with social cohesion, or harmony*. The starting point is the belief in individual freedom, but this natural desire is tempered by the recognition that social benevolence—concern for the welfare of others in society—is Sensible Self-Interest—a rational tactic for mutual survival and prosperity. The prime directive of Soft Libertarianism is freedom of thought, speech, and action, limited by respect for others' rights to the same freedom and the need to maintain social harmony.

Thus, for instance, the state should not allow individuals the freedom to breed an army of psychopathic Nazi storm troopers through Superbiology, because such behavior might just threaten social harmony. But generally, the first principle of the state would be to wish freedom on people, rather than seeking to control their behavior, beliefs, and attitudes—the latter being the underlying tendency of both Conservative and Liberal authoritarianism—the premodern Religious Right and the postmodern PC (politically correct) Left.

Soft Libertarianism could be seen as the recognition of the need to maintain a homeostatic balance between the self-interest of Social Darwinism and the social concern of Socialist Darwinism, based on a biopsychological understanding of human beings as evolved, biological entities whose biogenetic survival program includes both the neurochemical inducements to cooperate we call feelings, and the ability of the brain to logically deduce the optimum tactic for survival we call rational thought. In short, Soft Libertarianism is the belief in the benevolent individualism of Sensible Self-Interest.

In contrast to the conservative, the Soft Libertarian shares the Liberal belief in universal human progress. The Soft Libertarian, too, would like to see the continual enhancement of well-being for all human beings, not just the fortunate few—but not through the diktats of an authoritarian state! My response, then, to the central political critique of bioenhancement—that the result will be increased inequality—is a standard defense of free-market politics, *crucially, tempered* by state intervention where necessary to ensure social cohesion. Thus, I begin from the belief in individualism, but reject the extreme phobia against *all* things statist, as illogically self-defeating.

Applying the values of Hard and Soft Libertarianism in practice: while the Hard Libertarian might see state funding to provide, say, laptops for every schoolchild, as an imposition on the freedom of the individual taxpayer, the Soft Libertarian would view such an investment as Sensible Self-Interest, since well-educated children possessing the high-tech skills required in the modern workplace are far less likely to end up in jail or on welfare.

Dichotomal Thinking

> Despite the disparity of their beliefs, extreme libertarianism and repression share the same icy soul.
>
> Mary Riddell, *Observer* (March 27, 2005)

One of the greatest faults in political thinking is the assumption that social problems are only solvable on an either/or basis, by one or other of two opposing tactics represented by left and right of the political equation. This is an example of a tendency I call **Dichotomal Thinking**—the inability to see beyond polar opposites when evaluating ideas. Dichotomal Thinking replaces rational thought with unthinking orthodoxism—the knee-jerk response of the meme-washed disciple, inseminated with a rigidly preordained set of attitudes, which may never be challenged.

In politics, Dichotomal Thinking is the inability to conceive of the very possibility that one's opponent may possess any values of validity or virtue whatsoever. But a logical analysis of the two main dimensions of political belief, "tough or tender" and "elitism or egalitarianism," reveals them to be utterly false dichotomies. For in life as in politics, no one is—or can possibly survive by being—*wholly* tough or tender; *100 percent* selfless; or *absolutely* selfish. Life just isn't like that.

Social Homeostasis: Bispect Politics

The Meme War between opposing ideologies of left and right is based on a false dichotomy which serves only to create division in society. Political dichotomization may be resolved through a recognition of the biological basis of politics.

Here's how it may be done.

Politics is the art of social organization. "Man is a social animal" because cooperation benefits mutual survival.

A human being is an organism dependent upon its environment for survival. The two elements determining an organism's survivability are the fitness of the individual organism itself and the fitness of its environment to support life. Thus, man has a dual nature. He is both an autonomous entity and dependent upon the environment of which he is a part. This is an example of the natural principle I call Bispectism, which states that all things can be viewed in two equally important ways—as a configuration of distinct parts, or as a functioning whole.

Life is a process of self-regulatory adjustment between extreme states. Organisms automatically and continually adjust themselves to an ever-changing environment in order to maintain a state of dynamic balance—the process known as homeostasis. Failure to maintain a state of dynamic balance, both internally and between organism and environment, threatens the survival of both. If human beings fail to impose themselves on their environment sufficiently, they will become extinct. If they impose themselves too much on their environment, they will destroy nature, and thus, as a part of nature, themselves. In accordance with the natural principle of Bispectism, the needs of both species and environment, "parts" and "whole" must be recognized. This is the essence of ecological thinking.

Applied at the social level, the principle of Bispectism asserts the interdependence of individual and society. The needs of the individual are equal in importance to the needs of the society as a whole, because one can't exist apart from the other. The primary need of the individual is the right to live free from harm. The primary need of society as a whole is cohesion, harmony, or social stability.

The two main political ideologies concentrate their priorities on one side or the other of our dual nature as a biological species. The Right focuses on the freedom of individuals from dependence on the state (the "parts making up the whole") summed up by the term *self-reliance*. The Left focuses on the needs of society as a whole, summed up by the term *social concern*.

But the needs of the individual are equal to the needs of the society as a whole, because an organism cannot exist apart from its environment. The need for individual freedom is equal to the need for social cohesion. Politics is the art of balancing the two elements upon which survival depends—the individual and the society. Extreme politics of right or left lead to imbalance by focusing excessively on individualism or collectivism—the needs of the individual, or of the group, at the expense of the other.

A rational politics based on our biological nature would seek to balance the need for both individual freedom *and* social stability. The existence of such a politics would depend upon the willingness of politicians to recognize the principle of Bispectism applied to politics—the complementarity of the needs for individual freedom and social cohesion.

Freedom, prosperity, and social harmony could best be attained by following the principle of *maximum individual freedom compatible with social cohesion.* Such a **Bispect Politics** recognizes man's twin nature as an animal programmed for survival through self-interest, and a social animal that recognizes the mutual survival value in cooperation—benevolence as Sensible Self-Interest.

Politics should be regarded as a homeostatic adjustment mechanism, which maintains a society in a state of dynamic balance between extremes. The need for individual freedom is equal to the need for social cohesion, because the individual cannot survive independent from the society in which he lives. **Donne's Dictum**, "No man is an island," is a truism precisely because it is true. Self-interest and social concern are not incompatible, but two sides of a single coin. The future of politics lies in **Progressive Centrism**—a sensible middle ground between extremes of left and right.

The age of political extremism should be declared dead. Politics is the art of fine-tuning.

NOTES

1. Oxford English Dictionary, version 3.0 (Oxford: Oxford University Press, 2002), *bodypolitic.* Hereafter referred to as OED.
2. Plato, *The Republic*, ed. C. D. C. Reeve (Oxford: Oxford World's Classics, 2004).
3. Aristotle, *Politics*, trans. Trevor J. Saunders (London: Penguin, 1981).
4. OED, *state.*
5. Thomas Hobbes, *Leviathan*, ed. J. C. A. Gaskin (1651; repr., Oxford: Oxford World's Classics, 1998).
6. Camille Paglia, *Sex, Art, and American Culture* (London: Penguin, 1993).
7. OED, *corporation.*
8. Ibid., *bodypolitic.*
9. William Shakespeare, *Henry VIII*, in *Complete Works of Shakespeare*, ed. W. G. Clark and W. Aldis Wright (New York: Nelson Doubleday), 24.12.
10. Alexander Pope, *The Dunciad*, ed. Valerie Rumbold (Harlow, Essex, UK: Longman, 1999).
11. Encyclopaedia Britannica, Millennium Second Edition, CD-Rom, s.v. "Roman Empire."
12. Tobias Smollett, *The Adventures of Ferdinand, Count Fathom*, ed. Paul-Gabriel Boucé (1753; repr., London: Penguin, 1990).
13. Benito Mussolini, *The Doctrine of Fascism*, http://www.worldfuturefund.org/wffmaster/Reading/Germany/mussolini.htm.
14. OED, *corporate.*
15. Ibid., *corporative.*
16. Anna Bramwell, *Ecology in the 20th Century: A History* (New Haven, CT: Yale University Press, 1989).
17. OED, *soviet.*

18. Plato, *The Republic*, bk. 2.

19. Aldous Huxley, *Brave New World* (1932; repr., London: Flamingo, 1994).

20. OED, *corporational.*

21. Jean-Jacques Rousseau, *The Social Contract*, trans. Maurice Cranston (1762; repr., London: Penguin, 1968).

22. Ibid.

23. Aristotle, *The Politics*.

24. Ibid.

25. Ibid.

26. Winston Churchill, *Hansard,* November 11, 1947.

27. Julian Huxley, in *Man and His Future*, ed. Gordon E. W. Wolstenholme (London: Churchill, 1963).

28. Abraham H. Maslow, *The Farther Reaches of Human Nature* (London: Arkana, 1993).

29. Theodore Dalrymple, "We Don't Want No Education," *City Journal* (Winter 1995), http://www.city-journal.org/html/5_1_oh_to_be.html.

30. Huxley, *Brave New World.*

31. Jean-François Lyotard, *The Lyotard Reader*, ed. Andrew Benjamin (Cambridge: Blackwell, 1998).

32. Friedrich Von Hayek, *The Counter-Revolution of Science* (Glencoe, IL: Free Press, 1952).

33. Arnold Toynbee, *A Study of History*, pt. 3, rev. and abr. by Arnold Toynbee and Jane Caplan (Oxford: Oxford University Press, 1972); and Oswald Spengler, *The Decline of the West* (Oxford: Oxford University Press, 1991).

34. Camille Paglia, *Sex, Art, and American Culture* (New York: Vintage, 1992).

35. Ibid.

36. John Stuart Mill, *On Liberty* (Oxford: Oxford World's Classics, 1998).

Chapter 12
NEURAESTHETICS
Evolutionary Neuropsychology of Art

> What is beneficial is beautiful, what is harmful is ugly.
> —Plato, *The Republic*

Aesthetics is the attempt to answer the question, What is the nature of art and beauty?

Here, I outline an **Evolutionary Neuropsychology of Art**, defining art, beauty, design, and creativity in terms of their evolved benefits to survival and well-being. I propose a new aesthetic of **Up Art!** imbued with the euphoric **Oh, Wow! Factor**, and predict the emergence of a new breed of twenty-first-century artists, the **New Leonardians**, who combine the romantic passion for human transcendence with the classical logic of scientific rationality, in an uplifting spirit of **technoromanticism**, or **neuromanticism**.

SUMMARY OF TRANSHUMANIST AESTHETICS

Art is an extension of our evolved toolmaking facility—the practice of making things for the pleasure of perceiving them, rather than their practical use-value. The elements of art are power tools for ever-increasing survivability and well-being. Imagination is pattern making in the mind. Design is systematic pattern making. Creativity is practical pattern making. Through Designer Evolution, man is destined to become the creative artist of his own self-design.

The ideal art is an **Up! Art** imbued with the **Oh, Wow! Factor**, inspiring the awe and wonder of the **technosublime**. In the twenty-first century, a renewed belief in science and technology will produce a new breed of technophile artist-philosopher-scientists—the **New Leonardians**—who embrace the positive aesthetic of **Techno** or **neuromanticism**, and an ethic of **New Existentialism**—the recognition that life is here for the living, the pursuit of joy and beauty preferable to the willingness to wallow in the misery of misanthropic fear and loathing, futility, and despair. **The Three English Hs** of the late twentieth century provide a model of inspiration for the artists of the twenty-first century seeking to go their own way against the prevailing tide of miserabilism and self-mockery.

Art

The hushed tones of the whitewashed art gallery reveal its contemporary function as a secular church, but art evolved originally for its survival value.

Art is an extension of the toolmaking facility evolved for the survival benefits afforded by the ability to "make things."

Art is artifice—the products of man as opposed to nature. To be an artist is to make things from the materials of nature, primarily for the pleasure of perceiving them, as opposed to their use-value.

Art survived the evolutionary journey by conferring three significant benefits to our genetic programming for survival and reproduction:

1. Behavioral conditioning: The social function of art is the reinforcement of group cohesion through the dissemination of a metameme—a totalized world picture serving to guide behavior in the interests of survival.
2. Mood regulation: For the individual, art serves as a homeostatic regulation mechanism, helping to maintain the body in a state of equilibrium conducive to maximum survivability. In short, art serves to improve our moods—making us happy when sad, lively when tired, relaxed when stressed, brave when fearful—and aroused when wishing to mate.
3. Sexual advertisement: For the artist, art serves as a powerful advertisement for sexual availability. Here art is to man as plumage is to the peacock, birdsong to the bird—a mating call, carrying the message "I'm sexually available—and just look what I can do!" Hence that disappointingly common (but not ubiquitous) phenomenon, the artist who produces his best work in the most sexually fertile period of his life—his youth—and utterly fails to repeat the same quality in later years.

Beauty and Survival

Beauty means "the quality of that which we find attractive." We are programmed to find attractive that which helps us survive and reproduce. The pleasure of beauty is nature's reward for behavior conducive to survival.

Aesthetic appreciation is the ability to find things beautiful—an evolved, neurochemical conditioning mechanism inducing attraction to objects of benefit to survival and reproductive success. We find living nature beautiful because we are dependent upon her for our survival. We find feces unpleasant because our bodies recognize it to be an unhygienic waste product, not good for us to touch or eat. Not many people except postmodern artists find feces, death, or disease beautiful, and when they do, we might reasonably regard it as psychopathological.

Beauty and Harmony

The world is a metapattern of information. Human beings are imbued with a capacity for pattern recognition—the ability both to perceive patterns and to distinguish those parts of a pattern beneficial to survival and well-being. Nature rewards pattern recognition with a surge of neurochemically induced pleasure. The pleasure which ensues automatically from the ability to discriminate between life-enhancing and life-threatening parts in the patterns of nature is called aesthetic appreciation—or the ability to perceive beauty. Aesthetic appreciation is the pleasure of pattern-recognition.

Harmony is order, regularity, or *balance* within a pattern. Life is a condition of homeostatic *balance*. The cause of death is *im*balance—a cancer in the body is a group of rogue cells growing abnormally out of proportion to the rest (as a cancer in *society* is a group of rogue *memes* growing abnormally out of proportion to the rest). Harmony produces pleasure by signifying the stable order and balance that is the basic condition of life. The pleasure of harmony is nature's neurochemical reward for attraction to things which display dynamic balance, signifying good health. The ability to recognize harmony between parts in a pattern is the basis of aesthetic appreciation. Beauty is the attraction to health-signifying harmony. The instinctive pleasurable response to the perception of harmony is a faculty evolved and preserved by natural selection because it encourages the attraction to objects beneficial to survival and reproduction. A harmonious object is one in which all parts combine in an orderly, cohesive manner for maximum functionality—and hence of benefit to survival. Aesthetic appreciation involves discriminating between things according to their harmoniousness—that is, their manner of combining into pleasing patterns, which induce pleasure through the perception of order and balance, signifying health, and therefore survivability.

Human beings find the sound of total dissonance unpleasant because it signifies imbalance or disorder, and therefore ill-health. A preference for dissonance over harmony could be considered a masochistic pleasure in pain. A preference for Masochistic Art might be called **disphilia**—the love of dissonance, disorder, imbalance, or chaos. Disphilia is characteristic of an Opioid Neurotype addicted to the tranquilizing chemicals secreted by the body when harmed. (This is the *reason* for the efficacy of acupuncture—sticking needles into the body results in the secretion of pain-relieving endorphins in self-defense. Nothing to do with mystical "chi" energy.) Disphilia is an expression of Thanatos—the unconscious desire for self-destruction in the absence of joy in living.

Beauty and Complexity: The Mathematical Sublime

Complexity is a pattern with many parts. To perceive complexity is to distinguish between different parts of a complicated, but ordered pattern—an ability beneficial to our survival. The more complex the perceived pattern, the greater its potential value to survival. The ability to predict the occurrence of a tsunami by identifying complex causal connections within the patterns of nature is likely to save many lives. On a simpler level of complexity, the ability to distinguish between fresh and rotten food can mean the difference between life and salmonella. Life itself is a state of dynamic, organized complexity. Living things are complex, dynamic patterns of activity. We appreciate living bodies more than dead bodies by virtue of their greater complexity. Life is not homogenous but heterogeneous. Hence, most of us appreciate Michelangelo's *David* more than the block of stone from which it was sculpted.

Aesthetic appreciation is proportional to the ability to perceive complexities in patterns of ordered complexity—one aspect of the general power of the brain we call intelligence. The child is less intelligent than the adult because the brain has not yet developed sufficiently to allow for the recognition, storage, and processing of great amounts of information. A cat might be happy staring at a flashing colored light, while an intelligent human being may take greater pleasure in recognizing the beautiful complexity of a snowflake, a painting by Leonardo, a fugue by Bach, a mathematical equation—or a harmonious philosophical system. The philosopher Immaneul Kant spoke of the "mathematical sublime" as the appreciation of objects of immense complexity, such as an infinite mathematical equation. The pleasure of the mathematical sublime is the pleasure afforded by nature for the appreciation of complexity.

The inability to recognize any difference in aesthetic value between the ceiling of the Sistine Chapel and an unmade bed derives from the inability to appreciate organized, harmonious complexity.

Creativity

Imagination is pattern making in the mind. Creativity is pattern making in the world, a facility preserved by natural selection for the survival value of making things. A cave painting, a language, an axe, a house, a computer are all examples of creative pattern making in the interests of survival. The pleasure of creativity is nature's reward for behavior which increases our ability to survive.

In the process of creative thought, the brain experiments by making "educated guesses"—trying out connections between events which may not obviously appear to be causally connected. Sometimes the experiment in creativity works, sometimes not. The result of a series of successful educated guesses is a new pattern of connections which did not formerly exist—a creative act.

Creativity involves small diversions from a regular pattern. Too few diversions from an existing pattern results in the boredom of predictability critics call "unoriginality." Too many diversions from an existing pattern results in the boredom of incomprehensibility critics call "obscurity." The secret of good art lies in the ability to know when best to divert from a regular pattern to excite interest without destroying the harmoniousness of the pattern as a whole.

The greatest artists attain a perfect balance between innovation and consolidation. An artwork containing nothing new or surprising is of little interest—"We've heard it all before." Conversely, too much new information and the work is simply incomprehensible. The message for those wishing to attract the public, whether artist, philosopher, advertising agency, or politician, is to make your products, or memes, different—but not too different—sufficiently new to be of interest, but not so unique that they cannot be understood. For instance, not "the right-wing socialist party," but "New Labor."

The ability to make *new* patterns instead of simply repeating old ones is the difference between the *practical skill* of the pianist able to play Rachmaninov's Third Piano Concerto with brilliant manual dexterity—because his rational brain has learned the preexisting pattern (the score)—and the genius of Rachmaninov himself, whose facility for creative thought allowed him to construct a pattern which did not exist before. Originality of pattern is the difference between good art and cheap imitation.

An important aspect of creativity rarely mentioned is the importance of bravery. The ability to think the impossible, the unthinkable, the strange, the unusual, is the prerequisite for the creative artist. To try out new patterns—new ideas—increases the risk of appearing foolish and feeling humiliated if the experiment doesn't come off. In art, the courage to face ridicule is the difference between genius and mediocrity. The greatest creative thinkers and artists, from Leonardo to Einstein, possessed the audacity to do what they had to do entirely without fear of what other people might say. But then, the brilliant Roman poet Horace said it long ago: "The power of daring anything their fancy suggests has always been conceded to the painter and the poet."[1]

Design

Design is systematic creativity. To design means to make according to a plan, rather than to improvise. Design implies forethought—the ability to plan ahead, anticipate, recognize possible connections between events, and connect them together to produce a pattern—the object of design. Design is the opposite of the Dionysian. Rather than emotional impulsivity, design requires the Apollonian characteristics of intellect, order, reason, and patience.

The word *design* derives from the Latin meaning "to mark off." To design means to pick something out and mark it off as separate from other things—the act of distinguishing one thing from another. In short, discrimination is the essence of creativity. The ability to recognize some things as more useful, valuable, beautiful, or interesting than others is the basis of creativity in artist, designer, scientist, and philosopher alike.

A "designed" product is one that involves much thought and planning in its creation—as opposed to being thrown together quickly to make a profit. Design is an aspect of intelligence, requiring inhibition—the ability of the mind to inhibit instincts. In addition to creativity, the artist requires the ability to inhibit the impulse to go down to the pub and leave his symphony unfinished.

The capacity for intelligent design is ultimately what makes us human—better able to control our own destiny than our ape ancestors incapable of forethought and will—the ability of the mind to think ahead and control behavior in its own interests.

The ability to plan ahead is the essence of intellectual design. There is a famous quote by Vincent van Gogh from his letters to Theo: "The thing has already taken form in my mind before I start it. The first attempts are absolutely unbearable. I say this because I want you to know that if you see something in what I am doing it is not by accident but because of real direction and purpose."[2]

Design.

Michelangelo would sit for hours beneath the ceiling of the Sistine Chapel, merely looking and thinking—planning what he had to do. He might make one or two stokes of the brush in a day, and then go home.

Design.

Designers in all walks of life should be revered by society—for their craft represents the peak of evolutionary activity—the ability of the evolved human brain to create ever new and ever more complex patterns out of the stuff of the world.

Creative Evolution

Evolution, like art, is a creative process of pattern making. In evolution by mutation and natural selection, small changes to the existing pattern of an organism are either maintained or discarded by nature. The new, evolved organism is a slight but significant variation on the previous type. Without variation there is no creative evolution. Too much variation and the new organism will not survive in its environment.

The difference between human creativity and the creativity of nature is only a question of degree and materials. Nature uses chemicals to create the art of life, while man uses letters (literature/drama), notes (music), bricks and concrete, metal or plastic (architecture, sculpture).

A piece of music is composed from sequences of notes—a language of sounds. A human being is composed from sequences of DNA—a language of chemicals. When Darwinian evolution is replaced by Designer Evolution, man will be viewed as an artist, taking over the unfinished work of nature. Designer Evolution is the ability of a species to creatively direct its own evolution. In the twenty-first century, bioenhancement of the body and mind will be viewed as a creative act. Designer Evolution is the Creative Evolution of the human species. Evolution is an ever-expanding collage—an artwork to which all are free to contribute. Life as Art. Evolution as creative. Man as free. Such is the transhumanist vision. Through Designer Evolution, man is destined to become the creative artist of his own self-design.

Designer Evolution

Designer Evolution—the ability to redesign our own bodies and minds through Superbiology—is the ultimate in creative design.

Just as a top fashion designer differs from an amateur in a greater ability to recognize and enhance the beauty of the physical form by constructing new patterns from fabrics, so Designer Evolution by humankind will differ from Darwinian evolution by nature, in the ability of human beings, as creative designers, to recognize and enhance their *own* physical form, by manipulating *biological* material—the human body. Designer Evolution means the ability to recognize and enhance those aspects of our own biological makeup we consider capable of improvement, in pursuit of ever-increasing survivability and well-being.

If the design of fabrics to enhance the body is deemed virtuous, why should it not be deemed virtuous to design enhancements of the physical body itself? A recent TV show allowed members of the public the chance to undergo total makeovers to transform the way they look, involving everything from a change in clothes and hairstyle to comprehensive aesthetic surgery. The results were genuinely moving. Ugly ducklings were transformed into swans, and timid depressives discovered a newfound confidence. Only the most hard-hearted ideologue could possibly object to increasing someone's confidence by improving the way they look.

Yet plenty do. "We should learn to accept ourselves as we are," say the bio-Luddites. According to long-term bio-Luddite Jeremy Rifkin, the real goal of body makeovers is to achieve "Some socially mediated norm of acceptability . . . not a projection of one's own unique inner being onto the world—a celebration of communion—but, rather, an attempt to remake oneself to 'fit in' to the world."[3]

Oh, right. Not, then, the goal stated by the individuals themselves—to help them to look better, so they feel better, and make themselves, their partners, friends, family, and colleagues happier. If so, it certainly worked. Truly "a celebration of communion."

In Defense of Self-Design

Why should we not seek to redesign ourselves if we so wish? What right has anyone to criticize the desire of another to look the best they can? And if it's fine to design clothes to enhance the body, what's wrong with redesigning the body itself through Superbiology?

Mr. Rifkin says that the alteration of one's genome is "not art but artifice . . . merely a technological prescription . . . a set of choices purchased in the marketplace."[4] He seems to think that anything you have to pay money for is worthless. But composers pay for manuscript paper. Artists pay for paints. Why shouldn't people pay to enhance their bodies through biotechnology as through exercise—they pay to go to the gym—and why shouldn't they see it as a loving act of artistic self-design?

What exactly is wrong with regarding the body as creative material ready to be carefully enhanced—as bodybuilders and aesthetic surgeons do already! To regard one's body as a potential work of art is not to remove its dignity but to respect it. To see oneself and one's species as artworks in the making is not to devalue life but to venerate it. To see life as a process of creative self-design is the greatest honor one can bestow upon it. The artist no more feels contempt for his materials and work than a mother feels contempt for her child. The attempt to creatively enhance the self is the product of joy, love, and self-respect.

There can be nothing bad about the desire to be the *best* one can be. So let us reclaim our respect and admiration for designers, and choose to be artists of self-design.

Postmodern Art

"So What?" Art, Emperor's New Clothes Syndrome, and the ET Test

Most contemporary art is so banal it might be called **So What? Art**—an art of so little creativity, emotional power, or intellectual stimulation that it invariably fails to hold the attention for more than a few seconds, before the viewer thinks, So what? and moves on.

The ease with which the public is taken in by the unskilled, worthless tomfoolery of most postmodern art is such a distressing contemporary phenomenon I think it requires its own terminology. I call it the **Emperor's New Clothes Syndrome**. ENCS is a mass psychological disorder resulting from the tendency of individuals in a Bodypolitic to succumb en masse to meme control through meme washing by those who shout loudest for their attention.

As an antidote, I suggest what I call the **ET Test**—a method of evaluating ideas or behavior by asking the question, What would a highly evolved, intelligent extraterrestrial think of it? Observing our culture today, in such depths of postmodern nihilism that one of its most talked about artworks is the presentation of an unkempt bed, one may be forgiven for thinking that an alien superintelligence would be likely to respond with the extraterrestrial equivalent of the phrase, "Beam me up, Scotty, this planet sucks."

MISERABILIST ART

It is symptomatic of the postmodern anticulture of nihilism that joyous, sunny, or as I call it, **Up! Art**, is almost always regarded as inferior to **Miserabilist Art**. In the postmodern anticulture of nihilism, Up! Art is dismissed as lightweight kitsch, while Miserabilist Art is applauded as a "profound statement on the human condition."

Miserabilist Art is essentially a solace—a consolation for unhappiness. But just as one seeks no medicine if one is healthy, one seeks no solace if one is happy. No one could be fool enough to deny the value of sorrowful art. Men like symphonies of sorrowful songs because *Homo sapiens* is a sorrowful creature. But the belief that Miserabilist Art is *superior* to **Sunny Art** is just as absurd as the claim that unhappiness is superior to sadness, or illness to good health. Just as one prefers to be happy, not sad, healthy not sick, so Up! Art must logically be regarded as *better* than Miserabilist Art, where "better" means "of greater benefit to well-being." Since everyone prefers to be happy, not sad, it follows that emotionally uplifting Up! Art must ultimately be seen as superior to pessimistic, gloomy Miserabilist Art. The question must be asked, If what you seek is happiness, why expose yourself to—or produce—nihilistic art? The best art must logically be considered that which is uplifting, positive, and life-affirming, because that which we seek most of all is happiness, health, and life.

To which one may reply, How many of us expose ourselves to nothing but the art of conviviality and good cheer? Quite so. But one doesn't have to live up to an ideal to endorse it—no one condemns the dying smoker as a hypocrite for discouraging the habit in others. Personally, I can endure little else but a colorful **Up! Art** of joy, awe, and wonder (David Hockney being the world's greatest living exponent). Yet, when the mood arises, I suspect I shall go on listening to—and making—miserabilist *music* for as long as the miserabilist mood states exist. And I see nothing hypocritical in doing so. Rather, I'm expressing the attitude I call Positive Realism—the belief in seeking to *transcend* the misery and suffering inherent in the human condition, while at the same time acknowledging the existence of suffering as a present-day fact of life.

My point is not that we should "outlaw expressions of unhappiness" in a sort of Hitlerian frenzy of moral idealism, but that we should be *big* enough—*logical* enough, to recognize that the tactic of voluntarily drenching ourselves in miserable memes is the artistic equivalent of drowning our sorrows in alcohol or heroin (better put, "booze or smack," for the severity of the street names tells a tale).

As drug addiction is self-medication for the "dis-ease" of misery, so, too, an addiction to miserabilism is a solace for the suffering of life. Nietzsche got it right, when seeking to justify his addiction to the intoxicating music of Wagner: "If one wants to get free from an unendurable pressure one needs hashish. Very well, I needed Wagner."[5] He duly cured his addiction to what he called "romantic pessimism" in favor of music "cheerful and profound like an afternoon in October," saying, with just a touch of irony, "I would not know how to get on without Rossini."[6]

Van Gogh and the Neurochemistry of Inspiration

Those who fall for the myth that great art requires psychological disturbance are greatly deluded. First, there is the obvious fact that many of the world's greatest artists possessed wholly genial temperaments, and consequently produced plenty of gloriously Sunny Art—Mozart being the finest example. Second, no one creates art in a state of clinical depression. No one does *anything* much in a state of clinical depression, because its behavioral effect is the slowing down of the faculties. Depression is what happens when your batteries go flat.

The dazzling color, rhythm, and intensity of a classic van Gogh is not the work of a man in the throes of clinical (that is, deep) depression—a state characterized by the grinding down of mental and physical faculties. Rather, works such as *Starry Night* were clearly produced in a state of exultant euphoria—the mood artists call "inspiration." Like all mood states, "inspiration" is the product of certain chemical changes in the brain—surges in the level of neurotransmitters such as dopamine and serotonin. When artists talk of the need to wait for "artistic inspiration," they are not describing a touch from the hand of God, or the Greek muses, but a sudden rise in dopamine and serotonin levels, characteristic of the neurochemically imbalanced victim of "bipolar disorder"—contemporary eumeme for "manic depression."

Like many creative artists, van Gogh was surely a manic depressive who made the most of his periods of hyperelation to produce his work. The feverish speed at which he painted supports this view. Van Gogh worked as fast as his hand and eye would allow, knowing he did not have long to create before the Bliss Chemicals sank back into their receptors, plunging him back into a state of gloom, only to escape by stunning himself with absinthe, in a desperate effort to kick-start the depleted, serotonin-fueled engine of the brain, until the next period of endogenous, neurochemically induced elation began, and the cycle continued . . .

Van Gogh may be an extreme case, but the same applies, to a lesser degree, to all of us. Why should we condemn ourselves to unpleasant mood states imposed by our neurochemistry, when eugoics may allow us to replace uncontrollably negative moods with positive states of well-being? Why should we limit ourselves to feeling pleasurable emotions only in circumstances dictated to us by the selfish genes that ultimately have it in for us? Why should we not, instead, liberate ourselves, by learning how to produce positive emotions at will—not through "positive thinking" alone—the deluded dream of the New Age therapist—but through the new technology opening up in front of our eyes, the controlled neuroenhancement of the mind—eugoics.

It wasn't depression that van Gogh required in order to produce his work—it was elation. Creativity requires a jump-start, not flat batteries. And such a jump-start is now becoming possible—a neurochemical jump-start—through Superbiology.

Against Miserabilism

The myth of the romantic artist endures, by which states of emotional distress are held to be superior, privileged, and conducive to great art. But there is nothing brave, heroic, or glorious in depression or anxiety—they are nothing but disease states, as undesirable as cancer. No one should have to suffer the misery of an overly depressive, aggressive, shy, or nervous temperament through no fault of his own. Only the most callous of aesthetes would wish a wretched life on an artist in order to enjoy the products of his suffering. The idea that we should seek to preserve mental illness because it might throw up some interesting art, irrespective of the artist's well-being, is the moral equivalent of throwing Christians to the lions to provide jolly good entertainment.

As a Soft Libertarian, I am naturally averse to the idea of state censorship, in the arts as in elsewhere. So long as I harm no one in the process, I see no logical reason why I should permit the state to dictate what I am allowed to think, feel, do, or say. And yet—I cannot help wishing that artists who spread the negamemes of nihilism and misery in the name of "self-expression" would go and do something useful instead—like nursing or bricklaying, for instance.

The impulse to make miserabilist or nihilistic art is symptomatic of the Opioid Neurotype—a neurobiologically based temperament which predisposes to low mood states, probably caused by deficient levels of the neurohumors dopamine and serotonin. Miserabilist Art—or miserabilist attitudes in general—are not an objective depiction of "The State We're In," but the product of a neurochemically induced, temperamental tendency to see the world with a jaundiced eye—in which the glass is always half empty. Miserabilism is the universalization of a subjective state of unhappiness which carries the morally dubious message, "Be unhappy—see the world like me."

The belief that suffering is good for the soul is an attempt to make a virtue out of a necessity. Perhaps if negative-minded artists were to take a course of eugoics they may find their moods raised, and their art becoming more positive, uplifting, and thus socially useful.

Neurotypes predisposed to negative mood states are naturally attracted to art which expresses the feeling they know best—misery. Only when Superbiology allows us to eradicate our psychophysiological states of misery and despair will the miserabilist aesthete see the value in art which expresses joie de vivre. The solace of shared misery will continue to be sought through art until misery itself is cured by Superbiology.

Up! Art: Fully Firing on All Four Cylinders

Art is about perception. The way we perceive the world affects our behavior within it. The manner in which a society perceives its own culture is influenced by the nature of the artifacts it produces. Therein lies the importance of art, and the moral responsibility of the artist. For if happiness is preferable to unhappiness, and encouragement morally superior to disparagement, then logically, the *best* art must be that which uplifts and inspires, not that which depresses and deflates.

Human beings have four basic faculties—sensing, feeling, thinking, and perceiving—all of which are pleasurable to use because they increase our ability to survive. The best type of art is that which stimulates all four faculties, inducing the optimum state of well-being I call **Fully Firing on All Four Cylinders**. Such an art, if it existed, would be equally sensual, emotive, intelligent, and uplifting, and offer a positive, awe-inspiring vision of the world.

Instead of an art which depresses, deflates, or "shocks the bourgeois out of its complacency," let us seek to create a positive **Up! Art** which inspires and uplifts—an art which sees humanity not as some depraved animal, but as an intelligent species embarked upon an *upward* journey toward the betterment of the human condition.

Let us develop a new sensibility for the twenty-first century—an art made by and for those who consider positivity and joy a moral imperative for an evolving species.

The New Leonardians

> In the Renaissance, many artists were also scientists. I think that separation is now slowly beginning to disappear.
> —David Hockney, *That's the Way I See It* (1993)

> We are witnessing the emergence of a new type of artist, the artist/scientist/researcher.
> —Christine Paul, curator, Whitney Museum of American Art, *Guardian* (January 14, 2003)

George Bernard Shaw once said that the artists of the future will be artist-philosophers. More likely, they will be artist-*scientist*-philosophers. In the twenty-first century the ongoing explosion in popular technology will result in the development of new, syncretic art forms, which bridge the gap between art and science. We shall witness the emergence of a new breed of artists who express a confident technophilia and a newfound sense of wonder at the world. After the original Renaissance Humanist, we might call them the **New Leonardians**. These New Renaissance Men—and Women—will seek to identify and express the harmonious vision of the world emerging from the New Science of wholeness and order, through an art of intelligence and wonder. The New Leonardians will reunite the divided realms of art and science, and create a **New Music of the Spheres**—an uplifting depiction of the world as a creatively evolving unity-in-diversity, with humanity at its helm.

The true artists of our own time are the *scientists* who have discovered the extraordinary beauty of nature at all levels—from the structure of DNA to the paisley-patterned fractal; from the neural net of the brain, to the superstrings of the cosmos. *Here* is the material for the budding twenty-first-century artist! The possibilities of employing cutting-edge technology in art are endless.

The American multimedia artist Natasha Vita-More is the first artist brave enough to fully engage with the idea of redesigning the human body itself as a work of art. Therein lies one avenue for twenty-first-century art. Others include the "Light Artists" such as James Turrell, whose displays of fluorescent light in natural spaces are an awe-inspiring synthesis of technology and nature; or London-based husband-and-wife team Rob and Nick Carter, whose joyous experiments in "painting with light" signals a change in aesthetic, away from that of the miserabilist Chapman Brothers Grim of the *fin de siecle* twentieth century, to a bright, positive, postironic vision of the world (now seen as a place in which to experience the genuine joy of living—as opposed to the dubious pleasures of self-flagellation for the sins of humanity). Then there are the computer artists who mirror the transition from Darwinian to directed or Designer Evolution, by combining randomly evolving cyberorganisms with the selection of the most aesthetically pleasing—by the artist. And let us not forget the amazing work of the cutting-edge interior designers, with their sleek, modern, hi-tech furniture, atmospheric lightscapes, and "digital wallpaper."

How strange that so many young artists today still think they can compete with the vibrant, dazzling, multicolored, luminescent world of computer-based digital design that permeates everyday culture in the modern world, from "FX"-based megamovies and Web designs, to Sunday Supplements—with dull paint on canvas! Apart from David Hockney—the last great painter—paint on canvas is dead, a lifeless anachronism in a digital world of pulsating, electronic color.

The Oh, Wow! Factor and the Technosublime

Mahler once wrote that a symphony should "contain the world." I agree with this sentiment. The best or highest forms of art communicate a sense of awe and wonder at the beauty and magnificence of the universe of which we are all an intrinsic part, through works of ambition, skill, and magnificence—as opposed to, say, the expression of psychosexual anhedonia symbolized by two fried eggs and a kebab.

The human capacity for awe and wonder is nature's reward for appreciating the grandeur of life. Awe and wonder is an evolved, neurochemical conditioning mechanism positively reinforcing joy and fascination with the world as an aid to survival—for a species that found no pleasure in the environment upon which its survival depends would find itself lacking the will to survive.

The ability of the best art to inspire awe and wonder I call the **Oh, Wow! Factor**. It is a quality lacking in most contemporary art, but one I predict humankind will discover again in the twenty-first century.

The feeling of awe and wonder in the face of the "unimaginably Big" in nature is called "the sublime." The concept of the sublime was central to the dominant metameme of the nineteenth century—romanticism. Essentially, the romantic impulse is the human desire for *transcendence*, whether political (through revolution), or psychological (through religion, art, or science). When the sublime is the product of technology, such as the mind-boggling, global complexity of the Internet, we might speak of the **technological sublime**.

In the twenty-first century, the romantic impulse for human transcendence will shift from the political to the scientific realm, as we begin to acknowledge and appreciate the miraculous technowonderland of the modern world—product of our human hands—and seek ways in which to overcome the biological limitations of the human condition, through the new technology opening up in front of our eyes.

Technoromanticism and Neuromanticism

In the years ahead, a new breed of artists raised in the technowonderland of the modern world will begin to fashion a new art of awe and wonder at the majesty of both the natural world, and the world of technology—the **technosphere**. Technology and transcendence will combine in a new aesthetic for the DNAge —**Technoromanticism**.

Romanticism is the belief in imagination, passion, and transcendence by the mind. "Techno" derives from the Greek for "skill," "craft," or "art." By Technoromanticism I mean the combination of imagination—the ability to construct new patterns in the mind—and practical creativity, the ability of the mind to realize those patterns in the material world.

Technoromanticism differs from nineteenth-century romanticism in its practicality. It rejects the idle dreaming we call "fantasy" for *practical* creativity—the construction of new *things* for the benefit of survivability and well-being, rather than the escape into a virtual reality of *mental* constructs, common to the idle, romantic dreamer and psychotic madman alike.

Imagination is the ability of the conscious brain to construct new "virtual patterns" in the mind. *Creative* imagination is the ability to *realize* the virtual patterns constructed in the mind, in the material world of rocks and stones and trees. A vivid imagination is of little use to society without creativity—the practical ability to translate a mental design into a *product* of design—or work of art. Technoromanticism is the aesthetic basis of the belief in Designer Evolution—the passionate belief in the inevitability of human transcendence from the limitations of the human condition—not through the messianic memes of the mystics, or megalomaniacs of the Bodypolitic—but through the miracles of modern technology in the modern world, and the power of the conscious, self-governing, human mind.

Thus, when the spirit of technoromantic transcendence focuses on the belief in the enhancement of the conscious mind, through eugoics or Supergenics, we might call it **neuromanticism**. In a magazine piece of 1986, "The Neuromantics," Norman Spinrad argued that so-called cyberpunk science-fiction writers such as William Gibson, author of the novel *Neuromancer,* might be called "neuromantics," their work being "a fusion of the romantic impulse with science and technology."[7] Neuromanticism in this sense is a transhumanist attitude combining the romantic passion for human transcendence with the classical logic of scientific rationality. For the neuromantic, the transcendence of human limitations through scientific and technological advance is regarded as a creative act.

Thus, Transhumanist art rekindles the Promethean spirit of the modern age by passionately embracing the technowonderland of the modern world.

Techoromanticism and neuromanticism are twenty-first-century alternatives to the woolly minded emotionalism of romantic idealism, the "cold" sterility of scientific rationalism, and the life-devaluation of postmodern nihilism.

The Three English Hs

Most artists today have been conditioned by the high priests and priestesses of postmodernism to believe that negativity is aesthetically superior to positivity, euphobia to euphilia (the love of feeling bad above the love of feeling good).

Today's young generation of artists might benefit from reflecting on the work of those twentieth-century figures who failed to conform to the prevalent cynical view of humanity.

As just one example, I suggest the triumvirate I call the **Three English Hs**—David Hockney, Patrick Heron, and Howard Hodgkin—three late-twentieth-century colorists whose work shines like a beacon of beauty and joie de vivre in a sea of postmodern misery and despair.

The aesthetic I see in the Three English Hs is here expressed by David Hockney: "If you see the world as beautiful, thrilling, and mysterious, as I do, then you feel quite alive; I like that. I know there are people who can't see the world like that, who feel despair, who obviously cannot see much in the world, cannot see some immediate beauty. If they did, they wouldn't feel the despair. I see that part of my job is to show that art can alleviate despair."[8]

No surprise that Mr. Hockney found it necessary to move from England to California to find the inspiration he needed for his Sunny Art.

Needless to say, the Three English Hs are not as well regarded as the conceptual "Brit Pop Artists" with their fashionable artlessness and general *mal de vivre*. But all convey the message that the world is something of immense beauty, that life may produce ecstatic joy, and that the skill and patience required for human design is something to value, not deride.

I suggest this might be a preferable message for humanity than the view of the human condition emanating from so many of the postmodern antiartists, summed up in that tired old adage: "Life's a bitch and then you die."

But then, negs will be negs.

The New Existentialists

The postmodern nihilists, like the twentieth-century existentialists, and nineteenth-century romantics before them, assumed their depression was telling them the truth about the world. But a negative outlook on life is often the product of a neurochemical imbalance, which irrationally distorts one's attitudes to those of futility, misery, and despair. Romanticism without classicism fails because its mood-soaked exponents are the victims of their own negative emotions.

The goal of the proposed philosophical movement the English author Colin Wilson called a **New Existentialism** is to enhance the quality of consciousness, instead of accepting negative mood states as a given—an objective evaluation of the world: "The New Existentialism concentrates . . . upon the problem of life devaluation. It suggests mental disciplines through which this waste of freedom can be averted."[9]

In the twenty-first century, let us hope that a new breed of positive-minded artists—we might call them the **New Existentialists**—will begin to create an art of life, joy, and color in the recognition that, in the end, life is here to be lived—not merely suffered or endured.

As an inspiration, they could do worse than to look back at the greatest artist of the late twentieth century; not (as fashion dictates to the unthinking) those well-known painters of sadomasochistic fantasies and ugly nudes, but rather, the creator of a glorious Technicolor world of joy and wonder—David Hockney.

Hockney's world is not the lilac-pasteled dream of the New Age chocolate box artists, with their bucolic fantasies of a unicorned utopia, but the rational attempt to paint a positive vision of the world increasingly informed by a fascination with modern science: "In Euclidian geometry there are perfect shapes, perfect forms, like the cube. In fractal geometry the definition of a form is much closer to nature. We would be surprised if our lungs were a perfect cube, but in fractal geometry they can be seen as perfect forms. I find that absolutely engrossing because it is another way of seeing. And not only that; it is, at the same time, another way of feeling."[10]

The enthusiasm expressed in this quote is the enthusiasm for life itself. It is the joy of being conscious—of being able—not merely to *perceive* the wonder of the world around us, but also, increasingly, to *understand* it. For today it is the scientists, not the artists, who hear the beauty of the **New Music of the Spheres**.

It seems the cliché is true, that the real visionaries are never appreciated until history catches up with their way of looking at the world. History will come to regard the postmodern nihilists as the last gasp of the suffocated "soul" in the dying embers of the twentieth century. The future lies with those who, like David Hockney, possess the imagination, intellect, independence, and above all, the *will* to look around them and see not a world at war, but a world of wonder.

From the Art of the Ridiculous to the Art of the Technosublime

So let us reject the postmodern antiart of nihilism and embrace instead a sensual, emotional, intellectual, and uplifting art, which stimulates all faculties without resorting to the childish shock tactics of social taboo-breaking for its effects.

Let us rebuke the high priests and priestesses of postmodern nihilism and build a world to be proud of—a world of dignity and confidence, brimming with fabulous technology, sparkling with vitality, creativity, and pride.

Instead of abandoning self-belief and slipping quietly into the murky waters of postmodern cynicism, let us make ourselves a species worthy of communication with "other" intelligent life-forms—a world to which we could invite our extraterrestrial friends, without shame.

The postmodern anticulture of nihilism has had its fifteen minutes of fame.

Let us reject the postmodern art of the ridiculous for the transhumanist art of the technosublime.

NOTES

1. Horace, *The Complete Odes and Epodes*, trans. David West (Oxford: Oxford World's Classics, 2000).
2. Albert Lubin, *Stranger On the Earth (*New York: Holt, Rinehart & Winston, 1972).

3. Jeremy Rifkin, *The Biotech Century* (London: Victor Gollancz, 1998).
4. Jeremy Rifkin, "Dazzled by the Science," *Guardian*, January 14, 2003.
5. Friedrich Nietzsche, *Ecce Homo* (1888; repr., London: Penguin, 1979).
6. Ibid.
7. Norman Spinrad, "The Neuromantics," *Isaac Asimov's Science Fiction Magazine*, May 1986.
8. David Hockney, *That's The Way I See It* (London: Thames and Hudson, 1993).
9. Colin Wilson, *Introduction to the New Existentialism* (London: Hutchinson, 1966), p. 180.
10. Hockney, *That's The Way I See It.*

Chapter 13
NETAPHYSICS
The New Music of the Spheres

> Today the God hypothesis has ceased to be scientifically tenable . . . and its abandonment often brings a deep sense of relief. Many people assert that this abandonment of the God hypothesis means the abandonment of all religion and all moral sanctions. This is simply not true. But it does mean, once our relief at jettisoning an outdated piece of ideological furniture is over, *that we must construct some thing to take its place.* (Italics added)
> —Julian Huxley, *Religion without Revelation* (1927)

> Science offers the boldest metaphysics of the age. . . . Preferring a search for objective reality over revelation is another way of satisfying religious hunger. . . . We will in time close in on objective truth. While this happens, ignorance-based metaphysics will back away step by step, like a vampire before the lifted cross.
> —E. O. Wilson, *Consilience* (1998)

Metaphysics, and its contemporary scientific equivalent, cosmology, asks the question, What is the overall nature of the world?

By **Netaphysics**—or the **New Music of the Spheres**—I mean a new cosmology emerging from all areas of science in which the world is seen as a harmonious, evolving network of interacting parts, interconnected like strands in a spider's web. Such a view offers a positive alternative to theistic, pantheistic (New Age), and postmodern metaphysics, which either deny the blunt reality of death, or the ability of scientific rationality to explain the world at all.

I divide my metaphysics into three sections:

Theistic Metaphysics:	The Evolutionary Neuropsychology of Religion
Pantheistic Metaphysics:	Demystifying Mysticism
Transhumanist Metaphysics:	Netaphysics, or the New Music of the Spheres

Here I assert the naturalistic basis of the religious concepts of God, soul, spirit, spirituality, and Heaven and Hell in terms of **Evolutionary Neuropsychology**. I introduce a new position on the Ontological Question—**agnoskepticism** (rational doubt without certainty). And in preference to theistic and pantheistic metaphysics, I advocate the harmonious cosmology emerging from the new science I call **Netaphysics** or the **New Music of the Spheres**, which I outline in terms of reductionistic and systems science, and in its potential benefit both to the individual (**Neurotheology** or **"Mind of God" Theology**) and society (**Internetics**).

SUMMARY OF TRANSHUMANIST METAPHYSICS

The universe is neither a waiting room for heaven (theism), undifferentiated mystical oneness (New Age pantheism), or an unknowable, irrational chaos (postmodernism), but a magnificent metapattern of complexifying information. There is harmonious order in the cosmos. The emerging New Science of wholeness and order replaces both the childish nursery rhymes of theistic credulism, and the atonal dissonance of postmodern chaos with a **New Music of the Spheres**. In the new world picture emerging from science, nature is viewed as an evolutionary process of information complexification—a living, cosmic computer becoming aware of itself through its conscious aspect—the self-recognizing biocomputer we call the human mind destined for the future job of nature's reprogrammer. The **technosphere** resulting from the prevalence of advanced technocommunications in the modern world, in particular by the Internet, is leading to the evolution of a **cybermind**—a cybernetic extension of the human nervous system. Through an ever-expanding network of autonomously thinking, individual minds, communicating around the world like neurons in the neural net of a human brain, the species may increasingly come to regard itself as an interdependently functioning, conscious entity, or worldmind. If the cybermind can escape the dominance of the ultimate authoritarian Bodypolitic—the World State—it may serve to reduce conflict and increase cooperation without destroying individuality and creativity, in the evolving technowonderland of the twenty-first century (**Internetics**).

THEISTIC METAPHYSICS

THE EVOLUTIONARY NEUROPSYCHOLOGY OF RELIGION

> The final decisive edge enjoyed by scientific naturalism will come from its capacity to explain traditional religion, its chief competitor, as a wholly material phenomenon.
> —E. O. Wilson, *Consilience* (1998)

> Thereupon, in the year 2006 or 2026, some new Nietzsche will step forward to announce: "The self is dead"—except that being prone to the poetic, like Nietzsche I, he will probably say: "The soul is dead." He will say that he is merely bringing the news, the news of the greatest event of the millennium: "The soul, that last refuge of values, is dead, because educated people no longer believe it exists."
> —Tom Wolfe, *Sorry but Your Soul Just Died* (1996)

The Grand Delusion

No rational thinker could accept the claims of religious metaphysics. The essence of the theistic metameme might be stated thus:

The material world was created by an omnipotent ghost, inhabiting an invisible, undetectable dimension. Inside each human being resides a smaller ghost, called "the soul," which operates the physical body and will be rewarded by the big ghost with immortality in the invisible world on the death of the physical body it inhabits—but only if it worships the big ghost, and is kind to others. Otherwise, it will be sent to live forever in another invisible world inhabited by an evil ghost where it is unpleasantly hot.

No one has ever identified these spirit worlds of heaven or hell, and no two people agree on exactly what might go on there. No one has ever identified such a thing as a "soul"; rather, certain forms of human behavior have been subscribed to such an entity—much as the sound of thunder is sometimes ascribed by children to God moving the furniture.

Since no one has ever identified any of these phenomena—invisible beings, souls, heaven, hell, and the like—one must consider the question of why the human mind has conceived of and believed in such things to the extent that they have comprised the basic worldview of the majority of human beings throughout recorded history. The brute answer is that theists believe in the existence of something beyond the real world because they do not, unconsciously, regard this one as good enough to believe in. The real world of biological organisms is positively overflowing with the horrors of death, disease, evil, war, blood, sweat, and tears. Belief in a magical spirit world of pure bliss, and a self which defies death to live there forever, has simply proved far too great a temptation for the majority of humankind to resist.

The Grand Delusion is nature's memetic means of ensuring that humans do not confront the horror of their true condition—a brief, hard life followed by eternal nothingness. The religious metameme was invented through fear and preserved by natural selection for its ability to dispel existentialist anxiety about the inevitability of death.

The theistic metameme could properly be called "occultism"—a word originally meaning "hidden knowledge"—the knowledge in question being that of God, soul, heaven, and hell. This knowledge is certainly extremely well hidden, since no one at all has ever been able to produce it.

To theists, the prevalence of religious faith around the world is proof enough of God's existence. But the popularity of religion is no more an indicator of the existence of God than the popularity of Christmas is an indicator of the existence of Santa Claus.

There is no good reason for inventing invisible things. All religious phenomena are perfectly explicable in terms of evolutionary neuropsychology.

Religion

Religion is a type of meme map—a software program for the human brain, containing information as to the nature of the world and instructions on how best to live in it.

When spread among a tribe, a religion may be called a metameme—a shared belief system. A religion is a type of metameme based on faith, defined as hope without evidence.

Prophets, priests, and holy men are those afforded the social function of writing and uploading religious software into as many minds as possible, thence to continually update the program, in order to maintain its popularity.

The purpose of spreading a religious metameme is fourfold:

- To alleviate the fear of death
- To increase the tolerance of suffering in life
- To cement group solidarity
- To condition for benevolence through the promise of heaven and fear of hell

God

"God" is the anthropomorphization of that which humanity longs for most, but has so far been unable to attain: the power to avoid dying.

The word *God* means "that which will enable me to cheat death," imagined to be a superhuman being inhabiting another dimension.

"God" is the reification of the abstract quality of absolute power over life and death.

Religions which associate the ability to save them from death with strength, conceive of God as absolute power, and thus require total submission to his authority.

Religions which associate being saved from death with a benevolent mother conceive of God as love, and thus value gentleness and compassion.

Religions which associate being saved from death by forgetting one exists at all conceive of God as nothingness and value passivity and quietude.

People who think all this is a load of nonsense, and we should set out to cure death through science, are called transhumanists.

Soul

That which religion calls the "soul" is the reification of love—an evolved, neurochemical capacity for fellow-feeling imagined to be an immortal entity inhabiting the body.

The invention of the soul was a bribe by the founders of religion to encourage benevolence. For if the only part of us which survives death is that part capable of love (the immortal soul, which leaves the body), it follows that we must love to ensure immortality—quite an incentive to be good! Today we no longer require such a bribe. We recognize that benevolence is a rational survival tactic, an evolved predisposition to sociability in the interests of mutual survival.

When Nietzsche said, "God is dead," he meant to prophesize the imminent secularization of the modern world. Today, one might say, too, **"The soul is dead"**—not because we do not believe in love—quite the opposite—but because we no longer need to believe there is an immortal ghost living inside us in order to appreciate the *value* of love. We recognize that benevolence is common sense. "Goodwill to all men" is a rational tactic for mutual survival and well-being.

We no longer need God in order to be good—though a suicide bomber needs him to be bad.

Spirit

The word "spirit" derives from the Greek *pneuma*, meaning "breath." The original meaning of the word, then, is "breath of life"—life force, principle, or energy. To religions, "spirit" is the human personality regarded as an invisible, undetectable immaterial entity living within the body, which (if we please God), survives the death of our physical bodies to live forever.

However, the word *spirit* is also used to describe particular human characteristics, quite apart from any specifically religious inferences. In this sense, there are four attributes of "spirit":

1. Spirit as Physical Vigor	("spirited")
2. Spirit as Emotional Elation	("high spirits")
3. Spirit as Mental Courage	("brave spirit")
4. Spirit as Morality	("the spirit is willing")

All of these characteristics are biologically based. There is no need to posit the existence of a ghost-within to account for them. We can, however, begin to correlate them with specific neurochemicals:

Physical vigor is the product of adrenaline and androgens—evolved to assist hunting and mating.

Emotional elation is the product of dopamine—an evolved pleasure reward for survival behavior.

Courage is the ability to withstand (i.e., "not respond to") stimuli—the product of dominance-raising serotonin.

Emotional empathy is the product of oxytocin—evolved to assist child nurturing.

So the highest spirits result from high levels of neurohumors such as adrenaline, androgens, dopamine, serotonin, and oxytocin. Together, they produce a neurotype which exhibits the behavioral traits associated with the adjective *spirited*—liveliness, good cheer, bravery, and nobility of mind, or "human decency."

Conversely, the lowest spirits result from low levels of androgens (submissive behavior), dopamine (depression/inertia), serotonin (hypersensitivity), and oxytocin (emotional coldness). Together, they produce a neurotype prone to depression, anxiety, and anhedonia (inability to feel pleasure or love).

In short, the word "spirit" refers not to an invisible ghost inhabiting the body, but to neurochemically based dimensions of personality.

Spirituality

Spirituality is . . .

- A neurochemically induced feeling of universal love
- An extension of a neurochemical pleasure reward evolved to encourage benevolence, cooperation, and pair bonding in the interests of survival and reproduction
- Love divorced from any specific object of perception or thought, and thus experienced as a feeling of empathy for all things
- A pleasurable, neurochemically mediated feeling of "goodwill to all men"
- An evolved capacity for universal benevolence
- A feeling of loving everything, indiscriminately, caused by a surge of

chemicals in the brain—whether in response to meditation, drugs, or as a by-product of falling in love

The neurochemical basis of spirituality is probably a combination of oxytocin and serotonin, combining a maternal instinct for nurturing with a feeling of personal well-being, and absence of fear of others, resulting from a calm, unreactive nervous system.

Like the other varieties of the neurochemical attraction we call love—sexual, romantic, and maternal—spirituality is pleasurable, it is valuable, it is wonderful—but there is nothing mystical about it.

So let us be spiritual enough to enhance the neurochemistry of spirituality—that we may learn the art of creating universal love on earth—through science.

Heaven and Hell

The concept of "heaven and hell" is a behavioral reinforcement mechanism, serving to condition the tribe for benevolent behavior through the promise of reward and fear of punishment in the fantasy world of the imagination.

The power of the heaven-and-hell meme derives from its appeal to the greatest of all human desires and fears: immortality and death. Heaven and hell are symbols of absolute pleasure and pain: Be good—and experience absolute pleasure in Heaven. Be bad—and experience absolute pain in Hell. Quite an incentive to be good—for a dumb species. In the words of Einstein: "Humanity would indeed be in a poor way if he had to be restrained by fear of punishment and hope of reward after death."[1]

Theism

The main objection to theism for those who need it is not that it is untrue (one accepts, with a sigh, the psychological need of some to believe certain untruths as a defense mechanism against anxiety), but rather, that such a belief is not compatible with the social acceptance of Superbiology. For if one considers the world, and humanity, to have been deliberately designed by God in his image, and God's will is to be accepted unquestioningly, it follows logically that one must consider the practice of altering human biology to be immoral. Where that leaves the theistic parents of a child born with a life-threatening disease is a matter only they can decide.

If you believe that a human being is essentially an immaterial entity that survives the death of the body to live forever in heaven, you are hardly likely to be interested in curing death or creating a heaven on earth. Indeed, you are more likely to adopt extremist ideologies and behavior born of a "throwaway" attitude toward this life, such as this, expressed by the radical Muslim cleric Abu Hansa:

"This world is unworthy. . . . [W]e're talking about the authority who can give you a new body once your body is eaten by maggots. . . . He's the only one who's gonna replace your body."[2]

Those who believe they shall go to another world when they die may be less concerned of the havoc they cause in this one, but this life is the only one which concerns *transhumanists*, for this is the only one we have.

The fact that many nontheists also adopt the theist's disapproving stance against Superbiology (based on a fundamentally theistic "intuition" that it is somehow "not right" to interfere in the working of nature) only increases our desire to counter illogical beliefs.

The greatest threat to humanity's continuing evolution is theistic opposition to Superbiology in the name of a belief system based on blind faith in the absence of evidence. In the words of James Watson: "The biggest advantage to believing in God is you don't have to understand anything, no physics, no biology, I wanted to understand."[3]

Belief in the Prometheus Drive—the will to continually extend knowledge and abilities—is ultimately incompatible with those religious doctrines which see human weakness in the face of an invisible creator as a moral good.

Agnosticism

While it is nonsensical to declare certainty of God's existence, it is equally illogical to profess complete ignorance on the matter—the position known as agnosticism.

Essentially, agnosticism is atheism for cowards. Most agnostics do not really have "no idea at all" as to the possible existence of God—they are disbelievers who won't admit it for fear of upsetting theists. The Darwinist T. H. Huxley invented the term to avoid offending his clerical friends!

A second type of agnostic is merely "hedging his bets"—afraid of offending God, just in case he *does* exist and won't let him into heaven. The bet-hedging agnostic appears strangely unable to make the obvious deduction that an omniscient deity who rewards benevolence with immortality is hardly likely to be fooled or impressed by one who consciously decides to "keep an open mind" on his creator's possible existence merely to guarantee entry into heaven!

But then, of course, He moves in mysterious ways.

Atheism

While agnosticism cannot be considered a logical response to the Ontological Question, atheism, too, must properly be rejected on the same grounds. For it is logically incorrect to claim absolute certainty as to the nonexistence of any kind of creative intelligence in the universe.

It is quite possible, for instance, that we live in a sort of virtual reality—a computer program called The World written by some superintelligent alien programmer. Such a **Geek God of Virtuality** may even be observing our evolutionary progress, and waiting for a greater level of intelligence to emerge before making itself known to us. Such was the basis of Stanley Kubrick's film *2001: A Space Odyssey*, based on a short story by Arthur C. Clarke, who shares the view that the rejection of absolute atheism is illogical: "One of my objections to religion is that it prevents the search for god, if there is one. . . . I have an open mind on the subject, if there's anything behind the universe. And I'm quite sympathetic with the views that there could be."[4]

The belief in God as a sort of cosmic computer programmer, who wrote and set running the program called life without intervening in its subsequent effects, is known as deism. A popular view among Enlightenment rationalists, deism enabled them to retain a token religious belief in the face of a suffering world, by claiming the existence of a noninterventionist God who just looks on—and laughs, perhaps.

While the creator of our world certainly could have been some form of extraterrestrial computer programmer, to make the dogmatic claim that this is definitely the case—to believe in and place one's hopes in the existence of such a being—is wholly illogical. For we simply don't know!

It is perfectly possible that other forms of intelligent life exist in the universe, and perfectly sensible that we should search for them, not merely out of a sense of existential loneliness, but because such an advanced intelligence may help us to further evolve in the effort to improve our condition. But let us not waste time waiting for someone to save us from our misery! We have done enough of that in the past.

No more waiting for God or ET! Let us take our destiny in our own hands. In the words of James Watson: "We are the products of our genes. No one else is going to take care of us or give us rules for how to behave, except ourselves."[5]

Agnoskepticism

> I don't have the evidence to prove that God doesn't exist, but I so strongly suspect he doesn't that I don't want to waste my time.
> —Isaac Asimov, *Free Inquiry*, Spring 1982

Having rejected theism, agnosticism, atheism, and deism, we require a new word to describe a truly rational response to the Ontological Question.

I propose the term **agnoskepticism**, a combination of agnosticism (ignorance) and skepticism (rational doubt). Agnoskepticism consists of three principles:

1. To assert the existence of an omnipotent, benevolent creator in a world of inevitable, universal suffering is logically incorrect.

2. The existence of some kind of creative intelligence is perfectly possible, but in the absence of supporting evidence it is wholly illogical to place one's hopes in the existence or assistance of such an entity.
3. The only logical attitude is to assume we are alone in the universe, and seek our own "salvation" (self-improvement) through the most powerful tools at our disposal: logic, science, technology, benevolence, and the effort of will.

Agnoskepticism—doubt without certainty, self-help over solace—is a rational attitude for a rational species.

The sensible agnoskeptic has little interest in converting theists to his beliefs. Religion is a psychological security blanket—remove it and you risk inflicting emotional collapse on the vulnerable. It is also the dominant source of instruction in benevolent behavior for many millions of people. One should not attempt to remove the security blanket of religion—*without seeking to replace it with an alternative source of ethics*.

Agnoskepticism may, though, prove attractive to those who currently subscribe to atheism or its half-hearted brother, agnosticism, and remain unimpressed by the alternatives on offer in the memetic marketplace of the modern world: Luddite eco-fundamentalism, postmodern nihilism, theistic credulism, or New Age mysticism . . .

* * *

PANTHEISTIC METAPHYSICS: DEMYSTIFYING MYSTICISM

> I daresay you haven't had much practice," said the Queen. "When I was your age, I always did it for half-an-hour a day. Why, sometimes I've believed as many as six impossible things before breakfast."
> —Lewis Carroll, *Alice's Adventures in Wonderland* (1865)

New Ageism

The dominance of traditional theism in the West has been challenged over the past fifty years or so by the emergence of the "alternative religion" of New Ageism.

Essentially, New Ageism is Eastern religion repackaged for the West—a potpourri of Buddhism, Taoism, and Hinduism with a dose of Protestant enlightened self-interest and a dash of Kabbalistic occultism, to spice it up.

Though New Agers pay lip service to science (by trying to link quantum physics with mysticism), essentially, New Ageism is Neo-Medievalism—a return to the world of rabbit's feet and garlic.

New Agers have created a memetic whirlpool of nonsensical beliefs. More and more impossible things are being believed before breakfast.

New Age metaphysics are monistic (the world as one) or pantheistic (nature as God), depending on how Green you prefer your New Age. In the gospel according to New Ageism, God is immanent—somewhere behind the reality of rocks and stones and trees is the "real" world of "spiritual oneness." However, New Ageism retains the concepts of "soul" and "spirit" from theism, because these twin ideas are based on the underlying belief in the existence of another world behind the material "façade" of this one. The real or "true" self and world are somewhere else—nobody seems quite sure where—perhaps at "the quantum level." The proposed immaterial world—"cosmic oneness," Heaven, Nirvana, "noumenon," or "seamless coat of the universe," depending on your spin, is regarded as a spiritual unity synonymous with God, while the "soul" is the personal aspect of that immaterial whole.

In short, according to the New Age metaphysic, True Reality is not that of rocks and stones and trees and breast cancer, but a cosmic spirit world to which an undetectable squatter in the body called the "soul" returns once disease has destroyed the body.

Oh, that's all right then. No need to bother curing cancer.

Quantum Mysticism

It has been fashionable among New Agers for the past thirty years or so (in fact, precisely since the publication of Fritjof Capra's book *The Tao of Physics*) to claim that quantum physics supports their metaphysical monism, by suggesting that "all is one at the metaphysical level."

At the subatomic level of matter, single particles merge into "packets" of energy known as "quanta." New Age mystics see quantum energy as evidence of the "spiritual unity" behind the material world of things.

What New Agers seem unable to appreciate is the fact that the subatomic, quantum level of matter is utterly irrelevant to our existence as physical entities in the macro-universe of concrete objects. For we simply do not live at the quantum level! Indeed, to enter it would mean to die.

Quantum Mystics say, "All is One" behind appearances because there are no discrete, distinct concrete objects at the quantum level of observation. But the world is also "one" in the womb, prenatally, before the development of a physically autonomous and consciousness self. The problem is, we don't live in the womb! Nor do we live at the quantum level! Instead, we live in a very beautiful, fascinating material world full of distinct things—and if we are wise, we will avoid drifting off into empty-headed, wishful thinking concerning the possible existence of another one in which we don't get cancer.

Knowledge of the nature of reality at the quantum level has no relevance whatsoever to the fact of our existence as distinct entities programmed to die against our will.

Mystical Scientism is just another attempt to hide from death, by thinking of it as going "back to the source." Transhumanists don't want to go back to the source—we want to cure death. And at least we are honest about it.

New Age Stasism

New Agers say "Abandon Promethean fantasies of endless progress. Live in the eternal here and now. All is one. Time and death are illusions."

But the world is not one. It is billions of organisms devouring one another in an effort to stay alive. Time and death are not illusions. We are born, we live, we die—usually in agony.

"Living in the here and now" is what children do before they grow up—another name for it is "irresponsibility." The basis of civilization is the ability to plan for the future we call forethought. It is one thing to consider the lilies of the field, but "pay no thought for the morrow" and you're likely to end up on the street.

New Age No-Selfism

New Agers say, "Reject Cartesian duality—mind and body are one."

But mind and body are not one—they are two. That is why we have two words to describe them.

The dichotomy between mind and body is not an illusion but a painful reminder of our own mortality.

The unity of mind and body is possible only at the expense of the mind—by attaining the unconsciousness of cows in a field.

Religions that seek to eradicate the mind-body dichotomy do so by seeking to empty the mind. Trying to solve the problem of mind-body duality by emptying the mind is like trying to eradicate death by not thinking about it.

Only by recognizing a problem exists can it be solved. Only by acknowledging suffering can it be healed. Mind and body are not one but two. Mind wants to live, the body is programmed to die. Ego and id are at war, and if we are to survive, ego must win.

New Age Anti-Egoism

For New Agers, "the ego" has replaced the devil as the symbol of supreme evil. This is probably because it's easy to identify the ego, while the devil has only ever been spotted in bad horror movies.

To New Agers, "the ego" is a pejorative term for the self—hence desiring an "ego boost" is considered bad, while "raising self-esteem" is good—even though the two phrases refer to precisely the same thing.

The "ego" is just "the self"— an embodied biocomputer evolved to the level of complexity at which it becomes able to recognize its own existence—the facility we call "consciousness."

It was the evolution of "the self" which made possible the development of civilization, through the exercise of the brain's highest capability—reason. The emergence of the rational mind enabled humanity to create, plan, and think ahead, instead of living in the present like animals with smaller, less complex brains.

We owe to "the grasping ego" the very existence of human civilization. Without it we would still be living in caves.

Some people want us to return there.

Egoism versus Egotism

The demonization of the ego derives largely from a common misunderstanding about the distinction between the concepts of egoism and egotism.

Egotism means conceited self-obsession and disregard for others. *Egoism*, on the other hand, means "the ethics of self-interest." Egoism is necessarily the basis of all moral codes. Human biology dictates that self-interest is the basis of morality, for we are programmed for self-preservation—the will to live is our very essence. Only the suicidal masochist lives by a code based on something other than self-interest—and then we call it psychopathy.

Egotism, on the other hand, is not synonymous with Sensible Self-Interest. Arrogant, inconsiderate behavior is detrimental to self-interest because it automatically induces hostility in others. Lack of consideration for others is the essence of Stupid Selfishness. Egotism is but a few steps from "evil"—the difference being that egotism is out for itself, while evil is deliberately out to harm others.

It is not the ego that is bad, but egotism; not rational egoism, but irrational inconsideration for others; not Sensible Self-Interest, but Stupid Selfishness.

Meditation

New Agers say, "Abandon striving. Go with the flow. Experience the blissful transcendence of the ego through meditation."

Meditation is the doze's first cousin. The benefits of meditation are those of an extra hour in bed with your eyes half-open. The aim of meditation is the emptying of the mind. Its main virtue is to stop the chattering classes thinking too much—about themselves.

The New Age belief in "transcending the ego" through meditation derives from Buddhism, which holds that suffering is the essence of life and can only be alleviated by reducing life functions to a minimum, through sitting still, doing nothing, and losing all desire for anything at all. This is rather like treating a wart by cutting off your finger, or responding to a danger by lying down and playing dead.

When the mystic "loses his ego," he does not "melt into the oneness of the universe"—he stops thinking. When he stops thinking, he stops striving. If humanity stops striving, suffering will continue—indefinitely.

Without self-hood, there is passivity. Where there is passivity there can be no change. Where there is no change—people perish.

Mysticism is a cowardly retreat from the awareness of self and death. You can't meditate your way out of mortality. New Age mysticism is just another attempt to hide from death by not thinking.

Neurotheology—The Bliss Chemical Theory

The attraction of mysticism is not hard to understand—after all, everyone loves a mystery. But there is nothing mystical about mysticism.

The "bliss of mystic experience" is simply the pleasurable feeling associated with a surge in neurochemicals released during meditation as nature's reward for the attainment of the perfect state of homeostasis conducive to survival—complete mental and physical relaxation. Such a state places the least strain on the body, both by "emptying the mind" and by reducing the need for food, urination, or defecation—behavioral drives which "wake up" the brain by signaling a state of "want."

In other words, sitting still and doing nothing helps the body to recuperate (as do more conventional forms of relaxation, like lying in a hammock), so nature rewards and encourages the practice by secreting a "bliss chemical"—probably the neurotransmitter serotonin.

The same experience is now widely available in pill form on the black market so one need no longer attain the lotus position to attain it. It's known as "ecstasy."

Mystic experience is not evidence of a metaphysical spirit world, but a pleasant feeling produced by a surge in serotonin levels—the side effect of an evolved neurochemical conditioning mechanism, which rewards and encourages both relaxation and empathy in the interests of increased survivability.

Nirvana is a neurochemical called serotonin.

Scientific Superconsciousness

Mystic experience, demystified, does not lose its potential value.

The feeling of universal empathy characteristic of mystic experience is beneficial to survival because benevolence is a rational survival tactic. "Spirituality" is the feeling of falling in love with the world. One is more likely to do well in life if one feels in love with the world than if one rails against its unfairness and brutality.

Humanity will inevitably learn to re-create "mystic experience" through control of our neurochemistry, without the need to live in a tree or endure the lotus position to attain it.

The twenty-first century will reject New Age enchantments for neuro-enhancement.

* * *

TRANSHUMANIST METAPHYSICS: THE NEW WORLD PICTURE

> With postmodernists to the left of us and diehard Darwinians to the right, the evolutionary perfectionism of a Teilhard de Chardin goes down like a tepid science fiction. Nonetheless our global civilization continues to bank on the revolutionary promise of progressive technological change, a quintessentially modern perspective.
> —Erik Davis, *Techgnosis* (1999)

Netaphysics: The New Music of the Spheres

While premodernists delude themselves with wishful thinking, and postmodernists mock a world without meaning, modern science is busy discovering a **New Music of the Spheres**.

In the New Music of the Spheres, evolution is regarded as an endless symphony of harmonious complexity—and humankind destined for the role of its new conductor.

FOUR EXPRESSIONS OF THE NEW MUSIC OF THE SPHERES
THE WORLD PICTURE EMERGING FROM THE NEW SCIENCE

1. Reductionistic Science
2. Systems Science
3. "Mind of God" Theology
4. Internetics

Here I briefly outline the basis of the emerging scientific world picture, concentrating on the field of Systems Science, before exploring its potentially positive impact on the individual ("Mind of God" Theology) and society (Internetics).

The Limitations of Twentieth-Century Science

The term *metaphysics* generally implies theories as to the world that stand beyond or contradict the scientific description of nature; hence the common association of metaphysics with religion, the supernatural, and corresponding idealistic philosophies.

Where science analyzes specific aspects of the world, metaphysics generally implies the attempt to describe the significance or purpose of the world as a whole, beyond mere descriptions of its structures, forms, or operations.

In short, metaphysics is the attempt to describe the meaning of life.

Contemporary science has replaced metaphysics with "cosmology"—the scientific description of the greater universe of planets, stars, and galaxies. But at what cost? For unlike metaphysics, cosmology offers no suggestions as to the significance or purpose of the world—that is, the "meaning of life."

By abandoning metaphysics, science has abandoned ethics—the question of what to do with one's life—for how can a philosopher suggest one way of living over another, if the world itself is without purpose or meaning.

The divorce of ethics from science may not have been a problem had the philosophers taken it up. But instead, twentieth-century philosophers abandoned the attempt to derive a meaning of life from an understanding of the natural world—ethics from metaphysics. An intellectual taboo was placed on the attainment of ethical principles from facts about nature. Ethics and nature were divorced.

But if one rejects both theistic metaphysics and scientific naturalism as a source of ethics—where is there to go—except to abandon ethics altogether.

As a result, we now live in a postmodern era characterized by the absence of belief in any universal significance to human existence—a world without purpose—without a "meaning of life" for those sufficiently hardy of mind to reject the theistic metameme.

Where meaning is lost, morality follows. For why be good in a meaningless world?

Postmodern nihilism threatens the very survival of the modern world, by eliminating the necessary basis of a moral code—a meaning to life.

The solution is not a return to the comforting delusion of theistic metaphysics, but the rediscovery of a *naturalistic* metaphysics—the function, purpose, and significance of man—regarded as a conscious aspect of an evolving universe.

The New Science

Modern man needs a world-picture that offers meaning. Such a world-picture is now beginning to emerge—from science.

Increasingly, the world described by the **New Science** is not one of randomness, irrationality, and dissonance—but of evolving complexity and harmony.

Where the **Old Science** depicted the world in terms of emotionally loaded negamemes such as *naked apes*, *survival machines*, *selfish genes*, *cosmic fluke*, and *chaos*, the New Science depicts the world in terms of emotionally uplifting memes such as *Elegant Logic*, *Spontaneous Order*, *Self-Organization*, *Emergence*, *Complexity*, *Sync*, *Nexus*, *Connected*, *Linked.*

Where the Old Science viewed the world as a meaningless, purposeless fluke governed by chance, the New Science sees human benevolence, intelligence, and the ever-increasing order and complexity of the cosmos as inevitable products of the innate directionality of evolution.

One need only glance through a list of popular science books published over the past fifteen years or so, to recognize the message of the New Science. Today, it is not *The Selfish Gene* and *The Naked Ape* but titles such as *The Elegant Universe* and *From Chaos to Order* that litter the bookshelves of popular science. Here are just a few of the popular science titles that have emerged during the last years of the twentieth century and opening years of the twenty-first, revealing the dramatic change in tone, from negative to positive:

The Elegant Universe (Brian Greene)
The Harmonious Universe: The Beauty and Unity of Scientific Understanding (Keith J. Laidler)
At Home in the Universe: The Search for the Laws of Self-Organization (John Archibald Wheeler)
Linked: How Everything Is Connected to Everything Else and What It Means (Albert-László Barabási)
Science and Sensibility: The Elegant Logic of the Universe (Keith J. Laidler)
Emergence: The Connected Lives of Ants, Brains, Cities, and Software (Steven Johnson)
Emergence: From Chaos to Order (John H. Holland)
Sync: The Emerging Science of Spontaneous Order (Steven Strogatz)
Nexus: Small Worlds and the Groundbreaking Science of Networks (Mark Buchanan)
The Origins of Order: Self-Organization and Selection in Evolution (Stuart A. Kauffman)
Hidden Order: How Adaptation Builds Complexity (John H. Holland)
The Moment of Complexity: Emerging Network Culture (Mark C. Taylor)

Get the (new world) picture?

Reductionistic and Systems Science

In the New World Picture (or metameme) emerging from modern science, the stuff of the world at every level, from subatomic to cosmic, is seen essentially as a dynamic, complexifying metapattern of information.

A pattern is a consistent arrangement of parts that may be communicated from one source to another. Information is a pattern which maintains its shape for long enough to be recognized by an information processor such as the human brain. A pattern consists of two elements: structure and form; the individual parts, which make up the structure, and the overall form of the pattern as a whole, resulting from the interaction of parts. Correspondingly, there are two different strains within the New Science working to construct the emerging metapattern of the world—Reductionistic Science and Systems Science.

Reductionistic Science

Corresponding to the "parts" aspect of the world, Reductionistic Science tells us we can *decipher* the metapattern of the cosmos in order to improve those parts of it which "let us down" (through death, disease, and biological limitation) in pursuit of ever-increasing survivability and well-being.

Systems Science

Corresponding to the "whole" aspect of the world, Systems Science tells us there is a fundamental order, harmony, and balance to the universe, seen as an evolving unity-in-diversity—an organized metapattern of information composed of interrelated parts; thus contradicting the theistic and postmodernist assertions that the world is unknowable through science.

Here I shall focus mainly on the latter: the newly emerging Systems Science in which the world at every level is viewed as a network of harmonious interconnections.

Through this brief outline of the New World Picture, I seek to demonstrate that the emerging depiction of the world according to twenty-first-century science is not that of a random, and therefore undecipherable chaos, but a complex, harmonious system; a network of interrelationships based on logically comprehensible processes. I call it the New Music of the Spheres.

The New Music of the Spheres offers a logical alternative both to religious metaphysics based on wishful thinking and to the claim of the postmodernists that the world is too irrational and chaotic to understand at all.

REDUCTIONISTIC SCIENCE

> Life is just bytes and bytes and bytes of digital information.
> —Richard Dawkins, *River Out of Eden* (1995)

Reductionistic Science involves the breakdown of complex patterns into their constituent parts, in order to determine their structure and mode of operation. This has been the basic methodology of science to date, its most recent triumph being the decoding of the human genome—the bits of information contained in strings of chemicals within living cells, which act as a chemical template for the construction of bodies. The apotheosis of reductionism came with the discovery that living things are constructed like computer programs, from long sequences of digital information. The word *digital*, derived from "digit" (Latin for "finger"), means discrete, or distinct—as opposed to continuous, or "analogue." Digital information consists of individual elements, as opposed to continuous amounts. A continuous sound wave is an analogue signal; when translated into bits of information, it becomes digital. "Digital" thus describes communicable patterns of distinct parts.

Both computer programs and human bodies are now recognized to consist of patterns of digital information—bytes on one hand, genes on the other. The digital patterns of bits or bytes of information that comprise computer programs are written in a "binary" language consisting of two numbers—O and 1—corresponding to the basic states, "on and off." The digital patterns of genes that make up bodies are written in a "quaternary" language consisting of four chemicals known by their initial letters—A, C, G, T. The Book of Life is written in patterns of digital information.

The full impact of the extraordinary realization that living things are constructed in precisely the same manner as computer programs has yet to find its way into the public consciousness. Its main significance here lies in the inference that life is governed by precise mathematical laws—as opposed to being random, or unknowable, as the pre- and postmodernists would both have you believe—either because the creator moves in mysterious ways (science deemed incapable of explaining tragedies, miracles, or wonders), or because nature operates irrationally (for to understand nature is to control her, and "Control = Domination = Bad").

The metaphysics of premodern theism and postmodern relativism are simply wrong. It is no longer valid to promote the view that the world is unknowable through reason. The New Science has firmly put an end to the intellectual validity of belief in the intrinsic irrationality of life and the incomprehensibility of the universe. Life is not irrational at all, but constructed on the most logical principles possible!

Critics of reductionism like to associate it with cold-hearted men in white lab coats. But no one seriously contemplating the extraordinarily complex, yet sublime harmoniousness of the process by which human beings are created could fail to be struck by a sense of awe and wonder at the miraculous *logic* of life, by which precisely ordered strings of chemicals in tiny molecules within our bodies provide the complete instructions to build Life in all its extraordinary complexity.

SYSTEMS SCIENCE

Complementary to Reductionistic Science is an emerging holistic or Systems Science.

A system is an organized arrangement of independent but interrelated parts forming a unified whole. A pattern is a *communicable* system—a system composed of information that can be copied from one source to another. Where Reductionistic Science analyzes the parts within the patterns of nature, Systems Science focuses on the connections within and between those patterns.

A systems view is becoming increasingly necessary in all areas of science, as scientists seek to identify how patterns in nature form and connect at different levels of operation. Together, Reductionistic and Systems science provide the methodology by which to construct a unified body of scientific knowledge—a comprehensive meme map of the megapattern of nature we call The World.

Here I offer merely the barest outline of the systems approach in the New Science, with apologies to scientists for my inevitable simplifications and errors. This is a book of philosophy, not science. What is significant in the present context is not the precise scientific accuracy of my descriptions, but the *psychological and ethical implications* of the emerging **New Science** of wholeness and order.

EIGHT EXAMPLES OF TWENTY-FIRST-CENTURY SYSTEMS SCIENCE

1. BIOLOGY:	SYSTEMS BIOLOGY
2. MATHEMATICS:	COMPLEXITY THEORY
3. SOCIAL SCIENCE:	NETWORK THEORY
4. CHEMISTRY:	DISSIPATIVE STRUCTURES
5. EVOLUTIONARY BIOLOGY:	CREATIVE EVOLUTION
6. BRAIN SCIENCE:	NEURAL NETS
7. COMPUTER SCIENCE:	SUPERCOMPUTERS
8. PHYSICS:	SUPERSTRING THEORY

1. Biology: Systems Biology ("Omics")

The identification of genes is only one step toward a complete understanding of the biological makeup of human beings. Now that the Human Genome has been decoded, we are entering a new phase of research: the attempt to understand complex biological systems. A New Biology is increasingly becoming necessary, which focuses beyond the level of the genes to the structures and forms into which they combine to produce a human being.

The term ***Systems Biology*** has emerged to describe the complex interrelationship between all aspects of the process of life: genes, proteins, metabolism, neurotransmitters, tissues, organs, organisms, the living environment, and the ecosystems on which they depend. Systems Biology consists of a dizzying variety of emerging disciplines known collectively as *Omics*, which involve the study of the various interacting levels of life. Genomics is the study of the genome—the complete pattern of genetic information, which provides the template to build a living body. Genomics is now being complemented by new disciplines such as *proteomics*, *metabolomics*, and *phenomics*, as scientists move from the study of genes, to the proteins they construct, to the activity within and between cells, and the effect of all structural and functional levels on the behavior of the resulting organism as a whole.

The end result of all this feverish activity in the New Biology will be the most comprehensive account of the biological nature of human beings ever produced. Such a development will represent the triumph of Systems Science—the analysis of the interconnected "nature of nature."

The significance of Systems Biology in the present context is the view of life it presents as a harmonious, integrated network of complex interconnections—as opposed to the worldview emanating from both pre- and postmodernists, both of whom see life as a sort of eternally unknowable mystery, unfathomable by biological science—for God and Marx move in mysterious ways.

ODC Systems

One of the most significant characteristics of the emerging Systems Science is the type of causal connections that need to be studied. To put it crassly, where genetics was linear—the study of one gene at a time—genomics is nonlinear—the study of complex effects resulting from the interactions of many genes, proteins, and metabolic processes.

In Systems Science, the interrelated network of information patterns that make up the world are subject to nonsequential connectivity. In other words, given a pattern consisting of parts A, B, C, and D—rather than a linear sequence of causal connections by which A affects B which affects C, like a single line of dominos—instead, parts A, B, C, and D may all affect one another, like strands in a vibrating spider's web.

The term *nonlinear* is generally used by scientists to describe such complex interactions. The difference between linear and nonlinear causal connections is the difference between a straight line and a pattern—a bootlace and a net.

*Non*linear, of course, is a negative. Thus, it tends to imply that something is amiss, not quite right, not as it should be—*not* ordered. But this is not the case. "Nonlinearity" is *not* "disorder" but, rather, *complex* order. The semantic difference is important—not to science or scientists perhaps, but to the way in which scientific knowledge is perceived by the general public.

And that is what is important here—not what scientists may think among themselves, in their own self-enclosed, self-referential world, but how the information produced by scientists is conveyed to the general public. For the public's perception of scientific knowledge is of great importance to both psychological and social well-being. The health of individual and society depends upon the metameme—the dominant worldview within a culture. *The message is the meme.* Metaphysics determines both psychology and ethics. Our understanding of the world determines both how we think of ourselves and how we propose to behave. There is likely to be a great difference in the psychological effect of between being told on the one hand that the operating system of nature is one of "determinative chaos," as nonlinear systems have also been described, and on the other, that nature operates according to rules of "harmonious complexity."

For this reason, I prefer to call "nonlinearity" by a different name—**Omni Directional Connectivity (ODC)**—meaning connections between events (things and processes) in complex, dynamic, web, or net-shaped systems—which appears to be the fundamental type of system operating in nature.

2. Ecology, Geometry, Mathematics: Complexity Theory

Patterns of Omni Directional Connectivity—or **ODC Systems**—are the basis of the emerging science of Complexity; the study of dynamic (that is, functional) systems in nature, from weather patterns to political economies, which operate by means of nonlinear connections, such that small initial changes in one part of the system may have large, unexpected effects on the system as a whole, which cannot be predicted because of the sheer complexity of the connections involved.

In this context, the word "complex" means something different from "complicated." Complex refers specifically to the property of nonlinear or ODC Systems, the behavior of which cannot be predicted from initial changes in the system.

The best-known example of a complex ODC system is the idea that the fluttering of a butterfly's wings on one side of the world might affect the weather pattern on the other (apparently the original example was not a butterfly but a seagull). This suggests, of course, that the whole of nature is one grand ODC System—better known as an *ecosystem.*

Thus, nature itself appears to be an ODC System—a complex, dynamic metapattern of interconnections producing an "exponential divergence of trajectories," by which activity in one part of the pattern spreads out quickly to many other parts.

Fractals

Omni Directional Connectivity operates in the formation of the dynamic geometric patterns known as fractals, produced when certain mathematical equations with geometric correlates are fed into a computer and left to run of their own accord.

The best-known fractal—the Mandelbrot Set—is a simple equation which produces extraordinarily organic-looking, "paisley-style" patterns of ever-increasing complexity, in which each individual segment, when magnified, mirrors the pattern of the whole—a property known as "self-similarity."

Fractals have been called "the geometry of nature," and indeed, one only has to observe the obvious resemblance of such shapes to organic life-forms to feel that here science has discovered the mathematical order behind the structures and forms of nature.

The fact that such patterns grow of their own accord once a simple initial equation is fed into a computer should be food for thought for those who doubt the possibility that the myriad forms of nature could have evolved by themselves. If a simple equation can produce forms of such marvelous complexity in a few minutes, think what millions of years of evolution could achieve!*

Common fractal structures in nature include clouds, mountains, rivers, and blood vessels. Compare the branchlike structures of the human lung with a tree, or a piece of broccoli—then consider the possibility of a grand computer program called Life as a set of infinite equations, left to work themselves out, forever producing ever more complex geometric structures, forms, and processes . . .

Suddenly, the vision of a future cosmology comes to mind, and with it, a new breed of scientist-artist-philosopher—**The New Pythagoreans** —in awe at the mathematical harmony and unity of the cosmos.

Such is the nature of the new paradigm emerging from science in the twenty-first century.

From Chaos to Complexity

Fractals were initially popularized under the name "Chaos Theory." But, again, the choice of terms is misleading. For chaotic phenomena are not chaotic at all. Rather, they are composed of very complex Omni Directional Connections, the precise development of which the human mind is not sufficiently capable of predicting, at this point in its evolution. The fact that the outcome of Omni Directional Connectivity is too complex to predict with absolute certainty (as in the case of changing weather patterns, for instance) does not mean that such processes are random or chaotic, but merely that the sheer complexity of casual connections they involve make a completely accurate prediction of their development beyond our present capabilities. It may well be that such predictions will require laws of probability rather than absolute certainty since there appears to be an element of *freedom* in matter.

---

*In July 2004, newly uncovered fossils from Newfoundland, Canada, revealed that the Earth's earliest, complex forms of life, first appearing on the ocean floor 575 million years ago, were fractal structures called "rangeomorphs." Scientists believe that their classic fractal structure of "branches from which branches grow" allowed simple genomes to assemble into larger organisms. (*New Scientist*, July 15, 2004)

Again, scientists prefer the term *randomness* to *freedom*. But what is randomness but the property of material entities which allows them to "do their own thing" without constraint? Just as you cannot predict with absolute certainty what another person is going to do next—because they are relatively free to do as they choose within the parameters set by current biological limitations—so, too, there is freedom in matter—the freedom of "flexibility within limits."

It is the property of freedom in matter which allows for the process of evolution. Evolution is the automatic ability to construct new, unexpected patterns in nature. Without the freedom of parts to alter their position within a pattern, there could be no change—and thus no evolution. Evolution exists because of the freedom of movement within nature.

Freedom is a prerequisite for creativity. Creativity is the ability to construct new, unexpected patterns in nature. No freedom equals no creativity. Evolution is a consequence of creative freedom.

Chaos is not chaos at all, but sublime complexity.

The message that the operating system of nature is one of chaos will have a very different psychological effect than the message that the world operates according to rules of complexity, even though they mean the same thing. So let us choose our terms carefully. Just as the Selfish Gene would benefit from the memetic enhancement of the Benevolent Gene, so let us speak not of Chaos, but of a New Science of Complexity.

3. Social Science: Network Theory

The idea that Complexity operates in human cultures has been applied under the name Network Theory.

In Network Theory, social systems are seen as dynamic networks of relationships. To solve a social problem, one should seek to identify the hubs—the parts of the system which have the greatest affect on the whole by virtue of their greater connectivity. To simplify, imagine a society as a tree. If you want to "get to the root of a problem" in society, you should go to the roots from whence all the parts branch out.

Essentially, then, this is a scientific retelling of the truisms that one must "deal with the problem at source," "get to the bottom of things," or "identify the cause, not the symptoms." Its main significance here is the recognition of complex Omni Directional Connectivity, such that all parts of a whole may affect one another—in unexpected ways.

Complexity Theory is increasingly being applied to the world of economics and business. One lesson for the head of a company implied by Complexity Theory is not to ignore the problems of even the least significant member of staff. For what starts with a disgruntled cloakroom attendant could end with a disgruntled workforce, a collapse in productivity, and a slump in profits—

through the operation of complex Omni-Directional causation. Essentially, Complexity Theory here is describing the manner in which *gossip spreads*!

Similarly, as we have seen recently, the actions of a single Rogue Trader may bring an entire merchant bank to financial ruin. And cancers, like Rogue Traders acting on their own, may be seen as the product of Rogue Genes, which destroy the body by affecting many parts at once as they grow—hence the difficulty in destroying them.

The message of Complexity Theory for society is that we cannot ignore the importance of associations, relationships, networks. If human beings are a part of a metapattern of harmonious complexity, it makes sense that we should act accordingly—by seeking to foster and maintain harmonious networks of associations—in our own best interests.

4. Chemistry: Open Systems—Spontaneous Order through Self-Organization

A reaction between certain inorganic chemicals in a petri dish results in the sudden appearance of beautiful, dynamic, geometric patterns of circles or spirals. The Belousov-Zhabinsky Reaction (named after the Russian chemists who discovered it) is an example of an Open System—a pattern in nature which maintains its existence by automatically changing its internal structure in response to the environment.

By contrast, "closed systems" are those which inevitably run down through loss of energy—like a clockwork toy—through the inability to interact with their environment in order to maintain internal energy—the essential quality of living things.

In short, life is an Open System. Where closed systems are subject to the law of entropy, gradually running down to motionless stasis, or death, Open Systems are subject to the opposite law of syntropy or extropy—the automatic ability to sustain their own existence by interacting with their environment, and altering their structure accordingly, where required for survival—a property known as self-organization. Self-organization is the ability of life to produce **Spontaneous Order**.

The significance of self-organizing Open Systems is the fact that it overthrows the previous scientific orthodoxy, by which all processes in nature lead inevitably to eventual dissolution—that is, that everything runs down, like clockwork—including the universe as a whole. This view—the law of entropy—is the scientific underpinning of the regular newspaper reports we receive reminding us that the universe is doomed to end in dust. Just in case we were starting to get too enthusiastic about life on this beautiful planet.

Open Systems are also known as Dissipative Systems because they paradoxically maintain their structure by losing or "dissipating" energy into the surrounding environment, at the same time as taking in new sources of energy (such as light, heat, and food). Dissipative Systems was the term chosen by their discoverer, Illya Prigogine, the Nobel Prize–winning, Russian-born, Belgian chemist whose 1977 book, *Order Out of Chaos*, is a seminal work in the emerging Systems Science. But note, again, the difference between the two terms, *Open* and *Dissipative*: the first positive, the second negative in its connotations, yet describing precisely the same thing. Which term paints the more uplifting picture of life?

Words matter. For he who holds the power is he who controls the meme.

5. Evolutionary Biology: Beyond Darwinian Evolution?

Even the neo-Darwinian gospel of evolution by random genetic mutations may not be immune from revision by the Systems Science of Spontaneous Order through self-organization.

In Darwinian biology, "adaptation" is the process whereby fortuitous spelling mistakes occur during the copying of an organism's genes from one body to another (reproduction), resulting in some new attribute in the new body, which makes it more likely to survive than the old. It follows that organisms structured in such a way as to make beneficial genetic mutations more likely to occur are more likely to survive. Thus, the gene pool will increasingly become dominated by organisms better able to change their biological makeup automatically, in ways beneficial to their survival. The animals best able to survive are those best able to capitalize on genetic spelling mistakes by automatically rearranging them into a new "word" that "makes sense." So the ability to evolve *itself* becomes subject to evolution. John Stewart, author of the book *Time's Arrow* (which argues that evolution is progressive), expresses the idea thus: "Genetic evolution is not random or entirely blind. Genetic systems are like nervous systems and brains—they have been structured and organized by evolution to enhance their ability to discover effective adaptations."[6]

As living things have evolved, they have become better and better at automatically restructuring themselves into new patterns, in response to genetic spelling mistakes. Spontaneous order through self-organization is nature's way of staying alive—without outside assistance.

How are organisms able to spontaneously respond to genetic spelling mistakes by reordering their internal structure? Perhaps by Omni Directional Connectivity—the ability of an organism to act as a single system, through an interdependency of parts, produced by complex connectivity operating in all directions at once—like vibrations in a spider's web.

Genetic mutations may not be *entirely* random, as presumed by the neo-Darwinists, but governed by Omni Directional Connectivity, such that one genetic spelling mistake occurring either spontaneously or in response to an environmental stimulus automatically results in simultaneous changes to other genes, producing a sudden shift in shape—the formation of a new structure—through a process of spontaneous multi-mutation.

Evolution may proceed not by the gradual accumulation of random mutations, but through the ability of living things to automatically *capitalize* on genetic spelling mistakes by spontaneously reordering their internal structure, like anagrams, to produce new words—new life-forms.

Creative Evolution by Self-Directed Mutation

The ability to create new patterns from the stuff of the world is the definition of creativity. The door thus opens for a theory of **Creative Evolution through Self-Directed Mutations**, made possibly by Omni Directional Connectivity. Creative Evolution in this sense is the theory that biological evolution proceeds not by the gradual accumulation of entirely random genetic mutations (as dictated by standard neo-Darwinian orthodoxy), but rather by the automatic restructuring of many gene sequences simultaneously in response to initial random or environmentally induced mutations.

Creative Evolution provides a mechanism for the evolution of new species in a relatively short time, lending weight to the theory of "Punctuated Equilibrium," which holds that species evolve not by gradual or incremental change, but by relatively sudden leaps—thus accounting for the so-called missing link—the absence in the fossil record of a species intermediate between man and ape.

Creative Evolution theory would represent a paradigm shift away from the view of evolution as a random process in a universe doomed to gradually run down by losing its internal order, to a new recognition of the underlying order and harmony of nature as a unified system. Such a view is now no longer the preserve of a few renegade heretics, but is "in the air": "It's part of a quiet paradigm revolution going on in biology, in which the radical randomness of Darwinism is being replaced by a much more scientific law-regulated emergence of life."[7]

The message of Systems Biology appears to be that evolution is not random at all. The parts of the world possess an innate tendency to spontaneously come together in the formation of ever greater wholes, ever larger patterns of complex organization. The phenomenon of self-organization in nature rebuts the notion of the world as something chaotic, directionless, and guaranteed to end in nothing. Instead, nature fights entropy with extropy—the spontaneous creation of order. The world is not falling apart, but constantly trying to "pull itself together."

The phenomenon of Spontaneous Order reveals nature to be a *creative artist*, making use of the material at its disposal, as Michelangelo made use of the flaws in a block of marble. Just as nature fashioned the primeval soup into man, and Michelangelo fashioned a block of stone into *David*, so transhumanity is destined for the job of fashioning nature into ever more beautiful forms. By learning to creatively enhance our own genome through Supergenics, we shall be acting not *against* nature, but in accordance with its modus operandi. Human beings are part of the artistry of an evolving nature. As nature becomes aware of itself through the human mind, it becomes able to direct its own creative evolution. Our task is to act in accordance with nature in its role of evolutionary complexification—and *against* it in its other role—that of destroying us, through aging, disease, and death.

Nature may be a creative artist, but man is catching up fast. Once we have learned all we can, the apprentice will become the new master.

6. Brain Science: Neural Nets

Omni Directional Connectivity also appears to be responsible for the incredible processing power of the human brain.

The brain is now believed to be structured like a "neural net," processing information "in parallel" rather than sequentially—in other words, by means of Omni Directional Connectivity, allowing processing power of a level unattainable by simple, *linear* causal connections alone.

The significance of neural net theory lies in the idea that brain cells operate simultaneously as a harmonious whole, rather than distinct sequences—or disconnected lines of thought. The parallel with our everyday experience of thinking is immediately apparent. A single line of thought unconnected to any others is a perfect description of the psychopathology we call "obsession," manifesting as the illness today called "Obsessive Compulsive Disorder" (e.g., obsessive handwashing, checking locked doors, looking in mirrors, or avoiding cracks in the pavement), and at the worst extremes, in schizophrenia, often characterized by a fixation with some delusion—such as the idea that one is being controlled by aliens through the television (perhaps by association between the "virtual realities" of audiovisual hallucinations and those of the TV screen).

Neural net theory suggests something that the creative artist knows only too well—one's brain functions best when one is able to connect many trains of thought into a coherent whole, rather than being "stuck in a rut," fixated on a single train of thought, unable to broaden one's mental horizons for a bird's-eye view. Creativity is the construction of complete patterns. The greatest creative achievements require the ability of the mind to imagine (that is, mentally construct) a complete, complex pattern which does not yet exist (as in the quote by van Gogh: "The thing has already taken form in my mind before I start it": see chapter 12).

Neural net theory thus tends to support the hypothesis that the essence of creativity is the ability to produce completely *new* patterns, where "learning" is the ability to add new bits of information to *existing* patterns.

Designer Evolution represents the ultimate creative act of the human brain—the ability to imagine, design, and construct new patterns out of the human genome itself.

The brain has evolved to enable creative thought—so let us take advantage of the gift offered by evolution, and embark upon the ultimate creative task:

directing our own Creative Evolution, in the pursuit of ever-increasing survivability and well-being.

7. Computer Science: Supercomputers

Our relatively primitive computers have, until recently, processed information sequentially according to linear connectivity—as chains or lists of events. But the suggestion that the human brain operates by means of Omni Directional Connectivity has led cyberneticians to develop new computers which mirror the connectivity of the human brain, through "parallel processing"—essentially, joining up several computers to form a network equivalent to the neural net of the human brain. The power of the emerging "Supercomputers" by far transcends that which was attainable by linear processing.

The significance of Supercomputers here is the message that technology is progressing at an extraordinary rate *by mirroring the design of nature itself.*

Science and technology is the practice of examining how nature designs the world, then learning to do the job ourselves—better. Technology is increasingly becoming indistinguishable from biology, as we recognize that we can make much more complex machines from the stuff of life itself (dynamic, open, self-organizing, self-sustaining systems capable of Spontaneous Order) than from the relatively inert, dead matter of closed systems such as metals and plastics. The science writer Kevin Kelly has devoted a book to this idea, proposing that we are entering a new, "neo-biological" age, in which "Technos is enlivened by Bios."[8] I call it the DNAge. Mnemonically, it spells out Dawn of a New Age. The real New Age will be an age not of channelers and crystals, but of biology and technology combined to enhance the human condition.

8. Cosmology: Superstrings—The Theory of Everything

The ultimate expression of Systems Science is the ongoing attempt of physicists to complete the work begun by Newton and Einstein, by uniting quantum theory and the theory of relativity, in the construction of a so-called Grand Unified Theory of Everything (TOE).

Where Newton's theory of gravity united the worlds of heaven and earth, and Einstein's theory of relativity united space and time, the Theory of Everything attempts to unite the quantum and cosmological levels—that is, the very small with the very big.

The most popular attempt to produce a Theory of Everything has attracted much interest beyond the world of speculative or pure physics. According to Superstring Theory, matter consists of tiny, looped, vibrating strings. Just as different length musical strings produce different notes by vibrating at different frequencies, so different vibrations in Superstrings produce particles with different properties. The result is a vision of the world as a cosmic Net, producing order and harmony at every level—from subatomic particles to entire galaxies.

The view of a harmonious cosmos composed of vibrating strings naturally lends itself by mental association to the ancient, medieval, and Renaissance world picture of an ordered world of correspondences perhaps most associated with the mystical mathematician Pythagoras. In the world of correspondences (or harmonious relationships), the universe is regarded as a finely tuned instrument. The distances between planets are governed by mathematical proportions, analogous to the harmonics of musical strings, producing an exquisitely harmonious "Music of the Spheres," for those sufficiently sensitive to hear it. Pythagoras claimed to have heard the Music of the Spheres during an "out-of-body experience"—perhaps after overdoing the mandrake during a night on the town.

It must be stressed that the Theory of Everything is an untestable hypothesis, and a pretty way-out one at that, since according to the theory, we live not in a universe, but a "multiverse" of many different undetectable dimensions. The most popular version of TOE—known as M theory (apparently standing for "matrix"—meaning "net," although no one appears to be quite sure)—proposes the existence of no fewer than eleven dimensions, our own lying on a membrane inside a greater universe. Apparently this is rather like living on the underside of the crust of a loaf of bread, the rest of which contain many parallel universes. I confess to complete ignorance as to why this must be the case for the theory to be consistent—and frankly, I have little interest in finding out. My interest in the Theory of Everything lies in its psychosocial, or philosophical significance.

The Elegant Universe

First, the interest in Superstring Theory among the general public supports the assertion that human beings yearn for scientists to tell them that the world *makes sense*.

Second, the Theory of Everything demonstrates once again that Modern Science is not, repeat **not**, describing a world of unknowable chaos, but quite the opposite: over and over again, the natural structures, forms, and processes being described in every area of science depict a world not of chaos, but of exquisite complexity—of cosmic order and harmony. Thus, in his book on superstring theory, suitably entitled *The Elegant Universe*, cosmologist Brian Greene asserts that the construction of a coherent Grand Theory of Everything "would provide an unshakable pillar of coherence forever assuring us that the universe is a comprehensible place."[9]

Increasingly, modern science is providing us with just such an assurance. The world is neither the irrational chaos depicted by the postmodernists nor the unfathomable mystery imagined by the theists. The new **Creative Science** is finding order and harmony everywhere it looks.

In the twenty-first century, while the task of applied science will be the attempt to enhance minds and bodies, the task of pure science will be the identification of the creative order in the universe.

In the Theory of Everything, the universe is perceived as a grand cosmic symphony—and humankind is destined for the job of its future conductor.

Bispectism: The Universality of the Complementarity Principle

It may appear at first sight that the Systems Science of Omni Directional Connectivity is incompatible with linear, Reductionistic Science, since one sees the world as a web or network of correspondences, the other as linear strings of digital information. But reductionism and holism are simply the two basic ways we can observe any aspect of the world—as parts or wholes.

The Complementarity Principle in quantum physics states that matter at its smallest level oscillates between two forms: discrete particles, and waves of energy composed of *packets* of particles known as "quanta." A complete knowledge of any phenomenon at this level requires a description of both its particle and wave aspects.

The Theory of Everything is based on the belief that the laws of nature operate at *every* level, from subatomic particles to galaxies. The Complementarity Principle may therefore also be applicable not only at the subatomic level of matter, but at *every* level. The complementarity of parts and wholes in *any* structure may be a basic principle of nature. *All* things may have "particle and wave" aspects. I have called this principle not reductionism or holism, but **Bispectism**—defined as the natural principle that every phenomenon consists of a whole composed of an arrangement of parts, equal consideration of both aspects being required for a comprehensive understanding of its nature.

Reductionistic Science examines the parts that make up the whole, holistic or Systems Science focuses on the system as a whole. Neither is a more accurate way of looking at the world—rather, the most complete picture is that which combines the two.

Parts and wholes represent the two fundamental aspects of reality. The world is a metapattern composed of parts—"things" are both "entities in themselves," and parts of greater wholes. A "thing" is a pattern of information composed of parts—or bits of information. All things can be viewed from two equally significant aspects—as a configuration of identifiable parts or as a functioning whole; as individual parts within a pattern or as a complete pattern; as things *in themselves* or in terms of the relationship of things to other things. In other words, a full understanding of *any*thing requires a full consideration of both of its aspects—parts and wholes.

Bispectism in Transhumanist Philosophy

The principle of Bispectism unifies scientific reductionism and holism by asserting that there is no contradiction between analyzing parts of the world and identifying whole systems. Bispectism as a universal principle governing the structure of the universe may be applied to every area of science and every level of the world—from subatomic particles, to human societies, to ecosystems, to the cosmos as a whole. It underlines all areas of my transhumanist philosophy, thus:

- PSYCHOLOGY: Neuromotive Psychology asserts the complementarity of the Prometheus Drive for Individuation and the Orpheus Drive for Integration.
- ETHICS: The Nurethics of Sensible Self-Interest asserts the complementarity of self-interest and benevolence.
- POLITICS: Bispect Politics asserts the complementarity of the needs for individual freedom and social harmony.
- METAPHYSICS: The New World Picture emanating from modern science asserts the complementarity of parts and whole in physical structures, expressed in the combination of Reductionistic and Systems Science.

The new world of correspondences affirms the unity of microcosm and macrocosm, man and universe, painting a picture of man as an integral aspect of a cosmos perceived as an ordered whole. In the New Music of the Spheres, the universe is a cosmic symphony of elegant harmonies—an evolving metapattern of harmonious complexity. By contrast, the two dominant metamemes in the modern world—postmodernism and theism—view the world as either random-chaotic (because knowledge = power = tyranny) or beyond rational understanding (for its creator moves in mysterious ways).

The significance of the new paradigm to transhumanism, then, is the implication on the one hand that the world is an ordered, beautifully complex system (Systems Science), and on the other, that the structures of life are potentially *improvable* through an identification and rearrangement of specific information patterns within the metapattern of Life (Reductionistic Science). These two ideas—the beautiful, ordered complexity of the world, and our ability and right to improve upon it where desired in our own interests of ever-increasing survivability and well-being—are the bedrock of transhumanism. They correspond to the two most essential aspects of any practical philosophy—the attempt to understand the overall nature of the world and the attempt to determine our purpose in it—metaphysics and ethics.

The conclusion—we should embrace the positive, rational metaphysic of order and harmony increasingly offered us by the New Science, abandoning both the childish nursery rhymes of theism and the harsh dissonance of postmodern chaos—for the New Music of the Spheres.

Next, I shall explore the potentially positive impact that a recognition of the new world picture might have on the individual and society in turn. I shall argue that a recognition of the New Music of the Spheres may serve to satisfy the impulse for a sense of psychological integration with the world I have called the Orpheus Drive—the instinctive desire of a conscious entity to feel itself *at home in the world.*

"MIND OF GOD" THEOLOGY

> Philosophy is the product of wonder.
> —Alfred North Whitehead, *Nature and Life* (1934)

Abandoning superstition does not require losing a sense of wonder at the world.

In the worldview emerging from the new science, man and nature are regarded as microcosmic and macrocosmic computers, processing the metapattern of information that is The World.

In this sense, nature could be regarded metaphorically as a Universal Mind, gradually becoming aware of its own existence, through its conscious aspect—the self-recognizing human brain. And a rational contemplation of the universe could be regarded as a form of communication between the human mind and "the mind of the universe"—that is, "the absolute information content of the universe regarded as a cosmic computer becoming conscious of its own existence."

It would not be entirely unreasonable to reduce the metaphorical concept of "The World as an evolving information processor, becoming self-conscious through recognition of its own nature by its conscious aspect, the human mind," to a single word—*God.* When the physicist Stephen Hawking concluded his best-selling book *A Brief History of Time* with the statement that a future scientific Theory of Everything would allow us to "know the mind of God," he was using the word in this sense—as a metaphor meaning "the world as a cosmic information processor"— or World-Mind. Thus, the sense of awe and wonder experienced in contemplation of the ordered complexity of nature through a rational knowledge of her forms, structures, and operations might be called **"Mind of God" Theology**. It is a phenomenon well known to scientists, both old and new, from Einstein to Richard Dawkins!

> If something is in me which can be called religious then it is the unbounded admiration for the structure of the world so far as our science can reveal it.[10]

> The feeling of awed wonder that science can give us is one of the highest experiences of which the human psyche is capable. It is a deep aesthetic passion to rank with the finest that music and poetry can deliver.[11]

It is such a sense of wonder at the world which draws many a scientist to science.

The central significance of "Mind of God" Theology is the idea that one may feel a positive sense of emotional affinity with the world, or nature, both by understanding her structures, forms, and processes and by identifying with the natural process of evolution of which we are a conscious aspect.

Such an idea was expounded in the 1950s and '60s by the biologist Julian Huxley (brother of Aldous), who coined the term *transhumanism* in the sense I use it here, as a synonym for his philosophy of "evolutionary humanism":

> As a result of a thousand million years of evolution, the universe is becoming conscious of itself, becoming able to understand something of its past history and its possible future. This cosmic self-awareness is being realized in only one tiny fragment of the universe—in a few of us human beings.[12]

This perception of psychological unity between self and universe is the essence of "Mind of God" Theology. It is the impulse of a conscious entity to feel *at home in the world.* I have called it the Orpheus Drive, or the Will to Love—where to love is to empathize—to identify Self with Other, I with That.

This is the real basis of the "religious impulse"—not the desire to worship a creator, but the recognition of both the wonder and interdependence of things—the fact that nothing exists in absolute isolation, for the world is an evolving Unity-in-Diversity.

The word *spirituality* describes the neurochemically induced feeling of empathy and benevolence we call love, extended beyond its original evolved function (attraction to that which is beneficial to our survival and reproduction) to *all* objects of perception. Spirituality is a falling in love with the world.

"Mind of God" Theology is a "spiritual" attitude for those who recognize the sublime wonder of the world, but not the existence of an invisible, undetectable ruler who desires to be worshiped for making it. In "Mind of God" Theology, there is no irrationality, and no bowing and scraping, just a deeply held feeling of awe and wonder at the magnificent order of the cosmos of which we are a conscious aspect.

"Mind of God" Theology is a "spiritual" attitude for an intelligent, rational, evolving species.

For those averse to the metaphorical use of religious terminology, "Mind of God" Theology might equally be called **Neurotheology**—the view of the human mind as an embodied, microcosmic information processor able to recognize and empathize with the macrocosmic information processor that is The World.

Take your pick, for the rose smells as sweet.

INTERNETICS

> Our new electric technology is not an extension of our bodies but of our central nervous systems. . . .
> —Marshall McLuhan, *Understanding Media* (1964)

A future social expression of the Orpheus Drive for a sense of cognitive integration with the world may be found in the phenomenon I call **Internetics**, defined as "the sociological effect of the Internet."

Underlying "Mind of God" Theology is the view of the human mind and nature as micro- and macrocosmic information processors, evolution as a sort of cosmic neural Net, becoming conscious of itself through the evolving (complexifying) human mind.

The concept of nature as a brain becoming conscious of itself as its neural network expands through the activity of evolving human minds is clearly analogous to the phenomenon of the Internet. The growth of the Internet is comparable to the expanding neural network of electrochemical impulses in a growing human brain. The vast pool of individual human minds, spreading and sharing information through the ever-increasing network of globally communicated electronic information that is the World Wide Web, may be thought of as an evolving global brain.

The distinction between the Internet as hardware—a network of computers—and the World Wide Web as software—its information content (the global language of "hypertext" used to access the data stored in the Net) has been made clear by Tim Berners-Lee, the inventor of the World Wide Web. If the network of computers we call the Internet is a global brain, then the World Wide Web (its information content perceived by self-recognizing human brains) could be viewed as its emergent mind. At a certain point of neural complexification, self-recognition becomes possible—and an unconsciously functioning brain becomes a conscious thinking mind. As the collective content of the mind of the species, the Web could be regarded metaphorically as a **cybermind** emerging through the synthesis of human and computer brains.

The exponential growth of the World Wide Web denotes the evolutionary emergence of the **cybermind**.

The Cybermind

In the **New World of Correspondences**, the Internet as cybermind mirrors the neural net of the human mind, as the macrocosm mirrors the microcosm.

The significance of the cybermind is twofold:

Metaphysically—or cosmologically—the evolution of the cybermind mirrors the view of evolution as a process of complexification—the spontaneous emergence of ever more complex patterns in nature.

Ethically—the cybermind signifies the potential psychological unification of the human species through ever-increasing planetary communication. In this sense, the transhuman of the future—*Homo cyberneticus*—may be regarded as *Homo sapiens* cognitively evolved to operate as a cybermind in the interests of the species as a whole, rather than that of his own particular tribe—the unfortunate tendency of his evolutionary predecessor.

Thus, Internetics signifies the idea of the Net as a technological extension of the human brain and nervous system: human minds as neurons in an evolving cybermind, and the consequent potential for the psychological evolution and unification of the species.

The concept of the Internet as a cybernetic extension of the human nervous system was predicted with unnerving accuracy by Marshall McLuhan in his seminal book *Understanding Media*. But McLuhan was writing about the new *TV Age* emerging in the 1950s. The true power of Internetics will only become fully realizable in the twenty-first century, as we witness the full emergence of the power and influence of the Net—the greatest wonder of the modern world.

Gradually, through the Internet, the entire pool of human knowledge is being transferred from physical to virtual reality. Before long, the cybermind will constitute a virtual library containing the sum total of the human meme map. Like the Oracle at Delphi, the Net will eventually be able to respond to any question we ask it ("Mirror, mirror on the wall . . ."). The Magic Mirror is not as far off as you might think. Already, for a small fee, Google allows us to pose any question we like to a team of researchers.

As we enter the twenty-first century, the mind of the species is no longer scatterbrained. The cybermind is getting its head together, and starting to think . . .

The Danger of the Cybermind as Bodypolitic

The cybermind mirrors the concept of an evolving "World-Mind" central to the nineteenth-century German romantic philosophers of *Naturphilosophie* or "transcendental idealism," such as Fichte and Schelling. The movement reached its culmination in the ambitious metaphysical system of Hegel, who perceived human history itself as an evolutionary dialectic of *Geist*—a word significantly inferring both mind and spirit—or human mind and "World-Mind."

The emergence of the Internet lends itself to a new take on evolutionary "nature philosophy"—this time based not on extravagant metaphysical speculation, but on modern science. The Net is to transhumanism as *Geist* is to Hegel—the World-Mind becoming conscious of itself through humanity—this time, through technology.

Similarly, the potential *danger* in the concept of a global cybermind operating collectively for the common good is the same today as in the past—that it may come to negate the primacy of the autonomous individual, in favor of the society-as-a-whole—the Bodypolitic—*controlled by the state.*

Hegel identified his evolved World-Mind with the Prussian state of his day. Karl Marx duly materialized Hegel's philosophy, Dialectic of Spirit becoming Dialectical Materialism—the universal mind of the Absolute transformed into the International State of World Communism. Mussolini's Fascism—symbolized by the Roman emblem of an axe binding together a bundle of twigs—and Hitler's Third Reich, with its emphasis on "blood and soil," can be seen as national rather than international versions of the same idea—a mystical unification of parts into the whole. All of these totalitarian ideologies owe their origin to Rousseau's concept of the General Will: the idea that a universal human moral consensus justifies state-enforced collectivism—and the consequent subjugation of the freethinking individual.

How is the cybermind any different?

Protecting the Free Thought of the Cybermind

The cybermind is not a homogenous General Will, but a plurality of individual "Wills to Evolve." In short, the cybermind is the metaphysical equivalent of the political metaphor of the Neuropolitic.

Unlike the individual human mind, the cybermind has no self—no controller of consciousness—no ego, no "I." Thus, the cybermind exists only by virtue of the voluntary cooperative action of the individual minds of which it consists.

The ability of the cybermind to recognize itself through self-consciousness is only a metaphor, describing the ability of individual human minds connected by technology to perceive the nature of the world collectively, through shared knowledge—like cells in the neural net of a global brain.

While it remains relatively untainted by state control, the Net represents the apogee of freedom—the ability to spread one's thoughts, feelings, beliefs, and values around the world—at the click of a mouse. Uniquely, the Internet allows us to look at other individuals' points of view directly, rather than through the biased mouthpiece of the state, media, or corporate worlds.

However, if humanity should be so foolish—or sheepish—as to allow forces to take total control of the cybermind, it will find itself in the grip of the ultimate authoritarian Bodypolitic—complete planetary domination by the Defender of the Metameme. One World can mean both psychosocial unification and "nowhere to hide": privacy and individualism destroyed by Plato's "tyranny of democracy": a

Big Brother culture ruled by the mythical "General Will of the people"—euphemism for control by the powerful (the head of the global Bodypolitic).

As Marshall McLuhan pointed out, the emerging Global Village may have its downside—regression to a new tribalism. For in a small village, everyone is your neighbor, and when everyone is your neighbor, everyone knows your business. In the Global Village, the individualism of the big city may become the claustrophobic collectivism of the small tribe—this time, a planetary tribe, ruled by the Defenders of the Meme—whether state, media, commerce, or a stifling collaboration of them all.

If we are not careful to preserve our freedom, the cybermind may be infested by the ultimate virus of the mind—the killer that wipes out all other memes, to assume control of the planet, through complete domination of the metameme. If we fail to protect our freedom, Orwell's Newspeak, Ingsoc, and the Ministry of Truth will be realized with all the terrifying "banality of evil."

The evolution of the cybermind must not and need not negate the freedom of the individual. The Internet is and must remain a bastion of free thought. We must recognize the danger of allowing governments to control the content of the Net beyond the need to restrict incitement to violence (something which cannot be permitted in a civilized society). Let the cybermind be an *open* mind and think a billion thoughts (though hopefully, not *too* many impossible ones before breakfast). Let the people speak—not "as one," but "one by one."

The Net and the Neuropolitic: From Couch Potato to Cybersurfer

Having acknowledged its dangers, let us turn to the positive aspects of the Internet.

In the twenty-first century, the Net is destined to merge with and replace the TV as the central metaphor of Modern Life.

This can only be a positive development.

For half a century, television has served as an electronic soma—the tranquilizing drug of choice for millions—everybody's little helper. But increasing numbers are realizing that they no longer need or wish to be TV slaves, sitting goggle-eyed in front of Big Brother, mindlessly gorging on the junk-food diet offered by program (or mind) controllers more concerned with maximizing ratings by appealing to the basest instincts possible than with educating, uplifting, or inspiring.

In contrast to the mind-numbing, collectivist herd instinct that is TV viewing with its serflike willingness to be dictated to by program makers and advertisers, the Net is the epitome of individual choice. Far more is required of the Net Surfer than the ability to press the TV remote control. Unlike TV watching, surfing the Net requires intelligence and alertness. *You* choose where to go and what to do—not just from a selection of identikit channels, but millions of Web pages.

Want to know what's happening on the other side of the planet?—click for daily news updates on the subjects of your choice. Want information on any subject whatsoever?—click to browse a search engine. Want more information from specialists?—click to receive regular newsletters. Want to know what's going on in cutting scientific research or government reports?—click to download instantly. Want to contact anyone around the world?—click to send instant e-mail. Want to sell anything to anyone anywhere on the planet?—click to trade on round-the-clock world auctions. Want to buy something, anything, from anywhere on earth?—click to enter a global shopping mall. Want to know the meaning of a neologism you heard in the street, coined only yesterday?—click to search an online "Urban dictionary." Want to know what people around the world think about the news of the day?—click to read their responses on message boards. Want to find people with similar interests?—click to join newsgroups and communicate with like-minded people from all over the planet . . .

Where television *is* Big Brother—all citizens bowing down before the same God—the Internet is individualism in action. Where the TV is the realm of empty-headed celebrity worship, the Net is for people who think for themselves.

The Net is the symbol not of a Brave New World but a brave New Age of freedom and independence. Let us celebrate its wonders and the human ingenuity which made possible a free voice for the people of the world.

In the name of freedom, let the people speak.

The Net and the Technological Sublime

A new phenomenon is emerging: increasing numbers of people are recognizing the power of the Internet to induce a euphoric feeling of awe, wonder, and universal empathy.

Log in to a news-based Web site right now, and one can immediately browse through numerous beautiful color photographs taken all around the world over the past few days, by different photographers, across different continents, each capturing a moment in time—a moment of exhilaration, exasperation, or intensity in a life lived on this earth, at this time in history, here and now. Only the hardest of hearts could fail to be moved by the power of such a global celebration of life lived in all its myriad diversity and wonder all over the planet we call home.

The phenomenon of Net surfing as a sort of "spiritual experience"—inducing a surge of universal empathy, awe, and wonder at the vast, organized complexity of life of which we are each an intrinsic part—I call the **technological sublime**.

The philosopher Kant identified two types of sublime experience. The Mathematical Sublime is the appreciation of natural objects of such vastness that the mind cannot grasp their enormity, such as a starry night, the ocean, or the Grand Canyon. The Dynamic Sublime is the ability to contemplate dangerously powerful natural phenomena such as lightning, thunder, a stormy sea—but without fear—rather, by identifying oneself with the power of the cosmos.

As there is little to fear when surfing the Net (except sudden unexpected invitations to enlarge one's penis or observe naked teens), the **technosublime** may be considered a contemporary version of the Mathematical Sublime, a feeling of awe and wonder in contemplation of phenomena of unimaginable scale—namely, the sheer amount of information on the Web; the vast number of minds communicating simultaneously around the planet; the diversity of human life; and the unimaginable complexity of the technology involved in achieving such a modern miracle.

Science, too, invokes the sublime—that is the meaning of "Mind of God" Theology—the sense of awe and wonder experienced in contemplation of the miraculous complexity of nature. Thus we may equally speak of the **Sci-Tech Sublime**.

The technosublime, or Sci-Tech Sublime, is a spiritual experience suitable for a scientific age.

The Net and E-Phoria

In his book *Techgnosis*,[13] author Erik Davis explores the links between science, technology, and the "spiritual impulse" for meaning and wonder. The term *Techgnosis* neatly encapsulates the idea that new technology provides a kind of "secret knowledge" known as yet only to a few "initiates," thus mirroring the Gnostic sects of early Christianity (*gnosis* meaning secret knowledge).

The euphoria of the technosublime experienced when surfing the Net might be termed **e-phoria**. The e-phoria of the technosublime is a secular expression of the religious impulse—the desire for a sense of "belonging" to the universe—experienced in a technological age.

The instinct of a conscious entity to seek a sense of cognitive integration with the world I have called the Orpheus Drive for Integration. The e-phoria of the technosublime is an expression of the Orpheus Drive. Camille Paglia has expressed what I mean by the e-phoria of the technosublime perfectly:

> As a child of popular culture, teethed on the electronic media, I feel the neighborly nearness of nations, continents, planets. Wires, wires everywhere: our thoughts are beads on the endless chain of connectedness that is the cosmos. My metaphors are sports of a mysticism of things, a pagan crusade to save the phenomena.[14]

Precisely. A "mysticism of *things*." Such is the nature of *my* Techgnosis, my e-phoria, my technosublime.

It is, ultimately, a recognition of the beauty of This Life—lived Here and Now—

For I have no interest in another.

The Net as Humanist Heaven

It is impossible to escape the obvious parallels between the experience of cyber-surfing and religious metaphysics.

For what is the Internet but a humanist heaven—an immaterial realm in which mind communicates with mind without limitations of space, time, or the physical body?

Unlike a telephone call requiring use of the vocal chords, and thus greater awareness of the body, communication via the Internet feels experientially like a form of communication by thought alone, between disembodied minds in an immaterial realm—the "virtual reality" of Cyberspace. The feeling of disembodied mind communicating with disembodied mind in a realm transcending space and time is clearly analogous to the idea of life as an immaterial "spirit" in a virtual reality called "Heaven." With one great difference—unlike heaven, the Internet is a virtual reality which does not require the death of the body to experience its wonders.

Such is its potential power and influence that the Internet may come to form the basis of an unspoken "religious" attitude for a secular world—that is, religion in the *original* sense of the word—an attitude to life that binds people together.

Through its ability to link mind to mind around the world, the Internet may serve to increase the human instinct for cooperation rather than conflict. Such a statement may seem naïve in the face of the current political situation—a world at war, divided by geography, race, and religion. Could Internetics overcome our seemingly perpetual state of global conflict? Frankly, no. That would be asking too much. But the Internet may still help to improve the general level of human cooperation and benevolence—by changing the way we think about each other and ourselves—from the tribal to the global. There is nothing outlandish in the suggestion that the emergence of a vast number of individual minds, pooling information around the planet, may slowly begin to change the way we see the world from a continual territorial battleground between opposing tribes, to an integrated neural network of human minds, and that doing so will help to increase the general capacity for mutual understanding and tolerance.

For the essence of the Net is shared information—and that includes the sharing of beliefs and values. The more one understands the other person's point of view, the more tolerant and understanding one is likely to be. Hatred and aggression are often the result of an irrational fear of the new and unknown—an instinctive, primitive, self-defense mechanism—for the new and unknown is unpredictable, and unpredictability means danger. By allowing us to see those on the other side of the world as our neighbors, the Net may allow us to see our distant neighbors as our friends.

Perhaps we may come to realize the brotherhood of man not through religion or politics, but through technology—product of the rational mind in the modern world.

The Noosphere

> This externalization of our senses creates what de Chardin calls the "noosphere" or a technological brain for the world.
> —Marshall McLuhan, *The Gutenberg Galaxy* (1962)

The potential impact of the evolving cybermind was explored by the celebrated Russian geophysicist Vladimir Vernadsky in the 1920s. Vernadsky introduced the term *noosphere* (from the Greek *nous*, "mind" or "thought") as an extension of the geosphere and biosphere (the inorganic and biological levels of the world), to describe ever-expanding human consciousness as an emergent layer of "mind stuff" surrounding the globe, representing the next stage in the evolution of life on earth. In the last scientific paper written before his death, "Some Words about the Noosphere," Vernadsky optimistically listed the conditions he envisaged resulting from the evolution of the noosphere. The resulting shopping list of goals reads like a humanist political platform:

> Peopling of all the Earth;
> Abrupt transformation of the means of communication and commerce between different countries;
> Establishment of political and other ties between all the states of the Earth;
> Predominance of the geological role of man over other processes which take place in the biosphere;
> Expansion of the frontiers of the biosphere and Man's exit into the Cosmos;
> Industrial exploitation of new sources of energy;
> Equality of the people of all races and religions;
> Increase in the role of people's masses in the decisions on the questions of internal and foreign policy;
> Freedom of scientific thought and scientific search from the pressure of religious, philosophical, and political considerations, and the creation of the conditions, favorable for the free scientific thought, in social and state structure;
> Rise of the well-being of the world's people;
> Creation of a real possibility to exclude malnutrition, hunger, misery and to weaken the influence of the diseases;
> Rational transformation of the original nature of the Earth, with the purpose to make it capable to satisfy all material, aesthetic, and spiritual demands of mankind;
> Exclusion of wars from the life of society.[15]

In "The Biosphere and the Noosphere," an article written for *American Scientist* in January 1945, Vernadsky wrote, "We undergo not only a historical, but a planetary change as well. We live in a transition to the noosphere."[16]

But the noosphere was never to be associated with Vernadsky.

The Phenomenon of Man

The French Jesuit priest and amateur paleontologist Pierre Teilhard de Chardin had attended Vernadsky's lectures at the Sorbonne in 1927 with his friend and countryman, the philosopher Edouard Le Roy, who had written a book on the philosophy of Henri Bergson, known for his theory of Creative Evolution. The result was two papers by Le Roy published in Paris: "L'exigence idéaliste et le fait de l'évolution" (1927), and "Les origines humaines et l'évolution de l'intelligence" (1928), the third chapter of which was called "La noosphere et l'hominisation."

At the same time, Teilhard began writing his own book, developing the concept of the noosphere into a grand evolutionary theory.

But some memes remain recessive for a long time, waiting for the right moment to assume a position of dominance in the Meme Pool. It was to be some thirty years before the meme of the noosphere finally took off.

Authoritarian Bodypolitics of all persuasions sharing a dislike of heresy, Vernadsky's ideas were frowned upon by Communist Russia, while disapproval by the Catholic Church delayed publication of Teilhard's book until after his death in 1955.

Eventually, though, the meeting of minds between Vernadsky, Le Roy, and Teilhard resulted in a book which became a cult classic. In *The Phenomenon of Man*, Teilhard defined and elaborated the concept of the noosphere:

> Much more coherent and just as extensive as any preceding layer, it is really a new layer, the "thinking layer," which . . . has spread over and above the world of plants and animals. In other words, outside and above the biosphere there is the noosphere.[17]

In romantic prose, Teilhard presented an uplifting vision of the inevitable evolution of consciousness resulting from the increasing cerebralization of human life—a development which would eventually culminate in a state of absolute cosmic omniscience he called "the Omega Point."

The book was published in England in 1959, with an enthusiastic introduction by the evolutionist Julian Huxley, who had developed his own philosophy he called evolutionary humanism—or transhumanism. The postwar generation of idealistic youth that comprised the 1960s counterculture was naturally attracted to Teilhard's positive vision of evolutionary progress. *The Phenomenon of Man* helped make the concept of the "evolution of consciousness" popular with a generation.

The Net and the Noosphere

The world of science was not so enthusiastic.

Teilhard was soundly ridiculed as a prime purveyor of pseudoscience. The influential biologist Peter Medawar was particularly scathing in a famous 1961 book review, which had the effect of single-handedly destroying Teilhard's intellectual credibility with its dry wit—or supercilious sarcasm, according to one's memotype:

> It would have been a great disappointment to me had Vibration not somewhere made itself felt for all scientistic mystics either vibrate in person or find themselves resonant with cosmic vibrations; but I am happy to say that on page 266 Teilhard will be found to do so.[18]

However, forty years on, it is impossible to ignore the obvious correspondence between the concept of the noosphere as an evolving layer of mentality surrounding the planet—and the evolving global network of electronic information transmission that is the World Wide Web.

In short, the Net is the noosphere realized through technology.

In this regard, it is very interesting to take another look at Professor Medawar's review of *The Phenomenon of Man*. Taking away his temperamental aversion to romantic prose, Medawar's essential objection to Teilhard's book is the assertion that the noosphere "may be equated to 'information' or 'information content' in the sense that has been made reasonably precise by modern communication engineers. To equate it with consciousness, or to regard degree of consciousness as degree of information content, is one of the silly little metaphysical conceits I mentioned in an earlier paragraph."

But the equation of consciousness with information can no longer be regarded as a "silly little metaphysical conceit," for the brain is now generally recognized to be an organic information processor! The mind—as the active content of the brain—is, indeed, information! Pure information!

And through the emergence of the Internet, the idea that dramatic increases in human knowledge might be regarded as a layer of information surrounding the globe is not even a scientific metaphor—*it is a literal truth!*

Now look back at the Teilhardian vision condescendingly dismissed by Medawar as a "silly conceit": "to regard degree of consciousness as degree of information content."

Forty years on, it seems that Teilhard was right, Medawar was wrong.

The Technosphere and Neurosphere

Whatever its merits, the Teilhardian vision of conscious evolution has continued to attract a wide public fascination ever since. And today we are witnessing a new revival of interest—as a direct response to the emergence of the Internet.

In a 1996 speech to the Cato Institute, Louis Rossetto, founder and editor of America's popular technophile magazine *Wired*, announced "the dawn of the twenty-first century's digital civilization," by comparing the global rise of the Net to Teilhard's vision of "an electronic membrane covering the earth and wiring all humanity together in a single nervous system." Posting a message on the Internet, he went further: "What seems to be evolving is a global consciousness formed out of the discussions and feelings being shared by individuals connected to . . . computers. The more minds that connect, the more powerful this consciousness will be. For me this is the real digital revolution, not computers, not networks, but brains connecting to brains."[19] This is the phenomenon I call Internetics—the evolution of mind to cybermind through the miracle of technology—the World Wide Web.

And in the foreword to *The Biosphere and Noosphere Reader* in 1999, former president of the Soviet Union Mikhail Gorbachev expressed his sympathies with the Teilhardian vision: "We have reached the phase in cultural evolution where we must assume full responsibility for our power. . . . Knowing and reaching our fullest potential within the constraints of the biosphere must be the ultimate goal—the driving vision of the twenty-first century. And the noosphere concept suggests a philosophy for such a necessary balance."[20]

Noosphere, I think, is a rather ugly word. It also fails to distinguish between the "world of mind" intended by Vernadsky and Teilhard, and the phenomenon emerging specifically in our own time—the world of electronic information communication that is the Internet. Thus, rather than the Noosphere, I prefer to speak instead of two complementary concepts, the **Neurosphere** and the **Technosphere**.

The Neurosphere is the modern equivalent of the noosphere—an evolving **megameme**; a "layer" of cognitive information transmission surrounding the globe comparable to the geosphere and the biosphere (the geological and biological layers, respectively). The Neurosphere is made possible by the Technosphere; the combined *infostructure* of the Internet and other forms of modern technocommunications, such as satellite TV, cell phones, and fax.

The Technosphere is hardware, the Neurosphere is software. If the Technosphere is a cybernetic brain, the Neurosphere is a cybernetic mind. As they grow, the Technosphere and Neurosphere together signal the emergence of a **cybermind**; from an ever-expanding network of individual minds communicating around the world like neurons in a neural net, emerges something new—a self-recognizing, global brain—a "world-mind" facilitated by technology.

Earth becomes a planet populated by a species that increasingly thinks the same thoughts, at the same time—not by the force of a Bodypolitic imposed from above—but by the free will of *individual* minds, each acting independently in pursuit of ever-increasing survivability and well-being. Through the actions of a global multiplicity of individual Wills to Evolve, the Bodypolitic becomes the Neuropolitic. Fired by the Will to Evolve, the knowledge and abilities of the species increase beyond imagination. The cybermind becomes the "Mind of God"—awakening to its own existence. Marshall McLuhan put it thus:

> Today computers hold out the promise of a means of instant translation of any code or language into any other code or language. The computer, in short, promises by technology to create a condition of universal understanding and unity. The next logical step would seem to be, not to translate, but to by-pass languages in favor of a general cosmic consciousness.[21]

Netaphysics

The Internet—

Net-shaped interconnections in living systems—

Neural nets—

Network Theory—

The cosmic Net of Superstrings—

There appears to be something of a common thread here.

The New World Picture emerging from science finds a suitable metaphor in the image of the Net, a structure in which all parts interconnect such that altering one part affects the whole, by means of Omni Directional Connectivity.

Where the dominant metaphor of the world in the Industrial Age was that of a clockwork machine—wound up by God, the dominant metaphor for the DNAge will be that of the world as a neural Net—an evolving cybermind, synthesis of the organic and the technological, governed by the minds of individuals acting cooperatively through their own mutual interests, in pursuit of ever-increasing levels of survivability and well-being.

We might therefore call the emerging world picture **Netaphysics**—a term coined by Erik Davis in the conclusion of his book *Techgnosis*.

Netaphysics—the new scientific view of nature as a self-organizing meta-pattern of Omni Directional Connectivity—offers a positive vision of the world as a harmoniously integrated whole, and humankind its most highly evolved, conscious aspect.

Netaphysics—or the New Music of the Spheres—is a rational, scientific worldview for an alienated species that wishes to feel itself *at home* in the world once more.

NOTES

1. Albert Einstein, "Religion and Science," *New York Times Magazine*, November 9, 1930.

2. Abu Hansa, *James Whale Show*, Talk Radio, UK, February 4, 2003.

3. James Watson, Skeggs Lecture, Youngstown State University, Ohio, 2003.

4. Arthur C. Clarke, "Arthur C. Clarke Relishes His 2001," Associated Press, January 2, 2001.

5. James Watson, "DNA Pioneer Urges Gene Free-For-All," *Guardian*, April 9, 2003.

6. John Stewart, *Time's Arrow* (Australia: Chapman Press, 2000).

7. Harold Morowitz, "Chemistry Guides Evolution, Claims Theory," *New Scientist*, January 20, 2003.

8. Kevin Kelly, *Out of Control* (London: Fourth Estate, 1994).

9. Brian Greene, *The Elegant Universe* (London: Random House, 2000).

10. Albert Einstein, *The Human Side* (Princeton, NJ: Princeton University Press, 1979).

11. Richard Dawkins, *Unweaving the Rainbow* (Boston: Houghton Mifflin, 1998).

12. Julian Huxley, *Man and His Future*, ed. Gordon Wolstenholme (London: Churchill Ltd., 1967).

13. Erik Davis, *Techgnosis* (London: Serpent's Tail, 1999).

14. Camille Paglia, *Sex, Art, and American Culture* (New York: Vintage, 1992).

15. Vladimir Vernadsky, "Several Words about the Noosphere," *Uspekhi Sovremennoi Biologii* 18, no. 2 (1944): 113–20.

16. Vladimir Vernadsky, "The Biosphere and the Noosphere," *American Scientist* 33 (1945): 1–12.

17. Pierre Teilhard de Chardin, *The Phenomenon of Man* (New York: Harper, 1959).

18. Peter Medawar, *The Strange Case of the Spotted Mice* (Oxford: Oxford University Press, 1996).

19. Louis Rossetto, http://fusionanomaly.net/wired.html (May 12, 2005).

20. Mikhail Gorbachev, *The Biosphere and Noosphere Reader* (London: Routledge, 1999).

21. Marshall McLuhan, *Understanding Media* (New York: McGraw-Hill, 1964).

Epilogue
TRANSHUMANISM AS EVOLUTIONARY EXISTENTIALISM

Finally, I sum up transhumanism as a philosophy of life against death, and meaning against insignificance.

WHAT IS LIFE FOR?

The philosophy of transhumanism as I present it here is one that provides an answer to the question that often rears its head among those of a particular temperamental disposition. H. G. Wells expressed it thus: "Yes, you earn a living, you support a family, you love and hate, but—what do you *do*?"[1] In other words, is there nothing more to life than this? Is that all we are here for—to eat and sleep, work and play, fuck and fight? And what exactly is the point, if everything we are is snuffed out for eternity after three-score-years-and-ten (if we're lucky) of trouble and strife?

No nontheistic philosophy to date has been able to provide a satisfactory answer to the fundamental question—what is life *for*?

Existentialism said, bluntly, "Nothing—so just choose something you like and get on with it."

Hardly inspiring.

Postmodernism was even less so—"Life's a meaningless joke, so enter the absurd spirit of it all—eat, drink and be cynical, for tomorrow . . ."

And so the modern world became the postmodern world and the West slipped quietly into the deep dark waters of nihilism.

And this is the state of play at the dawn of the twenty-first century. Those who cannot accept the religious metameme, being presented with no positive alternative, are in danger of succumbing to postmodern nihilism, at a dreadful cost to both to individual and society alike.

But there is a way out of the impasse—a clear road ahead.

For the dawn of the DNAge opens the way forward for a brand-new version of existentialism—not the gloomy, angst-ridden, sermon-on-suffering of the old, but a positive, uplifting *New* Existentialism, suitable for the intelligent minds of the modern world in the twenty-first century.

EVOLUTIONARY EXISTENTIALISM

The Old Existentialism of Kierkegaard and Sartre took as its starting point the fundamental fact of one's own existence. Looking back to Descartes, the existentialist said, "I think, therefore I am. As a conscious agent, I must decide for myself why I'm here, and what I should do with my life."

The Transhumanist agrees. But unlike the Old Existentialist, he does not conclude that the world is objectively unknowable through reason, that whatever choice he makes is just as good as any other, so he may as well take a blind "leap of faith" into religion (Kierkegaard), or radical politics (Sartre), as a sort of empty gesture of defiance against a meaningless universe. Therein lies the relativistic path to the nihilism of the Postmodern Condition. Instead, armed with his four faculties and an attitude of judicious open-mindedness, the transhumanist investigates the world around him, and comes to the perfectly sensible conclusion that, all things considered, he actually rather *likes what he sees*.

For despite the obvious fact that by this point in our evolutionary journey, the semi-civilizations we have created have yet to conquer the major problems of global poverty and war (a truism that, contrary to the smug-superior nagging of the negs, concerns the transhumanist just as much as anyone else), he nevertheless observes around him something quite wonderful: a miraculous technowonderland, positively overflowing with fabulous products and services, endless opportunities for growth, development, self-expression, and joie de vivre. In short, he observes a modern world of plenty!

And further, he discovers that the scientists of the modern world are *not*, in fact, depicting the world of purposeless, directionless chaos that he has been meme-washed to accept by the twentieth-century Defenders of the Meme; but rather, a world of harmonious complexification—a magnificent, *unfinished*, evolutionary process of which he, like all human beings, is an intrinsic, self-recognizing (*conscious*) aspect.

He naturally concludes that the metaphor imprinted upon his mind by Defenders of the Meme, of humankind as nothing but an "insignificant twig on the cosmic bush of life," is in fact, quite untrue—that in the noble pursuit of knowledge through reason, aided by the power tools of science and technology, the collective activity of human beings could be more accurately described as *the mind of an evolving universe becoming conscious of itself.*

He thus finds himself more attracted to another scientific metaphor. For it appears that the universe itself is like a meme map—a cosmic computer, programmed with a single instruction, repeated over and over again for all eternity—"Evolve, evolve, evolve . . ."

And by following the instructions of the cosmic meme map through the ongoing pursuit of knowledge, abilities, and experience, it follows that the human journey can logically conclude in only one manner. Through the continual increase in knowledge pursued over centuries, the human family may ultimately come not only to *comprehend*, but to *empathize with* the mighty megapattern of the universe itself. At the end of the human journey, humankind may come to know the very "Mind of God"—reality in all its glory, thought and felt.

Such an understanding leads the transhumanist to a logical conclusion as to the best course of action in life. It can be stated in six words: "*Live to evolve, evolve to live.*"

The purpose of life is the continual expansion of abilities, in pursuit of ever-increasing survivability and well-being.

THE TRANSHUMANIST METAMEME

The philosophy of transhumanism outlined here represents a new, positive, Evolutionary Existentialism, in which human beings are not perceived as alone and isolated in a meaningless world, but conscious aspects of the ever-evolving, ever-complexifying megapattern of the universe, guided in their best course of action by a shared body of human knowledge, attained by hard work and effort of will—through science.

Such an attitude was expressed by the evolutionary biologist Edward O. Wilson, in his recent book, *Consilience*:

> We are entering a new era of existentialism. Not the old absurdist existentialism of Kierkegaard and Sartre, giving complete autonomy to the individual, but the concept that only unified learning, universally shared, makes accurate foresight and wise choice possible.[2]

Existentialist philosophers such as Kierkegaard and Nietzsche rejected the idea of philosophical "system building"—constructing a philosophy as one would a house, by systematically addressing each subject in the philosophical lexicon. But we are not all **system-phobic**. Transhumanism as outlined here could be seen as an attempt to construct a totalized philosophical system—or meme map—out of the principles of a new Evolutionary Existentialism.

The essence of the transhumanism meme map—the answer to the question, "Why am I here?"—is ultimately very simple. "*I am here to evolve*." And, by extension, because I am a human being of flesh and blood like any other, ultimately, the whole species is here to evolve, too, whether they recognize it or not. Yet, because we all differ in neurotype, life experience, and pressure from our own particular Bodypolitic (whether church, state, media, commerce, or peer group), there is little prospect of everyone on earth coming to a similar conclusion. Therefore, the only logical conclusion is to follow one's own path; to go one's own way—existentialist-style—*not* to force one's own meme map onto others whether they want it or not, but rather, to explain it (for no one likes to be alone), and to challenge the dominant metameme whenever the head of a Bodypolitic threatens to meme-wash the world into a program of self-destruction—whether that of the body—through death—or the mind—through the *living* death of No-Selfism, in the stultifying stasis of an authoritarian Bodypolitic.

A meme map is a communicated world picture, worldview, philosophy, or religion, thought of as a software program for the self-recognizing biocomputer we call the human mind, which serves to guide the behavior of the body. When a meme map spreads through a culture, it becomes a metameme—a socially influential meme map. The transhumanist metameme will be one among many in the memetic marketplace of the modern world in the twenty-first century. In the modern world (the free world, in which individuals are not prevented from thinking or doing as they wish so long as they refrain from harming others in the process), it will be up to individuals to "make up their own minds" as to whether they wish to adopt the *transhumanist* meme map, any of the *others* on offer—or alternatively, if they should have a couple of decades to spare—to construct one of their own. Members of other societies—those of the authoritarian Bodypolitics of this world—will not enjoy the same luxury. We should take care to ensure it is one we shall never relinquish.

WHERE HAVE WE COME FROM? WHY ARE WE HERE? WHERE ARE WE GOING?

So finally, after constructing the transhumanist meme map, we can return to the question posed at the very beginning of our journey—

A human being is a conscious, biological entity, thrown into the world, blind to its purpose, in possession of four faculties by which to answer the fundamental questions that naturally arise in its inquisitive mind:

How should I perceive the world?
What should I feel?

What should I think?
What should I do?

And now, armed with a new, comprehensive meme map of the world, the transhumanist is able to answer:

How should I perceive the world?

I perceive the natural world as a sublime miracle of awe and wonder, and the modern world as a technowonderland of human ingenuity.

What should I feel?

I feel the e-phoria of the technosublime. I strive to feel—not merely well—but *better* than well.

What should I think?

I think of myself primarily in two ways—as a unique individual, and as a member of the human species: both as self-governing individual and as one with society, species, and cosmos; Neuropolitic, cybermind, and the "Mind of God," united in the ever-evolving, ever-complexifying megapattern of the universe.

What should I do?

I shall live, not just for the moment, but for the evolution of self and species. In a spirit of benevolence and adventure, I shall *Live to Evolve, Evolve to Live!*

"WHERE IS THE DIRECTOR?"

It is in such a spirit that I offer my philosophy of transhumanism; as a blunt alternative to the nihilism of the postmodern age—as a hymn to humanity, freedom, progress, and the Will to Evolve.

For the New Music of the Spheres is no Symphony of Sorrowful Songs—but an eternal Ode to Joy.

And therein lies the ultimate purpose of this Transhumanist Manifesto. While recognizing the "tragedy of life" as one of biological limitation, I am saying NO to the tragic *view* of life in which man is perceived as a hopeless, worthless, debased creature in a meaningless universe. The tragic view of life is Thanatos in action—the final part of our genetic programming making itself heard as an unconscious death wish—the will to die.

Let us defy Thanatos. Let us replace the will to die with the Will to Grow and the Will to Love—Prometheus and Orpheus—the twin drives for Individuation and Integration, united in the Will to Evolve.

From Genethics to Nurethics, human to transhuman: Eros and Thanatos to Prometheus and Orpheus, let us abandon the will to die, and discover the Will to Evolve.

The original existentialist, Søren Kierkegaard, expressed his angst-ridden sentiments on the human condition, thus:

> Where am I? What is this thing called the world? Who is it who has lured me into the thing, and now leaves me here? Who am I? How did I come into the world? Why was I not consulted? And if I'm compelled to take part in this drama, where is the director? I want to see Him.[3]

"Where is the director?"

Transhumanists have found him: The curtains are up, and the first performance is about to begin:

Act 1, Scene 1; *Designer Evolution*
Scene: the modern world.
Characters: *Homo sapiens*
Plot: the evolution of the species
Director?—

Why, the directors are You and I!

MORE LIFE, PLEASE

And so we reach the end of our brief journey through the transhumanist meme map.

In the end, it's a simple story, a simple idea.

We are biological beings evolved by spontaneous genetic mutation. We have inherited a biological makeup that severely limits our potential. We are susceptible to the misery of disease—from the common cold to cancer. We age and die. Worst of all—we are acutely aware of the fact! Such is the meaning of "the tragedy of the human condition"—man is the only animal who knows he is doomed to die.

And therein lies the essence of transhumanism—the recognition that the war between mind and body is no illusion. Those who seek to abandon the mind for the safe ignorance of unconscious instinct will never see eye to eye with those who seek ever more mind; ever greater knowledge; ever wider experience. Transhumanists do not wish to be less conscious, but more.

And what of death? If life is a meaningless, purposeless fluke, then perhaps death can be justified. But if life is beautiful, wonderful, and purposeful—directed toward ever-increasing levels of survivability and well-being—well, then, we have our goal; our significance; our meaning—to maintain and extend life at all costs!

If one perceives the world as wonderful and life as purposeful, the goal of life is clear—*More, please.*

The Old Existentialism claimed to stand for freedom—existential freedom. But while the angst-ridden pessimism of the Old Existentialism followed naturally from the acceptance of death as a first principle, the positive vision of the new Evolutionary Existentialism signifies the belief in existential freedom from the chains of our biological limitations.

Through Superbiology we will soon be faced with the serious prospect of eradicating our mental and physical impairments and enhancing our capabilities to levels undreamed of by our ancestors. We will soon have the choice of transforming ourselves into the people we would like to be.

Let us seize the opportunity with open arms.

THE WILL TO EVOLVE

When Richard Dawkins concluded his book *The Selfish Gene* with the famous statement, "Let us wage war on the tyranny of the selfish replicators," he unwittingly signaled the dawn of the transhumanist age.

In essence, he was reiterating the insight of Cartesian dualism—the recognition that body and mind are really not "one" at all, but two very different phenomena. The body may want to self-destruct—but does the mind? No. Yet our genes insist upon it, against our will. Should we merely accept our fate, lie down and die at the approach of three-score-years-and-ten? Why should we not seek to reprogram the computer that has booked us down for self-destruction? Why should we leave our fate in the hands of a process of Darwinian Evolution which cares nothing for our survival?

Until nature is once again acknowledged as something to be improved upon rather than worshipped, admired but not idolized, humanity can never seriously alleviate the suffering at the heart of the human condition.

Only science, with its blunt acknowledgment of nature's tyranny, may allow us to conceive of the possibility that humankind should be set free, that suffering may finally be—not merely "alleviated"—but eradicated—and not in some other, illusory world of the imagination—but in This-Wonderful-Life.

Ultimately only technology can produce a truly radical transformation of self. Let us hope that the children of the DNAge are wise enough to reject those who preach a gospel of unreason and embrace the emerging wonders of a New Age of science with open arms.

If we want to, we can look ahead with confidence and determination, embracing the Will to Evolve beyond our present state of biological slavery, and striking out into the unknown with all the courage shown by our brave ancestors throughout history, who refused to accept the limitations of the present and struggled to find a better life—*more* life— in ever wider pastures. This pioneering spirit is the bedrock of transhumanism. It is the essence of the Will to Evolve—the innate impulse of human beings to strive—to struggle—to defy the odds—to be the *best* we can be.

The universe is our oyster once we dare to recognize our own remarkable abilities as human beings. Science and technology provide the most powerful tools in our continuing evolutionary journey—but the greatest requirement of all is the Will to Evolve.

NOTES

1. H. G. Wells, *Experiment in Autobiography* (New York: Macmillan, 1934).
2. Edward O. Wilson, *Consilience* (New York: Knopf, 1998).
3. Søren Kierkegaard, *Fear and Trembling*, trans. Alastair Hannay (London: Penguin Classics, 1986).

APPENDIX

THE BIO-LUDDITE ARGUMENTS SUMMARIZED

The main bio-Luddite arguments can be grouped into four types I call the Sanctity, Equality, Maleficence, and Incompetence arguments. I argue they are not rational arguments at all, but based on irrational fears I call respectively: hubraphobia, schadenfreude, malanthropy, and misanthropy.

1. THE AUTHORITY OF GOD/ SANCTITY OF NATURE/ INVIOLACY OF MAN ARGUMENT

"*God/Nature made man; He/She/it is cleverer/more powerful than us, therefore we shouldn't try to improve on the job ourselves.*"

This argument is based on *hubraphobia*: an illogical fear of hubris—punishment for challenging God or Nature. But we do so every time we take an aspirin, mow the lawn, or operate to keep a sick baby alive. Should we abandon *these* interventions in our God-given nature as well?

2. THE EQUALITY OF MAN ARGUMENT

"*Bioenhancement above the level of 'normal' health will be limited to the wealthy.*"

This attitude is based on *schadenfreude*—the disguised resentment of other people's happiness. Simultaneously equal access to all new technology could only be attained through a world communist dictatorship. Most of us have rejected that idea. In a free world, wealth and its privileges gradually spread throughout society. So, too, will enhanced genes spread within the gene pool, so *increasing* equality.

3. THE MALEFICENCE OF MAN ARGUMENT

"*Superbiology will be misused by dictators or mad scientists.*"

This argument is based on an attitude I call *malanthropy*—the belief in human wickedness, expressed in the idea that man is too dangerous to be trusted with power over his own destiny. We cannot base our ethics on fear of ourselves. If there are evil people in the world, we must have the courage and confidence to stop them from harming us.

4. THE INCOMPETENCE OF MAN ARGUMENT

"Human beings will only mess it up. They always do."

This argument is based on *misanthropy*—disguised contempt for one's own species. Had our ancestors shared this attitude, we would still be living in caves.

THREE LOGICALLY INCORRECT ARGUMENTS

One of the most powerful weapons in a Meme War, frequently used by bio-Luddites, Logically Incorrect Arguments (LIARs) are irrational arguments employed as verbal weapons by which to kill a meme—or destroy a potentially socially popular idea. Bio-Luddites typically employ a narrow range of arguments which do not stand up to logical analysis, but appeal to the public on an instinctive, emotional level, thus avoiding the need to engage in rational thought or debate. They might best be opposed by calmly demonstrating the absence of causal connections between ideas—or alternatively, by vociferously insisting on Logical Correctness. Here I describe three of the most insidious LIARs I call the Soundbite, Extremal, and Dichotomic Arguments.

1. The Soundbite Argument

"*A proposal is bad if it can be linked, however remotely, with an emotionally loaded, unpopular idea.*" For example, genetically enhanced food = "Frankenfood" = bad.

The Soundbite or Associational Argument is emotional scare-mongering to avoid rational debate; the attempt to destroy a meme (invalidate an idea) by association with another, emotionally loaded, unpopular meme, thus creating an emotional meme-link (automatic association in the public mind) between a meme and a negameme (unpopular, emotively threatening idea), without any attempt to demonstrate a causal connection between the two. Soundbite arguments are a powerful method of discrediting a meme in the eyes of the public, who tend to respond quickly and strongly to emotive words and phrases, thus bypassing rational analysis of the argument. Thus, for instance, the media turn public opinion against genetically enhanced crops (the only realistic way to "feed the world") with the continual repetition of the term *Frankenfoods*, and sensationalize the wholly compassionate attempt to prevent disease and disability in the newborn by endless repetition of the phrase "Designer Babies." The Soundbite Argument is useful for those who find it easier to convince by emotionalism than by rational debate.

We live in a "blip culture" overflowing with soundbites of information (a concept introduced by Alvin Toffler in his prophetic book *The Third Wave*). The modern mind must learn to judiciously select good memes (useful bits of information) and reject rogue memes (dangerous ideas) from the vast plenitude of junk memes in the meme pool. All the more reason, then, to condemn those who attempt to infest meme maps with irrational, destructive, memes through emotionally laden, associational arguments. There is nothing wrong with soundbites—this book is full of them! A good soundbite is one that condenses a rational argument or idea in a nutshell. The danger arises when soundbite criticism *replaces* rational argument.

2. The Extremal Argument

"*A proposal automatically denotes advocacy of its most extreme version possible, or the inevitability of the worst case scenario.*" For example, Superbiology means Eugenics.

Extremal means "furthest from the midpoint." The Extremal Argument is the attempt to destroy an argument by suggesting that its exponents are advocating the most extreme form it could possibly take. This is the basic methodology of meme control in totalitarian regimes, for it effectively bans all opposition entirely, by regarding any criticism whatsoever of current policy as extremist.

A variation on the Extremal Argument is the assertion that a benign proposal will inevitably lead to its worse case scenario, for example, that genetically enhanced food will inevitably lead to ecodisaster. Such an assertion is logically incorrect because there exists no such necessary causal connection to support it. Underlying the argument is a belief in "Murphy's Law"—that what can go wrong, will—an assertion both illogical and severely detrimental to the running of a society, for it effectively infers that "doing nothing" is the best policy, since any action is likely to result in disaster. In short, the Extremal Argument is irrational, pessimistic fatalism.

3. The Dichotomic Argument

"*Criticizing or supporting both sides of an argument means you don't know where you stand.*" For example, to criticize both state authoritarianism and lack of social concern means you're confused.

Dichotomizational thinking is the single most significant weakness in the majority of popular sociopolitical arguments. It is still generally assumed that the only intellectually valid position to take in an argument is "one side or the other"—that there must be one view that is wholly right, another that is wholly wrong. Hence one must be either a Conservative or a Liberal, and believe in toughness or tenderness. But the world is not so black and white. Opposites are useful abstractions, but extremes are intrinsically unstable in nature, lover of ecological balance. For life itself is homeostasis—a state of dynamic internal balance between extreme states. What is intrinsically *bad* for society is extremism born of the inability to recognize the universal principle of Bispectism. There are two sides to every argument, based on the needs of both parts and whole, individual, and society. Truth lies in a synthesis of opposites. The wise man sees the good and bad in both sides of the argument.

Let these be my final words.

GLOSSARY of neomemes

If one is constructing a new meme map—a communicable mental map of the world—one needs to invent new words, to describe new ideas. This glossary is confined to neomemes introduced for the first time in these pages, plus a few existing terms that need defining in the sense I use them here.

academowaffle. Private language employed exclusively in scientific/academic papers, pronounced with emphasis on third syllable—as in the pop song "It's a *Very*, Very (Mad World)."

Adrenaline Neurotype. In *Neurotypology*: part of the *Theory of the Four Neurohumors*. One of the *Four Neurotypes*; a neurochemically based dimension of personality characterized by a high predisposition for sensation-seeking and physical activity, with aggressive impulsivity at the extreme end of the scale; product of an overreactive sympathetic nervous system and a low basal cortical arousal state. A dominant adrenaline neurotype produces an emotionally unstable, "physical extrovert" (the equivalent of "choleric" in the ancient theory of the four humors). A genethic neurotype.

agnoskepticism. Rational alternative to atheism, agnosticism, and theism. Attitude of rational doubt without certainty. The existence of a creator both benevolent *and* omnipotent in a world of inevitable suffering is Logically Incorrect, thus invalidating the agnostic's claim to complete ignorance on the matter. The existence of *some sort* of creative intelligence is a perfectly logical possibility (see *Geek God of Virtuality*), thus negating the validity of atheism. But in the absence of evidence that such an entity either exists, wishes to make its presence known, or plays any part whatsoever in our lives, the only logical attitude is to *presume* we are alone until further notice, and to seek our *own* "salvation," that is, ever-increasing survivability and well-being, through benevolence and reason, aided by science and technology.

antimodernity. More accurate term for the *eumeme, postmodernism.*

Apocalyptic Scare-mongering Syndrome (ASS). Psychological disorder characterized by an irrational inability to prevent the impulse to run around shouting "the end is nigh," characteristic of *bio-Luddites.*

Apollonian Art. Cerebral art of balance, order, and harmony typically favored by the *Dopamine Neurotype.*

BEST centers. Predicted future clinics offering consumer access to "Biogenetic Enhancement Self-design Technologies," helping us to be the *best* we can be—through *Superbiology.*

Benevolent Gene. Proposed memetic engineering required to enhance the flawed meme of the Selfish Gene into a *Supermeme.* To survive, "selfish genes" require that the bodies they inhabit survive long enough to reproduce. Benevolence assists survival. Hence the evolution of neurochemically induced benevolent instincts—the neurochemistry of love. Hence, man is *not* "born selfish" (defined as self-interest *at the expense of others*) as stated in the *Dawkins Doctrine* (on page 3 of *The Selfish Gene*). See *Felt Morality.*

bio/neuro/genetic enhancement. Aka genetic manipulation/engineering. Means of enhancing body and mind through twenty-first-century biotechnology, as through exercise and education today.

bio-defense. Argument that innate biological predispositions provide a mitigating circumstance for criminal actions. Legal attempt to excuse a criminal act on the grounds of diminished responsibility due to biogenetic deficiency, for example, genetic predisposition to impulsivity or aggressiveness. My response: "Biology is no more of excuse to be bad than drunkenness excuses murder."

bioegalitarianism. Erroneous *experientialist* belief that everyone is born with equal potential abilities. Not to be confused with the attitude espoused here, *civil egalitarianism.*

bio-elitism. Preoccupation with judging groups or individuals according to alleged biological superiority. Rejected here in favor of *bio-empathy.*

bio-empathy. View that everyone is subject to biological limitations—we are all handicapped by nature. Death, disease, and fragility make (sometimes strange) bedfellows of us all. The emotional basis of the transhumanist desire to eradicate disease, defeat death, and enhance abilities.

bio-fatalism. View fostered, unwittingly or not, by *psycho-Darwinist psychoreductionists*; man as slave of unconscious or genetic impulses, incapable of freeing himself from his lowly condition. By stressing the power of unconscious id or selfish genes (take your pick), psycho-Darwinism fosters the fatalistic view that man is an unimprovable, "fallen" being. Bio-fatalism is the secularization of Original Sin.

biogenetic/neurogenetic. Adjectives sometimes used here as convenient umbrella terms for genes, neurotransmitters, and any other aspect of human biology that affect well-being or behavior.

bio-libertarianism. Belief in the freedom of the individual to seek mental and physical self-enhancement through Superbiology.

bio-Luddites. Opponents of *Superbiology* and *biopsychology.* Bio-Luddism is a hotbed of irrational fears I call, variously, malanthropy, hubraphobia, entelephobia, noo-phobia, sciphobia, technophobia, and euphobia: the fear of human evil, hubris, evolution, knowledge, science, technology, and feeling good about oneself and species.

biopsychology. The obvious umbrella term for the study of the evolved, biological basis of attitudes (beliefs/values) and behavior. However, the term (behavioral) *neuroscience* is currently ahead in the Meme War for term of choice to describe such a study, perhaps aptly, as it signifies the importance of the *brain and extended nervous system* as the central controller of behavior—the governor, or managing director of the body. Personally, I prefer *biopsychology* as a wider, more general, and immediately comprehensible umbrella term, and *Neuromotive Psychology* (or *evolutionary neuropsychology*) as my own attempted contribution—but you can't always fight the *metameme*.

Bispect Politics. Proposed political theory based on the natural principle I call *Bispectism*, the complementarity of parts and the whole, applicable to all areas of life. In politics, the complementarity of individual and social needs requires the maintenance of a state of dynamic balance between extreme states in the *neuropolitic* through a process of *social homeostasis*. Thus, "Political extremism should be declared dead. Politics is the art of fine tuning."

Bispectism. Proposed universal principle in nature. All things can be viewed from two equally important aspects—as a configuration of parts (structure), or as a functioning whole (form), a complete understanding requiring a full consideration of both aspects. Bispectism is not dualism. Where dualism asserts the existence of two distinct things, bispectism says each thing is perceivable in two ways—as parts or whole. The principle of bispectism can be applied to all areas of transhumanist philosophy. Ethically, bispectism explains why there are always two sides to every argument, whether personal, political, scientific, or philosophical—because there are two fundamental ways of viewing everything. It explains why arguments are not resolved—because of the failure to recognize that both sides are addressing and defending only one side of a double-sided argument; and it explains how to solve arguments, disputes, or problems—by finding a higher synthesis, or learning to "seeing both sides of the story."

Bliss Chemical Theory. In *Neurotheology*, part of *Neuromotive Psychology*, proposal that the "bliss of nirvana" experienced by the meditating mystic is the product of raised serotonin levels experienced as a neurochemical reward for the attainment of an ideal state of homeostatic balance conducive to maximum survival—complete relaxation.

bliss chemicals. The chemical correlate of pleasure; neurochemicals such as dopamine and serotonin, released as a pleasure reward for behavior conducive to survival and reproductive success. See *Bliss Chemical Theory*.

Bodypolitic. In the present context, society, social group, or "corporate body" ruled from above (by the head of the Bodypolitic), supposedly in the name of the "general will" of its members; for example, church, state, media, business, or "self-help" group.

Bodypolitics. Authoritarian or *marionette politics*. Control of individual behavior by the head of a corporate body, usually in the name of the "general will." Here rejected in favor of *neuropolitics*. For who decides the "general will"?

Candidism. Naïve optimism (originally that of the philosopher Leibnitz) satirized in

Voltaire's novella *Candide*, summed up by the phrase, "All's for the best in this best of all possible worlds."

chocolate box art. Unrefined (crass, gross, unsubtle), often sentimental (Opioid) art absurdly associated by postmodern artists with *any* art which gives pleasure, or expresses joie de vivre. See the *Three English Hs.*

civil egalitarianism. Belief espoused here in equal respect for all benevolent individuals, irrespective of abilities.

creative evolution through self-directed mutation. In evolutionary biology, the possibility, currently "in the air" among nonorthodox biologists, that evolution may proceed not by the gradual accumulation of random genetic mutations, but through the ability of living things to automatically capitalize on genetic "spelling mistakes" by spontaneous reordering their internal structure to produce "meaningful new words." In terms of aesthetics, the theory of creative evolution sees nature as a creative artist making use of the material at its disposal, as Michelangelo made use of the flaws in a block of marble, and humankind as its conscious aspect, ultimately destined for the job of fashioning the *megapattern* of nature into ever more beautiful forms.

creative science. The construction of a scientific theory regarded as a creative act. Scientists regarded as creative artists engaged in the construction of an accurate *metameme*—a comprehensive map of the world.

credulism. Credulity as a habitual behavior pattern. Scientific credulism is the willingness to believe anything scientists tell us—via the media.

cyber rage. State of technophobic hostility toward new technology not yet understood.

cybermind. Pool of autonomous, individual, self-governing minds communicating globally via the Internet, regarded metaphorically as neurons in the neural net of a human brain, forming a single mind, or global consciousness. The cybermind is to the *technosphere* as the human mind is to the brain—software and hardware respectively. Equally, the cybermind is to metaphysics as the *neuropolitic* is to politics—metaphorical planetary and political metaminds, composed of freely interacting, autonomous, individual minds, *without authoritarian control from above*.

cybernetics. In the present context, the science of human biological enhancement through the augmentation of body and mind with artificial parts, from pacemakers to computerized brain-enhancers. Just one aspect of twenty-first-century Superbiology, together with *eugoics*, *Supergenics*, and Nanomedicine.

cyberpolitics. Society governed by cybernetic or self-steering individuals (cybernetics; from Greek, *kubernetes*, steersman of a ship).

Dawkins Doctrine. View that humans are "born selfish" by virtue of their selfish genes (see page 3 of *The Selfish Gene*). In *Nurethics*, negated by the recognition of *Felt Morality*, the neurochemical basis of innate benevolence, goodwill, or fellow-feeling—the neurochemistry of love.

deathists. Those who seek to justify death.

Decreased Sensitivity Hypothesis. In *Neuromotive Psychology* (*biopsychology* or *evolutionary neuropsychology*), proposal that serotonin is an evolved pleasure reward for physical composure, psychological confidence, and social dominance, which functions by decreasing brain recognition of external stimuli, thus dampening the flight-fight response of the autonomic nervous system, and so reducing anxiety (that is, fear of the environment).

Defenders of the Megameme. Rulers of the world.

Defenders of the Meme. Propagandists. Influential members of a society who seek to protect the social dominance of their own world picture (*metameme*) through *meme control* attained by *meme washing*.

demoralization. Reduced capacity for benevolent behavior.

Designer Evolution. (Aka self-directed, volitional, willful, or superevolution.) The gradual replacement of Darwinian Evolution (by spontaneous mutation and natural selection), with evolution self-directed by human beings through genetic enhancement as a service available to to consumers in the modern world.

dichotomal thinking. "All or nothing" attitude criticized here, by which a philosophy or idea must always either be accepted wholesale or rejected as worthless. See *Bispectism*.

disphilia. Schadenfreude in action; delight in disparaging and discouraging others. In art, love of dissonance, disorder, imbalance, chaos characteristic of postmodern antiart; an expression of thanatos, the unconscious desire for death, third and final instruction in our genetic programming.

DNAge. Future civilization defined by the prevalence of *Superbiology*. Where preindustrial, agricultural society was defined by the manipulation of plants and animals through farming, and industrial society by the manipulation of metals to make machines, the DNAge (or *Gene Age*) will be defined by the ability to manipulate the human body itself through *Superbiology*.

dominant and recessive memes. Prevailing/as yet socially unpopular ideas, which define and hover beneath the surface of a society, respectively.

Donne's dictum. "No man is an island": a truism precisely because it is true. Early seventeenth-century English poet John Donne's famous maxim encapsulating the essence of *Sensible Self-Interest*. The individual cannot survive independently from the society in which he lives. Hence, self-interest and social concern are not incompatible, but complementary; two sides of a single coin.

Dopamine Neurotype. In *Neurotypology*; part of the *Theory of the Four Neurohumors*. One of the *Four Neurotypes*; a neurochemically based dimension of personality characterized by a high level of goal-driven mental activity. Dominance of the dopamine neurotype produces an emotionally stable, "mental extrovert" (the equivalent of "sanguine" in the ancient theory of the four humors). A *nurethic* neurotype.

eco-fundamentalism or **ecoism.** Extreme environmentalist ideology. Belief that nature is good, modern techno-industrial society bad.

Emperor's New Clothes Syndrome. Ease with which society is collectively conditioned to accept *negamemes*, for example, art which depresses or deflates.

entelephobia. Irrational fear of evolutionary progress, from Aristotle's term *entelechy*, the innate potential for growth and development in all living things.

e-phoria. Experience of euphoria in presence of the *sci-tech sublime* experienced when Net-surfing.

Epsilonia. Culture dominated by the attitudes of its least educated members, from *epsilons*, the lowest class in Aldous Huxley's *Brave New World*.

ET test. Evaluation of art/attitudes/behavior by imagining the response of an extraterrestrial superintelligence. Based on the proposal that humanity should endeavor to make itself worthy of communication with other intelligent species.

eugoics. Well-being enhancers. (From Greek, *eu*, well or good; *egoic*, pertaining to self;

hence "good self"). Drugs designed to improve well-being by altering levels of specific neurochemicals. More generally, any rational techniques for enhancing well-being, whether through neuropharmacology, *Supergenics*, or *Neuroanalysis.*

eumeme. Disturbing or dangerous idea disguised as harmless; for example, antimodernity masquerading as *postmodernism,* or "destruction" disguised as "deconstruction." Memetic equivalent of euphemism. Alternately, a good meme.

euphilia. Love of feeling good. Opposite of *disphilia.*

euphobia. Irrational fear of feeling good about oneself or species. Instinct for self-abasement. Product of Thanatos—the death drive in the service of the *selfish genes.*

Eutopia. "Good place" resulting from widespread voluntary use of *eugoics* (rational well-being enhancers, whether neuropharmacological or psychotherapeutic). Realistic alternative to utopia ("nowhere place").

evolution. The process of unfolding potential in matter, from the Latin to unroll, like a carpet, or a scroll. Evolution is complexification—a movement from the simple to the complex. The information implicit within an unrolled, patterned carpet or written scroll is *waiting* to be revealed. Neither can be considered to have fulfilled their proper function before the process of unrolling is complete (for what's the point of an unrolled carpet or scroll?). Darwinian evolution is biological evolution by spontaneous alterations to gene sequences preserved over generations for their benefits to survival. *Designer Evolution* is biological evolution by genetic enhancements *deliberately directed* through *Superbiology.*

Evolutionary Epistemology. Belief in the continual increase in accuracy and quantity of human knowledge over time—otherwise known as human progress. Logical alternative to *premodern* theistic absolutism (truth is in an old book) and *postmodern* relativism (if it works for you, it's true).

Evolutionary Ethics. Ethics based on appreciation of the benefit to humankind of evolution's fundamental mode of operation—complexification (because the complexification of human faculties increases survivability and well-being).

evolutionary existentialism. Belief in existential freedom from biological limitations, attained through *Superbiology.* New existentialism proposed here, transcending the angst-ridden pessimism of the old, which followed from the acceptance of death as a first principle.

evolutionary humanism. Alternative name for *transhumanism.*

evolutionary neuropsychology. Alternative name for *Neuromotive Psychology.*

experientialism. Belief that behavior is determined by life experience rather than biogenetic makeup. Term suggested here for its clarity, in preference to "nurturism" or "environmentalism." In the *New Biology*, behavior is seen as resulting from a *combination* of nature and nurture—biogenetic predispositions mediated by social conditioning. "Free will" is proportional to the ability of the mind to determine behavior for itself. See *Nurethics.*

Felt and Thought Morality. Central concept in *Nurethics*; the twin sources of benevolent morality. Felt Morality is nature's neurochemical reward for benevolent behavior conducive to survival, the source of our basic instinct for goodwill. Thought morality is *Sensible Self-interest*—an extension of the instinct for benevolence through the faculty of reason; the ability of the evolved rational mind to inhibit *stupidly selfish* asocial instincts through intelligent self-control. Thinking of others requires the ability to *think*. Consideration for others requires the ability to *consider*. Thought morality is the *logic of love*.

FO PAS. Fashionably Orthodox Position Among Scientists; acronym conveying criticism of a certain herd mentality operating within science, as in all *Bodypolitics*, in which challenging the current orthodoxy is deemed taboo, through threat of rejection from the tribe (i.e., the sack, ridicule, or a suddenly inability to find glowing blurbists among one's friends for one's new book).

from Candidism to (techno)can-do-ism. Phrase summing up rejection of naïve optimism in favor of *positive realism*—rational belief in our ability to gradually overcome acknowledged biological limitations through technology.

from Genethics to Nurethics. Phrase expressing the essence of transhumanist ethics. From behavior determined by genetic programming through instincts (or "predispositions") in the survival interests of the selfish genes, to that determined by the autonomous, human mind through reason, in its *own* interests, of ever-increasing survivability and well-being. Human evolution proceeds "from Genethics to Nurethics" as humanity takes over its own genetic programming. Genethics are instructions for human behavior programmed by the genes. Nurethics are instructions for behavior determined by the self-conscious, rational, thinking mind. Genethically, man is programmed to survive, reproduce, and self-destruct. Nurethically, he seeks to override the third instruction in his genetic programming—the command to self-destruct.

fully firing on all four cylinders. Super-optimal state of well-being, involving full stimulation of all faculties; feeling, thinking, sensation, and perception. The ideal state of well-being; the optimum functioning of all faculties.

Geek God of Virtuality. Phrase denoting the possibility that God is a computer programmer, the world a computer program, life lived in the simulation of a "virtual reality." A perfectly valid hypothesis, thus negating the validity of atheism in its wider sense, and leading to advocacy of *agnoskepticism*. Though the concept of a "geek God" may sound ironic, it in fact raises a serious point: if God is an extremely intelligent, logical, self-controlled being, shouldn't we seek to emulate his example?

General Neurotypes. Neurochemically based *dimensions* of personality, as opposed to personality *types* (*Personal Neurotypes*.) See *Neurotypes*.

genethic politics. Fascism and Communism; authoritarian politics based on primitive, aggressive survival instincts.

Grand Delusion. Belief that death is not death; the metaphysical basis of theism.

Great Pessimists. The influential thinkers, Schopenhauer, Nietzsche, Freud, and more recently Dawkins, whose memes, unwittingly or not, have served to promote the belief that behavior is motivated—not by the independent mind—but by forces beyond our control—whether the World as Will, Will to Power, Unconscious Id—or Selfish Genes.

groupthink. Inability to think for oneself, characteristic of *Bodypolitics*, as opposed to *Neuropolitics*.

Hard Libertarianism. See *Soft Libertarianism*.

Homo cyberneticus. The next stage in human evolution. *Homo sapiens* evolved to *con-*

sciously direct its own evolution, through *Superbiology*. From the Greek, *kubernetes*; steersman or pilot of a ship. Man the *steersman* of his own destiny.

hubraphobia. Irrational fear of hubris; punishment by God or nature for seeking to transcend human limits.

human progress. Increasing survivability, through the complexification of human faculties and the *power tools* they make possible.

immortologies. Technologies to cure the disease of aging.

impossibilism. Irrational tendency to automatically dismiss a difficult but desirable goal as unattainable.

increased intentionality hypothesis. In *Neuromotive Psychology;* theory that dopamine is an evolved pleasure reward for intentionality, or will—the ability of the mind to control the body.

Internetics. Phenomenon and study of the sociological effects of the Internet, here regarded positively as the eventual storehouse of the complete human *meme map*—the sum total of human knowledge; and as a potential source of increasing psychosocial unification. The Internet mirrors the neural net of the human brain as the macrocosm mirrors the microcosm; an evolving *cybermind*, becoming conscious of its own existence through a shared body of scientific knowledge.

judicious open-mindedness. Logical assessment of new scientific theories, however strange, rather than instant dismissal (*scientific orthodoxism*) or uncritical acceptance (*scientific credulism*). Judiciousness (soundness of judgment, discretion, good sense) is the ability to logically assess new information, rather than to accept or reject it on impulse, the latter tendency arising in a rigid, "closed-minded" meme map—product of a brain that has stopped growing.

junk memes. Meaningless or valueless ideas, spreading through a culture.

logic of love. Benevolence as a rational survival tactic. See *Thought Morality.*

logical correctness. Rational response to the authoritarian "political and moral correctness" of liberal left and religious right, in their attempt to impose their *metamemes* on the *meme culture* through *memetic invasion*.

malanthropy. (Latin, *malus*, bad; Greek, *anthropos*, man) Irrational belief in the innate wickedness of man; judgment that human beings are too evil to be trusted with power over their own destiny; belief that "fallen man" must be protected from himself—by the state.

marionette politics. Authoritarian *Bodypolitics* in which the heads of the body corporate pull the strings to control the public body.

megameme or **megapattern.** The world, universe, or cosmos, regarded as an ultimately comprehensible metapattern of information. Today's *metameme* will be tomorrow's megameme. See *"Mind of God" Theology*.

meme attack. The deliberate modification or replacement of a culturally popular word, phrase or idea, thus altering the attitude promoted by its original meaning, in order to control the *metameme* (prevailing social attitudes). For example, Nietzsche and Freud's meme attack on the concept of "sublimation" replaced its original, positive inference of "transcendence through refinement" with the negative, antihumanistic connotation of "a social outlet for brute animal instincts."

meme control by **meme washing.** Cultural thought/mind control through brainwashing.

The attempt by a *psychohacker* to coercively delete files from a *meme map*. The basic *Weapon of Memetic Destruction* in a *Meme War*.

meme culture. Culture defined in terms of its dominant *metamemes*.

meme map. Information content of the brain regarded as a software program for attitudes and behavior which can be copied from one brain to another, whether voluntarily—through communication, or coercively—through indoctrination (*meme washing*). Causally connected set of *memes*, uploaded into the biocomputer that is the human brain to guide behavior. A communicable mental map composed of a set of attitudes (beliefs and values), forming a totalized World Picture, belief system, philosophical system, or religion.

meme pool. Sum total of *memes* in a *meme culture*.

Meme War. Battle of ideas. Meme wars are ideological arguments for influence or control over the direction a culture will take, by influencing or controlling its prevailing language, attitudes, and behavior.

meme-link. Automatic association between two acausally connected ideas, placed in the mind of the *Bodypolitic* as a weapon in a *Meme War* to provoke fear and loathing, for example, genetically enhanced food = "Frankenfood" = bad; or *Supergenics* = *eugenics* = bad.

memes. Attitudes (beliefs and values) or behavior (habits) regarded as units of cultural communication or transmission between brains, as genes are units of biological transmission between bodies. An idea or habit is only a meme if communicated to others. The term was introduced into the meme pool in 1976 by Richard Dawkins, in his book *The Selfish Gene*.

memespeak. Language of memetics.

memetic determinism. Complete control of a culture by *Defenders of the Meme*.

memetic enhancement. Improvement of a meme, for example, by the insertion of the meme of *The Benevolent Gene* into that of the Selfish Gene.

memetic invasion. All-out attack on a popular world picture, worldview, belief system, ideology, or dominant metameme, by another, as yet unpopular, recessive metameme, in order to gain control of social attitudes and behavior.

memetic mutation. Introduction into the *meme pool* of a new or semantically altered word, phrase, or idea, causing a change in attitudes or behavior. Directed mutation is the basic *Weapon of Memetic Destruction* (WMD) in a *Meme War*. See *meme attack*, *memetic invasion*.

memetic reproduction. *Meme*, *meme map*, or *metameme* copied to other brains by means of open communication or *meme washing*.

memetics. Study of cultural information transmission as genetics is the study of *biological* information transmission. Seen here as a future social science. To understand the metameme is to understand the culture.

memocide. Deliberate eradication of a meme from a *meme pool*, or *metameme* from a *meme culture*, through a *meme attack*, or *memetic invasion*, in a *Meme War* for control of behavior within society.

memotype. Classification of individuals in terms of their *meme maps*, or attitudes.

meritocracy. Society which rewards individual effort and abilities, advocated here in preference to enforced elitism (autarchy or oligarchy: the tyranny of the one or the few) or egalitarianism (democracy as the tyranny of the "general will") by an authoritarian *Bodypolitic*. The *Neuropolitic* is a meritocracy, rather than autarchy, oligarchy, or democracy.

metameme. An individual *meme map* copied to sufficient number of brains that it affects behavior within the culture as a whole.

metamind. Any group of minds acting together without coercion; for example, the *neuropolitic* and *cybermind*, as opposed to the *bodypolitic* in which control is imposed from above—by its head.

metaphysical naturalism. Pantheistic view that nature is sacred, so whatever it does is good—like kill us.

Mind as Virus Metaphor. Analogy describing the mind-brain theory of *neurodualism* proposed here. Comparison of the mind's emergence through evolutionary complexification to a virus infesting the brain, seeking to assume control from the selfish genes which have the body programmed for self-destruction.

"Mind of God" Theology. Scientific spirituality for a New Age of reason, combining the theist's reverence for life, with the scientist's commitment to rationality. Awe and wonder experienced in the rational comprehension of the organized complexity of the universe, regarded metaphorically as a form of communication with a "universal mind," becoming self-aware through the self-recognizing human brain, its conscious aspect. Nature as cosmic computer, man as biocomputer; macrocosm united with microcosm in a *New World of Correspondences*. The expression "Mind of God" derives from the concluding words of Stephen Hawking's populist book on cosmology, *A Brief History of Time*, in which the laws of physics governing the universe are compared to the contents of a universal mind, cosmic meme map—or *megameme*.

miserabilism. Art or attitudes based on a subjective state of unhappiness, posing as an objective world picture. Regarded here as ultimately inferior to *Up! Art*, on the logical grounds that happiness is universally more desirable than unhappiness.

modern world. In the present context, the technologically advanced industrialized nations, originating with, and now extended way beyond European civilization, based on the values of *modernity*. The developed free world. See also *New Modernity*.

modernity. Belief in individual freedom, ongoing human progress through reason, science, and technology, originating in the European Renaissance, now extending globally in a *New Modernity.*

moral correctness. The equivalent of political correctness imposed by the Religious Right. See also *logical correctness*.

Mountain Lecture. The Sermon on the Mount viewed as a lecture in ethics by a brilliant moral philosopher.

negameme. Psychologically or socially destructive idea communicated to others so as to deleteriously affect attitudes and behavior.

negametameme. Negative world picture. See *bio-Luddites*.

negativism. Predisposition toward negative evaluations. Negativity as a habitual, temperamental predisposition.

negs. The habitually negative-minded.

neomeme. Potentially propagating neologism. A new word or idea capable of being copied to other *meme maps*.

Netaphysics. Or *New Music of the Spheres.* Twenty-first-century *New World Picture* emerging from the *New Science* depicting a harmonious cosmos composed of dynamic patterns of interrelated parts, maintaining order through self-organization, with an innate tendency toward complexification. Derived from Systems Science, an approach increasingly required in all areas of science, which focuses on interrelationships between parts in the patterns of nature (e.g., complexity and network theory; fractal geometry; dissipative structures; neural nets; superstrings). Netaphysics replaces the pessimistic *Postmodern World Picture* with a vision of the cosmos as a spontaneously evolving Unity-in-Diversity; a dynamic, self-sustaining, cosmic Net. The term *Netaphysics* was coined by Erik Davis in the final pages of his book *Techgnosis* to describe a "digital metaphysics" based on the World Wide Web—the equivalent of my concept *Internetics.*

Netropolis. "Internet-friendly" metropolis; culturally influential, often university-based city with a high proportion of young, creative "brainworkers" involved in Information Technology. Evidence of the cultural transition from a survival society dominated by *genethic neurotypes* to a *Self-Enhancement Society* dominated by *nurethic neurotypes.*

Neuroanalysis. *Neuromotive Psychology* potentially applied in clinical practice, consisting of three overlapping components: Cognitive (encouragement of the *Will to Evolve*); Experiential (development of a *Rational Plan for Creative Living*); and Neurological (*neuroenhancement* through *eugoics*).

Neurodualism. Or neuro-Cartesianism. Alternative to both Cartesian dualism and monistic *No-Selfism* on the mind-body question. The mind is the name the conscious brain gives to itself, once able to recognize its own existence via sensation and thought. The *Self* is the name the brain gives to the mind and body as a functioning entity. Contrary to the claims of the *New Age neurophilosophers*, the self is alive and well. To deny its existence is to denigrate the importance of reason and will—conscious control of behavior by the mind, rather than selfish genes or memes. Neurodualism does not imply that mind and brain are entirely distinct entities, but rather two aspects of a single phenomenon. The brain is a configuration of biological *parts*, the mind is the functioning brain as a *whole*, able to re-*cognize* its own existence (see also *bispectism*).

neuroenhancement. Improvement of brain function through *Superbiology.*

Neuroepistemology. Acquisition of knowledge viewed as a process of information retrieval by the evolved, self-recognizing, self-programming, embodied, organic information processor, or biocomputer that is the human mind, in its own interests of ever-increasing survivability and well-being.

neurohumors. Umbrella term for all classes of functionally active neurochemicals, including neurotransmitters, peptides, and hormones. See the *Theory of the Four Neurohumors.*

neuromanticism. Passionate belief in the transcendence of the mind—or neuroenhancement—through *Superbiology.*

Neuromotive Psychology. Or *evolutionary neuropsychology.* Proposed psychological model for the twenty-first century. Human beings are not robots unconsciously acting out a genetic drive for survival and reproduction, but conscious agents pursuing their own purpose; self-recognizing, embodied brains capable of taking control of their own behavior in their own interests. The evolution of consciousness—the embodied brain's recognition of its own existence—naturally results in the emergence of the *Will to Evolve*—the cognitive drive of a conscious entity to expand its knowledge and abilities in its *own* interests of ever-increasing survivability and well-being. But the complete "fulfillment of potential" envisaged in humanistic psychology is not attainable by a species subject to the biological limitations of *Homo sapiens*. "Positive thinking" alone cannot overcome our biogenetic deficits. In the twenty-first century, neuroenhancement through *Superbiology* will increasingly enable us to attain the optimum neurochemical balance for maximum psychological well-being. See also *Prometheus and Orpheus Drives*.

Neuropolitic. Independent-minded, nonauthoritarian alternative to the *Bodypolitic*. See *Neuropolitics*.

Neuropolitics. Political theory introduced here. In the Neuropolitic, behavior is determined not by an unthinking public body controlled by head of state or other corporate bodies (*Bodypolitic*), but by a multiplicity of autonomously thinking, individual minds, each operating out of *Sensible Self-Interest*, resulting in the automatic maintenance of *social homeostasis* in the *cultural* body, as the autonomic nervous system maintains a state of internal homeostatis in the *physical* body—by allowing the free activity of individual parts within limits required for its harmonious functioning as a whole. However, just as the mind must assist the body to perform its basic functions when the body breaks down (for instance, by administering medicine), so, too, the state must guide society where necessary to ensure social stability or cohesion. Hence, the harshness of laissez-faire Social Darwinism is avoided in Neuropolitics by the application of *Soft* rather than *Hard Libertarianism*—summed up by the dictum, "Maximum individual freedom compatible with social cohesion or harmony."

Neurosphere. See *Technosphere.*

Neurotheology. Alternative term for *"Mind of God" Theology*, for those disturbed by the use of religious metaphors in secular philosophies. The term has also been used to describe recent neuroscientific explanations of religious experience in terms of brain activity: visions and voices have been explained as audiovisual hallucinations resulting from damage to the parietal lobes of the brain; the mystic's experience of "ego-lessness" as the product of temporal lobe epilepsy reducing awareness of self. My own contribution to the field is the *bliss chemical theory* of mystic experience.

neurotranscendence. Naturalistic transcendence through the uplifting of the mind to a sense of psychological unity with the world. Satisfaction of *The Orpheus Drive* for Integration.

neurotypes. In *Neurotypology*, four neurochemically mediated dimensions of personality, proposed in the *Theory of the Four Neurohumors*. The habitual balance between the four *General Neurotypes* produces a unique *Personal Neurotype,* or personality type.

Neurotypology. Proposed study of neurochemically based dimensions of personality. See *neurotypes.*

New Age neurophilosophers. In *Neuromotive Psychology*, contemporary philosophers of the mind whose denial of the existence of the mind or self resembles the core belief of New Ageism.

New Ageism. Belief in numerous impossible things before breakfast.

New Alchemists. Individual, private exponents of experimental neuroenhancement by means of *eugoics* (well-being enhancing drugs), often attained through the Internet.

New Biology. (Aka New Genetics.) Modern, gene-based, technology-assisted biology. Also, in contrast to the *Old Biologism*, contemporary theoretical biology which recognizes the influence of both biological makeup *and* life experience (nature *and* nurture, temperament *and* character) on personality and behavior.

New Enlightenment. Predicted revival of science, art, and cultural optimism in the twenty-first century, in the light of the explosion of new scientific knowledge, particularly in the increasingly combined areas of biology and technology. A renewed faith in the ability of our species to comprehend the universe.

new invisible hand. In *Neuropolitics*, the "hand of nature" operating automatically through the laws of complexity to maintain order in a Neuropolitic without the requirement of extreme, centralized control.

New Leonardians. Predicted new breed of artists reuniting art and science in the twenty-first century, Renaissance-fashion, by combining a fascination for scientific knowledge and techniques, with a sense of awe and wonder at the natural world.

New Modernity. (Aka transmodernity, hypermodernity.) Technological and global extension of *modernity* beyond the confines of European culture in the twenty-first century. The New Modernity represents a renewal of belief in the inevitability of human progress, through reason, science, and technology, combined with the belief in individual freedom.

New Music of the Spheres. (Or *New World of Correspondences.*) Metaphor describing the *New World Picture* emerging from the *New Science*, inspired by superstring theory, a "Theory of Everything" in pure physics, which unites quantum and relativity theory (the very small and the very big levels of reality) by depicting the cosmos as a harmonious network of tiny vibrating strings—thus mirroring the Pythagorean worldview of mathematical relationships between planets, as between man and universe, microcosm and macrocosm.

New Prometheans. Those united in the pursuit of the knowledge by which to transcend the human condition.

New Renaissance. Predicted revival of humanistic self-belief in the twenty-first century out of the ashes of fin de siècle postmodern nihilism, resulting in a rebirth in the arts.

New Scientism. Predicted twenty-first-century revival of belief in science and technology as a source of human progress and explanation of the universe, in the light of dramatic advances in knowledge and techniques. Renewed respect for the scientific method of observation, hypothesis, and experimentation as the knowledge base responsible for the miraculous *technowonderland* of the *modern world.* Belief in science without the dogmatism and "cold" rationality of the *Old Scientism.* In the New Scientism, new scientific pronouncements are evaluated with *Judicious Open-Mindedness*, not blindly accepted (*Scientific Credulism*) or rejected (*Scientific Orthodoxism*); and "cold" scientific rationality is replaced by "warm logic"—logic mediated by feeling in the interests of maximum well-being.

New World Picture. Harmonious depiction of the world, or *metameme*, emerging from modern science in the twenty-first century. See *Netaphysics*, *New Music of the Spheres.*

9/11/2001. Date on which the New Age fantasy of the Aquarian Age dissolved.

99 percent fallacy. Attempt by *postmodern scientists* to debase humanity by reducing its level of significance in the scheme of things to no more than that of chimpanzees, based on the assertion that we share 99 percent of our genes. My response: It's the other 1 percent that makes the difference between an ape and Einstein, stupid.

No-Selfism. In *Neuromotive Psychology*, the denial of any meaningful distinction between mind and brain, and hence the nonexistence of the Self, characteristic of *New Age neurophilosophers*. Belief in abandoning the "illusion" of selfhood and allowing entities called memes to control our attitudes and behavior, independently of our will. Rejected here as music to the ears of authoritarians of all persuasions (church, state, media, commerce) who wish to control the public by imposing their own *metameme* through *meme washing* (cultural indoctrination).

Nurethics. Evolutionary ethics proposed here, consisting of two main elements, corresponding to the questions, "How should we live?" and "How should we treat one another?" Ethics are a software program for the evolved, self-recognizing biocomputer that is the human mind, which may allow us to replace *Genethics*, the genetic program for self-destruction, once the autonomous rational mind assumes responsibility for control of behavior from both *Selfish Genes* (*bio-authoritarianism*) and *negamemes* (the social authoritarianism of the *Bodypolitic*). The human species can and should take control of its own evolution, by developing the *Bio-Enhancement Self-design Technologies* (BEST) by which to overcome death, disease, and biological limitations. Humankind is the conscious aspect of a natural process of evolutionary complexification. The best way to live is to participate in the ongoing evolutionary journey by seeking to extend one's abilities throughout life. Morality requires no deity. We are not "born selfish" but imbued with the neurochemistry of love. Benevolence is *Sensible Self-Interest*—a rational survival tactic. "Good-will to all men" is the optimum tactic for mutual survival and well-being.

Oh, Wow! Factor. Awe and wonder in the presence of *Up! Art*, or in contemplation of the *Sci-tech Sublime*.

old biologism. Attitudes characteristic of nineteenth-century scientism, which played down the role of the environment (i.e., society) in contributing to social disadvantage, in favor of a biological determinism which categorized races, sexes, and classes as biologically distinct and unequal. Rejected here for the *bio-empathy* of the *New Biology.* In accordance with the principle of *Bispectism*, we should think of ourselves *primarily* in two ways—as unique individuals, and as members of a single species. Other self-definitions, such as those of sex, race, nation, politics, or preferred football team, are not *un*important, but *less* so.

old scientism. Overidealized belief in "the perfectibility of man" through the "Enlightenment Project of Modernity," characteristic of European Enlightenment thinkers, ending in the mid-twentieth century as a result of collapsed confidence in the light of consecutive world wars. Limited by dogmatism (rule by a technocratic elite), and an underestimation of the biological limitations to human perfectibility. See *New Scientism.*

Omni Directional Connectivity (ODC systems). The complex operation of causality in net- or web-type structures found in nature, producing an exponential growth of trajectories. Proposed alternative to the negative term, *nonlinear*.

Opioid Neurotype. In *Neurotypology*; part of the *Theory of the Four Neurohumors*. One of the *Four Neurotypes*; a neurochemically based dimension of personality characterized by a tendency toward anxiety and depression and the desire to avoid oversensory stimulation, product of an overreactive nervous system, and high basal cortical arousal state. A Dominance of the Opioid Neurotype produces an emotionally unstable, "physical introvert" (the equivalent of "melancholic" in the ancient theory of the four humors). A *genethic* neurotype.

Orpheus Drive. See *Prometheus Drive*.

Orphic Art. "Intuitive" art, which induces calm, relaxation, serenity, or peace of mind; typically attracting the *Serotonin Neurotype.*

orthodoxism. Rejection of all but the current scientific orthodoxy. Criticized here as narrow-minded and illogical. For where else do new scientific discoveries come from if not the willingness to think the unthinkable?

Personal Neurotype. Neurochemically influenced personality type resulting from a unique, habitual balance between all four dimensions of personality, or *General Neurotypes*. See *Four Neurotypes*.

philosophy. A type of computer software program designed by the self-recognizing, embodied biocomputer we call the human mind, in order to determine the nature of the world, and how best to live in it.

positive realism. Proposed transhumanist attitude to life; alternative to both optimism (a naïve, "rose-tinted" view of life) and pessimism (irrational negativity). Sober recognition of the existence of fundamental, universal, biologically based problems (death, disease, malevolence, the limitations of body and mind) combined with the logical prediction that they will gradually be overcome through science and technology. Belief in meliorism, the inevitability of gradual improvement in the human condition.

postmodern. Adjective defining the prevailing *metameme* in the modern world of the late twentieth century. To its advocates, postmodern is synonymous with "pluralistic," in defiance of the perceived "hegemony" of *modernity*; to the judicious analyst, postmodern is synonymous with "antimodern, antihumanistic, cynical, nihilistic, and ironic. See *postmodern irony*.

postmodern antiart. Art expressing a negative view of the world, whether that of nihilism (art as disguised death wish), *antagonism* (art as social critique), or *miserabilism* (expression of personal unhappiness).

postmodern irony. The inability to take anything seriously, leading inevitably to nihilism. Postmodernism is existentialism with an ironic sneer, instead of gloom.

postmodern relativism. Belief that knowledge, morality, and aesthetic appreciation are individually subjective, or relative to the culture from which they derive.

postmodern scientists. Contemporary scientists who promote a fourfold negative view of the world: Indeterminacy (evolution as directionless); Contingency (life an evolutionary fluke); Insignificance (man a naked ape); and Maleficence (ruled by selfish genes). I respond with the counterclaims of Complexification (evolution proceeds toward ever more complex forms, structures, and operations): Cerebralization (the human brain is a predictable outcome of evolutionary complexification); Ascendancy (*Homo sapiens* is the peak of evolutionary complexification on earth); and Beneficence (we are born with the neurochemistry of love).

Postmodern World Picture. *Negametameme* spread by *postmodern scientists*, depicting evolution as directionless, nature "pitilessly indifferent"; man a "cosmic fluke"; "twig on the bush of life"; and "naked ape"; the product of "selfish genes."

postmodernism. Attempt to replace *modernity* with epistemological, ethical, and cultural relativism, more accurately termed *antimodernity*.

power tools. Conventionally—electrically powered tools. Here, all tools that increase survivability and well-being.

practical versus speculative science. Science which produces clear evidence of causal connections between events that may be of practical use in the fight to cure disease (for example, "a mutation of gene CFTR on chromosome 7 causes cystic fibrosis"), versus science which claims to find causal links between events based on dubious statistical evidence, presented to the media as if proven (for example, "fizzy drinks cause throat cancer," due to an increase in both). Transhumanists place their faith in Practical Science.

premodernism. Rejection of *modernity* for religious dogma and/or a preference for less complex, nonprogressive, preindustrial cultures, characteristic of theists, New Agers, and eco-fundamentalists.

progressive centrism. In *Neuropolitics*, a sensible middle ground between the extremes of left and right. *Soft Libertarianism* is the politics of the sensible center.

progressive evolution. The argument espoused here, that evolution leads inevitably to improvement in the human condition: 1. Human progress means "increasing survivability"; 2. Evolution is complexification; 3. The evolutionary complexification of the human brain necessarily increases survivability, by increasing flexibility of response to the environment; 4. Therefore, human evolution is progressive.

progressive fallacy. Denial of evolutionary progress. Assertion by *postmodern scientists* that evolutionary progress is an illusion, refuted here, thus: 1. Progress = increased survivability; 2. Evolution = complexification; 3. Complexification = increased brainpower = increased survivability = human progress.

Prometheus and Orpheus Drives. In *Neuromotive Psychology*, cognitive or *nurethic* drives for Individuation (Will to Grow) and Integration (Will to Love) transcending the *genethic* Freudian drives of Eros and Thanatos (will to live/self-destruct), product of the *selfish genes*. Associated with the *Dopamine* and *Serotonin Neurotypes*, respectively.

psycho-Darwinism. In the present context, Freudian and neo-Darwinist views regarded as constituting a single model of human motivation. Freud's life and death drives of Eros and Thanatos (sex and aggression) = the neo-Darwinian "predispositions" to act in the service of genetic survival/reproduction (Eros) and self-destruction (Thanatos).

psychohacker. One who attempts to surreptitiously delete or alter files from others' *meme maps* in order to control their attitudes and behavior. See *meme washing*.

psychoreductionism. Theories which reduce human motivation to unconscious or genetic drives for survival and reproduction (*Genethics*), ignoring higher, cognitive drives (*Nurethics*) (where "higher" means "more complex" and therefore of greater potential benefit to survival).

pure sublime. Appreciation of subtle complexity, as opposed to crass superficiality, in art or nature; the difference between the ceiling of the Sistine Chapel and an unmade bed.

Rational Plans for Creative Living. In *Neuromotive Psychology*, part of *Neuroanalysis*; the identification of personal strengths and weaknesses, plus a commitment to develop the *Will to Evolve*, leading to the construction of a personal *meme map* for the attainment of life goals. As creativity is the ability to construct new patterns out of the stuff of the world, Creative Living is the ability to create new patterns out of the stuff of *life experience*. The opposite of Creative Living is being "stuck in a rut"—acting out the same old routines day after day—without growth, development, progress, or creativity. Creative Living is an attitude of openness to possibilities.

recessive memes. Ideas which remain dormant, hovering beneath the surface of a culture, awaiting a better time in which to prosper.

Reciprocal eugoics. Voluntary, mutual encouragement of ever-increasing well-being, characteristic of a *Self-Enhancement Society*.

reciprocal selfishness. What happens when you're Stupidly Selfish and lack consideration for others; they lack consideration for you, too. See *Nurethics*, *Sensible Self-Interest*, *Thought Morality*.

rogue memes. Socially unorthodox, potentially dangerous ideas, which threaten the survival of existing memes.

sapienism. Prejudice and discrimination against the human species. Attempt to denigrate *Homo sapiens*, the thinking animal, by downplaying the vast differences in ability between man and ape. Coterminous with sexism, racism, and the newly discovered psychological disorder, homophobia. Expression of antihumanistic misanthropy.

scientific credulism. Uncritical acceptance of new scientific theories handed to the public by the media.

scientific orthodoxism. Instant dismissal of any new theory which threatens the current scientific orthodoxy.

sciphiles/sciphilia. Lover of science/the love of science and technology, and belief in their power for the good of humanity.

sciphobes/sciphobia. Those displaying an irrational fear of science.

sci-tech sublime. See *technosublime*.

Self-Enhancement Society. Vision of society as a harmonious multiplicity of autonomous, self-governing, freethinking individuals, engaged in the voluntary, mutual encouragement of ever-increasing well-being. See *Reciprocal eugoics*.

Sensible Self-Interest versus Stupid Selfishness. Central concepts in *Nurethics*. The basis of *Thought Morality*. The brain's evolved capacity for reason enables us to recognize the rational basis for benevolent morality: "Good will to all men" is a rational tactic for mutual survival. Evil is stupid selfishness because malevolence attracts equal malevolence in return. See *reciprocal selfishness*.

Serotonin Neurotype. In *Neurotypology*; part of the *Theory of the Four Neurohumors*. One of the *Four Neurotypes*; a neurochemically based dimension of personality characterized by mental calmness, serenity, or peace of mind, product of an unreactive (i.e., strong) nervous system, and high basal cortical arousal state. A Dominance of the serotonin neurotype produces an emotionally stable, "mental introvert" (the equivalent of "phlegmatic" in the ancient theory of the four humors). A *nurethic* neurotype.

ship of the wise. Opposite of Ship of Fools. A Cybership bound for *Eutopia*.

So What? Art. Trivial, postmodern art of so little creativity, formal skill, emotional power, or intellectual stimulation that it fails to hold the attention for more than a few seconds, before the observer says "so what?" and moves on.

Social Homeostasis. Central concept in *Neuropolitics*; a healthy society requires self-regulation between political extremes, as a healthy body maintains life through self-regulation between *physical* extremes—the process known as homeostasis. See *Bispect Politics.*

Socialist Darwinism. Social concern regarded as an evolutionary adaptation benefiting mutual survival. The recognition that benevolence is both an evolved, neurochemically induced instinct, and a rational tactic for mutual survival, leading to the recognition of the need for a certain level of state regulation in favor of the disadvantaged. The *Nurethics* of *Sensible Self-Interest* applied to politics, summed up in the maxim, "Ignore the poor on Sunday, quiet a riot on Monday."

Soft Libertarianism. Belief in the ethical primacy of individual freedom of thought, speech, and action, tempered by the pragmatic recognition that a certain amount of restriction on individual freedom through state intervention is a practical necessity for the maintenance of social cohesion; expressed in the dictum, "Maximum individual freedom compatible with social harmony." A midpoint between total selflessness, attainable only under totalitarian rule, and the self-defeating egotism of *Hard Libertarianism* characterized by hostility to *any* state limitations on personal freedom, rejected here as *stupidly selfish.*

Star Trek Philosophy. Modern equivalent of the Greek myths. See *technomythology.*

stasism. From Greek, "stoppage of flow." Denial of evolutionary progress; belief that evolution is directionless; view that man is "going nowhere," promoted by postmodern scientists for their own, mysterious ends. Opposite of progressive or *Creative Evolution.* The logical response—evolution is complexification, and Designer Evolution is *self-directed* complexification.

Stupid Selfishness. Illogically self-defeating absence of thought or consideration for others. See *Sensible Self-Interest, Felt and Thought Morality.*

sunny art. Positive, uplifting art of joie de vivre. See *Up! Art.*

Superbiology. Umbrella term used here to describe the emerging biotechnological methods of curing disease, enhancing abilities, and extending life in the twenty-first century; including: *eugoics* (neuropharmacology); *Supergenics* (gene therapy/enhancement); *cybernetics* (artificial/computer implants); nanomedicine (molecular scale engineering, or nanotechnology, applied to the human body).

Supergenics. Predicted, twenty-first-century "consumer-controlled" genetic enhancement of body and mind through the alteration or addition of genes, to enhance well-being, expand abilities, and extend life; viewed as the next logical step in personal self-development after exercise, education, or aesthetic surgery. Genetic enhancement (improvement beyond the level of "normal" health) will gradually replace gene therapy (curing disorders), as diseases become preventable. Neurotheorist Zack Lynch has proposed the term *enablement* to cover both.

supermeme. Big Idea dominating a *meme map* or *metameme.*

system-phobia. Irrational aversion to philosophical system-building, characteristic of gloomy existentialists and their ironic successors, the postmodernists.

technofatalism. Irrational fear that technology will lead to doomsday scenarios.

technomythology. Or *Star Trek Philosophy.* Modern equivalent of the Greek myths, encapsulating the essence of *evolutionary humanism.* Belief in the ability of the species to unite in the ongoing pursuit of ever-greater knowledge and experience in a spirit of benevolence and adventure, symbolized by the vision of the starship *Enterprise* and her crew, "boldly going where none have gone before." In its combination of harmonious, multiethnic cooperation and enthusiastic technophilia, Star Trek Philosophy offers a positive model for the future of the species. As in the Greek myths, individual characters symbolize human faculties; thus, Captain Kirk represents physical and emotional will, Mr. Spock—logic; their perfect collaboration signifies the ideal synthesis of feeling, thinking, and willing.

technoromanticism. Transhumanist attitude combining the romantic passion for human transcendence with the logic of science and technology. Passionate belief in the transcendence of human limitations through science and technology (as opposed to religion or politics). Union of Dionysus and Apollo; feeling and thinking. Alternative to both naïve, romantic idealism, and "cold" scientific rationalism.

technosexual. Contemporary technophile aesthete combining a love of technology and stylish design. One who appreciates the *technowonderland* of the *modern world.*

Technosphere. Global technocommunications (Net, TV, radio, phone, fax) forming a level of cognitive information, mentality, thought, or "mind-stuff" surrounding the planet (the *Neurosphere*), comparable to the geosphere and biosphere (levels of inorganic and organic matter), signifying the union of human mind and technology. The Technosphere is hardware—a global brain; the Neurosphere is software—a global mind. Together, they represent Vernadsky and Teilhard de Chardin's concept of the "noosphere" (mind sphere) realized in the age of the Internet.

technosublime/sci-tech sublime. Scientific and technological sublime. Awe and wonder at the immensity, complexity, and interconnectedness of the world, appreciated through science or technology; for example, contemplation of the miracle of complexity by which simple sequences of DNA chemicals automatically construct the myriad forms of life (scientific sublime); or the vast storehouse of information and complexity of global communication that is the World Wide Web (technosublime).

technowonderland. The *modern world* viewed as a miracle of technological invention—instead of an oppressive ideology responsible for cultural imperialism.

Thanatosian Art. Romantic or decadent art typically attracting the *Opioid Neurotype*, which induces pleasure from emotional or physical pain, by the release of endogenous tranquilizers such as endorphins.

thanatosians. Deathists. Those who happily accept mortality.

theistic absolutism. Dogmatic belief in absolute, incontrovertible, eternal truth, characteristic of religions. See *evolutionary epistemology.*

Theory of the Four Neurohumors. Proposed theory of neurochemically based dimensions of personality, or *neurotypes*, mirroring the ancient biological personality theory of the four humors. See *Neurotypology.*

Thought Morality. In *Nurethics*, the evolved capacity for *Sensible Self-Interest*. See *Felt and Thought Morality.*

Three English Hs. In *Neuroaesthetics*, three late-twentieth-century colorists (David

Hockney, Patrick Heron, and Howard Hodgkin) suggested as a model for a positive *Up! Art* of the twenty-first century, for their willingness to swim against the proverbial tide of postmodern cynicism, with an uplifting, nonironic art of joie de vivre that has nothing in common with the *chocolate box art* derided by postmodernists.

Transhumanism. Evolutionary Humanism. Belief in transcending human limitations through reason, science, and technology. A hi-tech humanism for the *DNAge*. Humanism extended into the biological realm in the age of biotechnology. Positive futurism. Belief in self-design through *Superbiology*. Belief in *Creative Evolution*—evolution by self-design regarded as a creative act. Belief in a *New Modernity*—a renewal of the Enlightenment project of modernity in the *DNAge*. Belief in a *New Renaissance* in the arts and a *New Enlightenment* in science. Belief in the transition from Darwinian to *Designer Evolution*.

transnaturalism. Scientific naturalism (the belief in the objective scientific study of nature, regarded as all that exists), with an added ethical dimension, intended to refute the Naturalistic Fallacy (the assertion of the invalidity in seeking to infer values from facts about nature). Ethics must *necessarily* be based on nature, because human beings are biological entities, programmed to survive. Ethics can only be divorced from nature if one refutes the fact that self-preservation is an innate, universal value. As conscious aspects of nature, it is perfectly valid for human beings to base an ethics on the fundamental directionality of natural evolution—toward increasing complexification—while seeking to transcend the operations of nature in other ways—by eliminating disease, death, and the biological limitations of mind and body.

Up! Art. Or *sunny art*. Proposed transhumanist aesthetic. Art which seeks to uplift and inspire, rather than deflate, depress, disturb, or dumb down.

warm logic. In *Neuroepistemology*. Logic mediated by feeling. Proposal that science should not seek to reduce all experience to logic—rather, logic should be used to guide feelings. Rational decision making for increased emotional well-being, as opposed to the "cold logic" of thinking divorced from feeling.

Weapons of Memetic Destruction (WMD). Words or ideas intended to destroy the meaning or validity of other words or ideas, in order to attain *meme control* of a culture.

wheelchair analogy. Analogy signifying the difference between transhumanist politics and the conventional politics of left and right: Liberals say, "Let me buy you a new wheelchair"; Conservatives say, "Let me take away your chair to make you self-reliant"; transhumanists say, "Let me *cure* you that you may walk again, *through Superbiology*."

Will to Evolve. In *Neuromotive Psychology* and *Nurethics*, determination by the self-recognizing, embodied brain to expand its capacities in the interests of ever-increasing survivability and well-being, regarded as an emergent drive in a conscious entity. The Will to Grow (*Prometheus Drive*) + The Will to Love (*Orpheus Drive*) = *The Will to Evolve*.

INDEX